REAL-TIME SYSTEMS
Engineering and Applications

THE KLUWER INTERNATIONAL SERIES
IN ENGINEERING AND COMPUTER SCIENCE

REAL-TIME SYSTEMS

Consulting Editor

John A. Stankovic

REAL-TIME SYSTEMS
Engineering and Applications

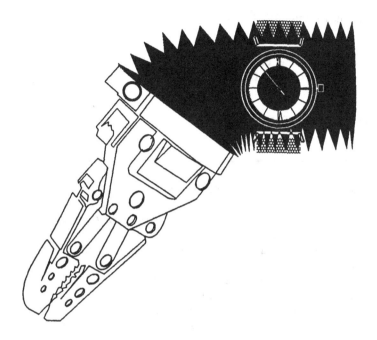

edited by

Michael Schiebe
University of Ulm

Saskia Pferrer
Fachhochschule Ulm

with a foreword by
R. Lauber

KLUWER ACADEMIC PUBLISHERS

Boston/Dordrecht/London

Distributors for North America:
Kluwer Academic Publishers
101 Philip Drive
Assinippi Park
Norwell, Massachusetts 02061 USA

Distributors for all other countries:
Kluwer Academic Publishers Group
Distribution Centre
Post Office Box 322
3300 AH Dordrecht, THE NETHERLANDS

Library of Congress Cataloging-in-Publication Data

Real-time systems engineering and applications / [edited] by Michael
    Schiebe, Saskia Pferrer.
        p.   cm. -- (The Kluwer international series in engineering and
    computer science ; SECS 167)
        Includes bibliographical references and index.
        ISBN 0-7923-9196-9 (alk. paper)
        1. Real-time data processing.   2. Systems engineering.
    I. Schiebe, Michael.   II. Pferrer, Saskia, 1947-   .   III. Series.
    QA76.54.R4295   1992
    004'.33--dc20   91-40129
                                                                      CIP

Contents

V.   EXAMPLES FOR APPLICATIONS

VI.   FUTURE DEVELOPMENTS

CONTRIBUTING AUTHORS

T. Bemmerl,
> *TU München, Inst. für Informatik, Arcisstr. 21, D-8000 München, Germany*

F. Demmelmeier,
> *SEP Elektronik GmbH, Ziegelstr. 1, D-8153 Weyarn, Germany*

K. Fischer,
> *Institut für Informatik, TU München, Orlean Str. 34, D-8000 München 80, Germany*

K. F. Gebhardt,
> *BA Stuttgart, Rotebühlplatz 41, D-7000 Stuttgart 1, Germany*

B. Glavina,
> *Institut für Informatik, TU München, Orlean Str. 34, D-8000 München 80, Germany*

E. Hagg,
> *Institut für Informatik, TU München, Orlean Str. 34, D-8000 München 80, Germany*

W. A. Halang,
> *Univ. Groningen, Dept. Comp. Science, Box 800, NL-9700 Groningen, The Netherlands*

B. Hoogeboom,
> *Univ. Groningen, Dept. Comp. Science, Box 800, NL-9700 Groningen, The Netherlands*

P. Hruschka,
> *GEI-Gesellschaft für Elektronische Informationsverarbeitung mbH, Pascalstr. 14, D-5100 Aachen, Germany*

A. Knoll,
> *TU Berlin, Institut für Technische Informatik, Franklinstr. 28-29, D-1000 Berlin 10, Germany*

K. P. Kratzer,
> *Fachhochschule Ulm, Prittwitzstr. 10, D-7900 Ulm, Germany*

D. Langer,
> *MBB Deutsche Aerospace, Postfach 801160, D-8000 München 80, Germany*

R. Lauber,
> *Institut fuer Regelungstechnik und Prozessautomatisierung, Pfaffenwaldring 47, D-7000 Stuttgart 80, Germany*

K. Mangold,
> *ATM Computer GmbH, Bücklestr. 1-5, D-7750 Konstanz, Germany*

S. Pferrer,

Fachochschule Ulm, Prittwitzstr. 10, D-7900 Ulm, Germany

J. Rauch,

MBB Deutsche Aerospace, Postfach 801160, D-8000 München 80, Germany

M. Rößler,

MBB Deutsche Aerospace, Postfach 801160, D-8000 München 80, Germany

H. Rzehak,

Universität der Bundeswehr München, Fakultät für Informatik, Werner-Heisenberg-Weg 39, D-8014 Neubiberg, Germany

M. Schiebe,

IAK Informatik, Universitaet Ulm, Einsteinallee 11, D-7900 Ulm, Germany

E. Schnieder,

Institut für Regelungs u. Automatisierungstechnik, Langer Kamp 8, D-3300 Braunschweig, Germany

G. Schrott,

Institut für Informatik, TU München, Orlean Str. 34, D-8000 München 80, Germany

J. Schweiger,

Institut für Informatik, TU München, Orlean Str. 34, D-8000 München 80, Germany

H.-J. Siegert,

Institut für Informatik, TU München, Orlean Str. 34, D-8000 München 80, Germany

A. D. Stoyenko

New Jersey Institute of Technology, Department of Computer and Information Science, Newark, NJ 07102, U.S.A.

H. Windauer,

Werum Datenverarbeitung, Erbstorfer Landstr. 14, D-2120 Lüneburg, Germany

FOREWORD

I am very pleased to write a foreword to this book, which is so timely and is authored by so many recognized experts in the evolving new discipline of "real-time computer science".

As a matter of fact, real-time aspects have been considered in the past only marginally - if at all - by the computer science community. This fact can easily be explained from the very definition of real-time systems as *those systems in which the correctness depends not only on the logical results of computations but also on the time at which the results are produced.* A mathematically oriented computer scientist is educated to deal with logics, algebra, etc.; thus, with time-independent computations. It is the engineer who finds himself involved in real-time problems. Therefore, it is not surprising that efforts to cope with real-time problems have been undertaken by application oriented institutions, such as software firms, control engineering departments of universities, etc. And, as well, it is not surprising that application-oriented thinking produced ad hoc results, but not a consistent and sound theory of real-time systems.

It is only recently that computer science institutions are becoming interested in real-time computing, realizing the significance of real-time systems and their increasing practical importance. Some indications of this interest are, for example, the foundation of a committee on real-time systems in the German computer science association (Gesellschaft für Informatik) two years ago, and the introduction of the journal *Real-Time Systems* by Kluwer Academic Publishers in 1989.

The two editors of this book, Dr. Schiebe and Dr. Pferrer, owe the merit of bringing together many well-known experts in the field of real-time systems, and, as well, to cover the most important areas in a consistent way. Starting with the historical background, the book presents theoretical foundations, continues with models and tools, with practical considerations and with examples for applications, and ends with future developments. It is thus a compendium of the actual state-of-the-art of real-time computer science and it certainly will help to teach real-time thinking to future computer scientists.

R. Lauber
Stuttgart, Germany

PREFACE

Now that real-time systems have been freed from the former squeeze between hardware limitations and cost-constraints, it becomes feasible to be concerned about a more systematic approach to the whole field of Real-Time Systems Engineering.

Why this book? Although quite a few books on the subject of Real-Time Systems Engineering already exist, they often reflect a very personal view of one author only. This book, in contrast, attempts to cast light on this rapidly developing discipline from diverse angles without harmonizing the different perspectives the authors may have. At the same time, it is organized systematically so that it can be used as a source book and a textbook at the same time. This book, however, is not a problem-solver or a cookbook. Instead, it makes suggestions of different kinds for novel approaches to Real-Time Systems Engineering, keeping in mind that the most dangerous phrase employed by computer people, as Grace Hopper (who retired as rear admiral of the US Navy in 1986 after working successfully more than 40 years on computers) once stated:

"But we've always done it that way".

(viz, writing programs in Assembler, avoiding programming tools, overlooking operating systems developments, denouncing safety as a personal matter...).
Hopefully, by presenting innovative concepts in this important field of real-time processing, this book will encourage the reader to reassess his own thinking and to even

"try it also another way".

M. Schiebe
University of Ulm

How this book is organized:

Part 1 presents an overview to the origins of real-time processing.

Part 2 contains material of predominantly theoretical character that will be new, if not somewhat alien, to most readers and thus may trigger some discussions.

Part 3 introduces useful modeling concepts and tools that may have escaped, e.g., the weathered engineer who has concentrated his energy on ploughing through major Assembler programs to be pressed into mini-memories of present and former generations of microcomputers.

Part 4 focuses on the more practical aspects of Real-Time Engineering and is meant to give an overview of the present state-of-the-art, both in hardware and software including related concepts in robotics.

Part 5 gives examples for novel and even controversial real-time applications that illustrate the present state-of-the-art.

Part 6 focuses on future developments, gives suggestions for future research activities and for an educational program.

The book is complemented by a Glossary that tries to define the terminology employed in Real-Time Systems Engineering.

Acknowledgements

This book was conceived during a Workshop on Real-Time Processing organized by the Editors at the University of Ulm, FRG, in 1989. This fact - together with the long-standing experiences in Europe with civil applications in real-time engineering - is responsible for the dominance of contributions by European engineers and computer scientists, a fact which may be new for the readers from overseas.

The Editors would like to thank all authors for their valuable contributions and in particular the consulting editors, Drs. W. A. Halang, A. Knoll, and O. Künzel for their work.

The lay-out and editing (performed on an ATARI ST with Calamus (c)) is always a painstaking effort and has thus required the fullest attention of Mrs. U. Richter to whom we are particularly grateful.

We thank Dr. R. Rüdel, Univ. of Ulm, FRG, for his encouragement in publishing this book, and Mr. R. Holland of Kluwer Academic Publishers for his invaluable help with all technical matters.

The Editors

REAL-TIME SYSTEMS
Engineering and Applications

1

THE ORIGINS OF REAL-TIME PROCESSING

M. Schiebe

University of Ulm

D-7900 ULM

Real-time systems engineering in the society

Real-time system engineering, a discipline between Engineering and Informatics, has certainly a great impact on everybody's life.

Interestingly, microprocessor-operated management units of, say, a microwave oven, a robot, and a heart pacemaker - which obviously are to different degrees important in our everyday life - may have more in common than one might assume: they may be operating on the same piece of hardware, may use some of the same interfaces and may have been developed on similar host computers. And, they have one more thing in common: they have to stay in synch with the outside world, which turns them into real-time systems.

Recently, micro-controllers have quietly started to revolutionize such jobs as controlling the motor management of even the simplest car, just as running whole factories or safety-sensitive parts of a nuclear power station. They are often "hidden" physically in a rapidly growing variety of unassuming pieces of machinery and do their job largely "transparently" to the user. Following current trends, the transparency goes so far as to completely mimic a piece of hardware long replaced by the microchip (e.g., an electronic clock exhibiting a pendulum or a computer display in an aircraft, imitating the presence of analog instruments).

However, if at fault, real-time systems may cause devastating damage, which is when they finally catch the public's eye. It is thus a surprise that as yet not enough has been done to develop the corresponding disciplines theoretically and practically world-wide in a consistent way (with few notable exceptions like [5]) and to educate students accordingly.

As one of the consequences, there is currently not widespread agreement on principles that should underlie a theory of real-time software engineering, not to speak of generally accepted reliability and safety standards. The field is further hampered by the lack of dedicated low-cost hardware because large-volume hardware has primarily been developed for purposes that are not time-bound in the strict sense of real-time processing.

Even a precise definition of the term *REAL-TIME* is still lacking as most authors of the related literature prefer their own definition of this term.

A look back into the history

In the face of these problems, it may be interesting to excavate historical facts that could explain some of the problems we still face and thus may help to overcome some of the controversies of which the theory of real-time engineering is by no means free.

As a reference model for the system design of the Universal Automatic Computer (UNIVAC) a decimal machine, based on the ideas for the EDVAC (Electronic Discrete Variable Computer), J. Presper Eckert and John Mauchly started to build a smaller and somewhat infamous computer, the BINAC (BINary Automatic Computer, shown together with its designer in Fig. 1). This project was by no means modest [1, 7]: the BINAC was impeded by the additional intention of being a feasability study for airborne computers, intended to navigate long-range missiles, at that time under development by Northrop Aircraft, an US manufacturer. It was as such planned to be the first real-time computer and the first US stored-program computer and the first computer to use Short Code, a forerunner of a high-level computer language (Interestingly, the first programmable computer was the Z3, built by Konrad Zuse 1941 in Germany who also developed the first high-level programming language "Plankalkül" [4,8] which had at least some impact on Algol). However, like many real-time projects to come, this one suffered from substantial budget overdrafts and was 15 month behind schedule when finished in 1949. The processor relying on 7000 electron tubes was most unreliable, too heavy and was left in a close-to-experimental state due to financial restraints. In spite of these drawbacks, BINAC also had something to offer [2]: contrary to most then exisiting computers, like the ENIAC and the above-mentioned UNIVAC that used decimal numbers, the BINAC employed the binary numbering system. It would hold 512 *31 bit words in a mercury delay line and, interestingly, had a dual processor architecture working in parallel to improve its reliability, a problem that has haunted the real-time world ever since. When it ran - on "sunny" days - the machine could perform 3500 additions and 1000 (integer) multiplications per second .

Figure 1: John V. Mauchly (left) and J. Presper Eckert and their BINAC

In order to direct a trainer-analyzer for military aircraft, in 1944 Jay W. Forrester (Fig. 2), then a young engineer at MIT, inaugurated the later famous WHIRLWIND project. This was the first functioning and reliable real-time computer. It started off as a master's thesis and ended in 1951, after a lot of toiling, with a true 16 bit real-time system with 2kB of fast magnetic-core memory (also invented by Forrester). Why did he choose a 16 bit word-length? By his collaborator, R. R. Everett's account

4

[4,8], the reason for this was: "...*our analysis of the programs we were interested showed us that 1000 words was tight and 2000 considerably better. That gave us 11 bits and we knew we needed at least 16 instructions; 32 sounded more reasonable and that gave us 5 bits more.*" Obviously, the popular 16 bit architecture did not enter the world in order to gain access to 16 address lines or to manipulate (short) integers.

Figure 2: Jay W. Forrester

Forrester's team that worked on this computer came up with many other firsts: the first computer display, the first character generator, the first vector generator, and the first light pen. However, they missed the invention of the interrupt technique, a pivotal idea that would allow one to deal with asynchronous events in an orderly way. Instead, untimely data were buffered in the WHIRLWIND and dealt with when there was time. Nevertheless, the WHIRLWIND showed its competence by computing interception courses for 48 (prop-driven) fighter planes and, no wonder, was as such the test-bed for SAGE (Semi-Automatic Ground Environment air-defense system). SAGE was another first, as it was the first computer network, meant to synchonize defense monitoring over the northern hemisphere. The SAGE could add two 16 bit words in two microseconds and multiply two integers in twenty [3]. The WHIRLWIND was taken out of service in 1959.

In 1960, at the beginning of the transistor age, the PDP (Programmable Data Processor) series of Digital Equipment Corporation (DEC) was introduced, the first minicomputers and the first computers designed for real-time jobs on a small scale. The PDP-1 would still cost $120.000, but the PDP-8, designed by Gordon Bell (Fig. 3), would already sell for under $10.000 and could be used in embedded systems. It was capable of handling interrupts and would revolutionize science, medicine but also manufacturing control. In 1970, Bell also designed the pdp-11 and thus introduced the bus concept allowing an easy extension of the computer. Together with the pdp-11 came RT-11 and later RSX11-M, two very successful Real-Time operating systems. Bell was also the father of DEC's VAX (Virtual Address Extension) and popularized, although not invented, the Ethernet as the backbone of local-area networks (the Ethernet was invented at Palo Alto Research Center (PARC) by Rank-Xerox engineers. - This is slightly digressing from the history of real-time engineering, as both, the VAX-VMS and the Ethernet, lack real-time capabilities).

At the beginning of the computer age, computers, of course, could not be miniaturized sufficiently and thus had only a few applications, mainly in mathematics and statistics. Therefore, the lack of a consistent theoretical backbone for something like a slowly emerging real-time technology did not matter much. This changed, however, in 1959 with the beginnings of Integrated Circuit Technology and in particular with the invention of the microprocessor by Intel in 1971. It was designed by Ted Hoff and Federico Faggin. Intel's 4004 would soon be followed by the popular 8080 which became for some time the market leader and thus the work-horse for a legacy of microcomputers, both in the PC-

Figure 3: Gordon Bell

market and in embedded systems. Ted Hoff can also be credited for the invention of the first signal processor in 1974 [8]. In the following years, when the microprocessors had finally entered the markets in volume, their reliability had encouraged engineers to incorporate them into increasingly more safety sensitive systems.

Ever since, a plethora of new microprocessors, organized in only very few families, has emerged. However, although designed for speedy execution, none of these were primarily designed for real-time applications. RISC-technology was not designed for real-time applications either, although it purported to be useful under such conditions.

This stormy development led to a mass market where a microcomputer, in the most trivial case, has now deteriorated to a subordinate part of a wedding card, playing away Mendelson's Overture, a deed that is of course accomplished in real-time.

The beginnings of real-time software

The WHIRLEWIND computer and their lot were of course difficult to program, when one considers that the first programmers always had to know their computer hardware inside out (a capability that some real-time programmers have preserved until now), to be able to get anything up and running. So something had to be done: A first attempt to simplify programming can be seen in the *Symbolic Language* of Alan Turing, a low-level language which ran on the Ferranti Mark I computer in Manchester, England, in 1951. The Symbolic Language can be regarded as the forerunner of the *Assembler* language which, of course, still requires that the programmer understands somewhat the working of the innards of his/her computer.

The schism between hardware and software came with the advent of high-level or application-oriented languages. After less successful predecessors (summarized in [4], like the Short Code or Grace Murray Hopper's first compiler, the A-0 written in 1952 for the UNIVAC), the conception of the first high level computer language, FORmula TRAnslation (FORTRAN 0 by John Backus of IBM), can be dated to 1956 and must be regarded as a milestone, at least for the IBM 704 on which it ran. This language allowed for the first time to regard the writing of programs as a discipline in its own right, distinct from the hardware on which the program would run. Unfortunately Backus and his team *"did not regard language design a difficult problem, merely a simple prelude to the real problem: designing a compiler which could produce efficient [binary] programs"* (quoted after [1]). This was feasable then because competing compilers were notoriously inefficient and would slow down the computer operations enormously. In spite of this, not many real-time programmers ever since have really trusted a compiler to be as efficient as a hand-coded Assembler program. Apart from other widely discussed conceptual defects that have ever since infested all of FORTRAN's versions and offsprings due to its original design deficiencies and that eventually led to the development of ALGOL and the ALGOL family of languages (which includes Pascal, Modula, Ada, and "C"), neither FORTRAN nor ALGOL were intended for real-time programming [4]. Unfortunately, the kernel of the first languages designed for real-time applications, Coral 66 and RTL/2 did not have real-time features either (cf. Chapter 4.1 for a more detailed discussion). Thus a development started that set standards that again turned out to be unfavourable for real-time programming.

From the early beginnings of computer design, the number of instructions a computer could execute or its clock rate were used to indicate the speed of a computer. This served the manufacturers as a marketing argument, somehow suggesting that *computers are so fast that they will easily master any given workload* (cf. Chapter 2.1 and 4.2). However, the concept of timeliness was not integrated into the design of computers with the result that it was left to trail and error when exactly a certain service can be had, even in most self-proclaimed real-time systems.

Although the BINAC, the first real-time computer, already had a dual-processor architecture and facilities to work in parallel, the impact of von Neumann's much cherished concepts, as written down in his "First Draft of Report on the EDVAC" in 1945, later dogmatized sequentialization both in the design of hardware and software, thereby reflecting more of von Neumann's pessimistic assessment of future progress beyond the technical state-of-the-art, than being an enlighted general concept for the computer sciences of the future [3,7]. Sequentialization was consequently realized in the EDVAC; but ironically, in von Neumann's "own" machine (the "JONNYVAC"), the arithmetic unit ran in parallel and thus pioneered what is now known as Instruction-Level Parallelism (ILP), a popular technique in modern microprocessor design. Nevertheless, the concept of sequentialization, both in hardware and software, became somewhat like a basic law for a long time to come, and this may have prevented early experiences in parallel programming and related techniques that might have been useful at present times, not the least in real-time computing.

Conclusions

After almost 50 years, Real-Time Systems Engineering still faces many challenges [6]: Major concepts that stood at the birth of the modern computer were unfavourable for the development of a general and consistent framework for real-time systems engineering. It is obviously difficult to match contrasting demands within one real-time computer language, such as timeliness vs. concurrency of resources or sequential vs. parallel program execution, or single vs. distributed processing, or fast response time vs. fault-tolerance. This situation requires major rethinking of prevailing dogmas and calls for efforts both in research and in education.

Unfortunately, over the years a "meddling-through" approach was often adopted, both with respect to the development of real-time hardware and real-time software, a late reflection of which are, for example, the many attempts to endow operating systems like UNIX with real-time properties. This should be given up in favor of a more systematic approach that does not sacrifice important long-term goals for futile, short-term successes that are, for example, sold under the lable of portability.

Do we face a crisis in resl-time systems engineering? In comparison to the well publicized software crisis, the present situation in real-time engineering is not so different. If it does not change, this will eventually block further progress and will prevent the establishment of quality and safety standards that are needed in a society that has to rely more and more on its real-time technology.

REFERENCES

[1] S. Augarten, *"Bit by Bit, An Illustrated History of Computers"* New York, Ticknor & Fields, 1984.

[2] H. M. Davies, "The Mathematical Machine", *Mathematics in the Modern World*, pp. 325-335, San Francisco, Freeman and Cie, 1963.

[3] R. R. Everett, "Whirlwind", in N. Metropolis, J. Howlett, G.-C. Rota, Eds., *"A History of Computing in the Twentieth Century"*, Academic Press, New York, 1980

[4] D. E. Knuth and L.T. Pardo, "The early developments of a programming language", in [4].

[5] R. Lauber, *"Prozessautomatisierung"*, Vol.1, 2nd Edition, Springer Verlag, Berlin-Heidelberg-New York, 1989.

[6] J. A. Stancovic, "Misconceptions about real-time computing: a serious problem for the next generation", *IEEE Computer*, Vol.21, no.10, pp 10-19, October 1988.

[7] N. B. Stern, *"From ENIAC to UNIVAC"*, Digital Press, Bedford, MA, USA, 1981.

[8] R. Slater, *"Portraits in Silicon"*, Cambridge, MA, USA, The MIT Press, 1989.

2

THE CONCEPT OF TIME IN THE SPECIFICATION OF REAL-TIME SYSTEMS

Boudewijn Hoogeboom and **Wolfgang A. Halang**
Rijksuniversiteit te Groningen
Vakgroep Informatica
NL-9700 AV Groningen/The Netherlands

INTRODUCTION

Everyone knows what time is, or at least how the word "time" is used in everyday language. Time is so much a part of our everyday experience that it has become a self-evident aspect of the world in which we live. One might expect that this familiarity with time would enhance the ability to relate it to the behavior of computing systems. In particular, the timing of input-output relations should pose no special problems. But a quick glance at the state of affairs in computer science tells us that, in sharp contrast to the alleged exactness of the discipline, there is little concern for descriptions of systems or programs that are exact with respect to time. Most of the attention devoted to time is directed toward speeding up the rate of data processing or developing time-efficient algorithms.

In 1966, the International Society for the Study of Time was founded in order to promote the study of time and to stimulate discussion and interdisciplinary meetings on the concept of time (Whitrow, [42]). While, contributions have been made from several disciplines, this activity has passed the computer science community without being noticed. In searching for publications in the area of computer science on the subject of time, one finds nothing of interest - with some minor exceptions (e.g., [1, 14, 30]). While time appears to be essential in the area of real-time systems, it has been largely neglected as an object of study.

As the development of real-time systems is changing from a craft to a discipline, the need for special tools and constructs to deal with timing constraints becomes apparent. So far, low-level mechanisms and ad hoc programming concepts have been used for this purpose, while some efforts have been made to extend existing languages and specification formalisms to include time. Of course, within real-time languages there are constructs that use time as a parameter, but an integrated approach to the expression and enforcement of timing constraints is still lacking.

For reasons mentioned above, we feel that reflection on the role of time in real-time systems is necessary. A clear understanding of the reasons why and of the manner in which time is involved in specification and implementation is a prerequisite for a sound methodology in this area. Our concern is with the informal part of development of real-time systems and with some basic assumptions underlying computer science. Suggestions will be made to bridge the gap between theoretical and practical approaches in design and specification of real-time systems.

This chapter is divided as follows: The following section introduces the concept of time as far as is relevant for our purpose. In section "Requirements for Real-Time Systems" real-time systems will be characterized by their general requirements. In connection with these requirements, the treatment of time in the field of computer science is investigated in the section on "The Primitiveness of Time Concept in Computer Science"; we will demonstrate that the role of time is neglected and even suppressed. In the next section ("A Different Approach: The Pragmatic Value of Common Sense Concepts of Time"), a different approach will be taken in order to develop a concept of time that is adequate for the specification and design of real-time systems. The implications for specification of real-time systems are outlined in the ensuing section, and in section "Time Slot Based Synchronization" a method is presented that demonstrates how time can be used for synchronization purposes. A summary of the main argument is given in the final section.

THE CONCEPT OF TIME

This section is intended as a general introduction to the concept of time. The word "concept", as used here, refers to a logical entity that is open to analysis. A concept has meaning by virtue of its relation with other concepts and by the way it is used in speech. Many concepts can readily be defined within some scientific framework, but the concept of time eludes easy definition; this seemingly innocent, self-evident concept is notoriously difficult to capture in a description.

The "Elimination" of Time in Philosophy and Physics

Although there has been much speculation about time in the field of philosophy, this has not resulted in a formulation comparable to the generally accepted framework describing spatial relations. Some outstanding examples demonstrate this:

- In his *Statics*, Archimedes analyzes the equilibrium of forces in machines, without reference to time.
- Euclidian geometry, which provides a consistent description of spatial relations, has for ages been accepted as a final statement. In contrast, no comparable formalization of time took place until the 18th century.
- In Newton's mechanics, time is a parameter in terms of which the relative displacement of material bodies may be described. But the dynamic equations that describe velocity and acceleration are equally valid if the direction of time is reversed, i.e., if t is replaced by $-t$. The one-directionality of time, the so-called "arrow of time", is thereby denied. Irreversible phenomena like heat propagation or biological processes are thus interpreted as macroscopic effects of reversible microscopic behavior.

A main characteristic of Newtonian dynamics is that, given the initial conditions of a closed dynamic system, the behavior of the system is fully determined by equations, both in the future and in the past. This amounts to an elimination of time, because the passage of time does not have effects that need to be incorporated for a complete account of dynamic behavior.

- With his theory of general relativity, Einstein presented a static, timeless view of the universe. His main thesis was that there is a limiting velocity for the propagation of all signals (c, the velocity of light in vacuo). This limiting velocity plays a fundamental role in his new concept of time. We can no longer define the absolute simultaneity of two distant events; simultaneity can be defined only in terms of a given reference frame. This meant a radical departure from the classical account of time. But in two respects, the theories of relativity remained well in line with the classical tradition:

- In the *Special Theory of Relativity*, the local time associated with each observer remained a reversible time.
- In the *General Theory of Relativity*, the static geometric character of time is often emphasized by the use of four-dimensional notations. Thus, events are represented (in the so-called Minkovski diagram) in space-time.

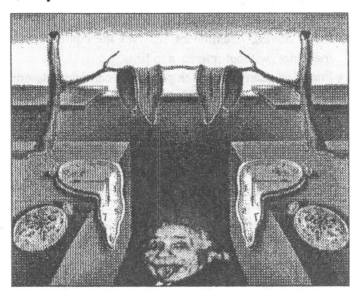

Hommage à Dali

Mathematical Formulation

The structure of time can be mathematically defined in terms of an ordered set. A few proposals to do so have been made, of which the most famous are by Russell [34] and Walker [39]. Their formulations have a different starting point, but their results are quite comparable (cf. [42], Ch. 4).

The basic properties of the temporal order are the subject of what may be called "mathematical metaphysics". The following principles have been proposed (cf. [1], p. 4):

First-order conditions:

- Transitivity: $\forall x, y, z : (x < y \land y < z) \rightarrow x < z$
- Irreflexivity: $\forall x : \neg (x < x)$

Together these imply asymmetry: $\forall x, y : x < y \rightarrow \neg (y < x)$

- Linearity (excludes branching time): $\forall x, y : x < y \lor x > y \lor x = y$

Existence principles:

- Succession: $\forall x : \exists y : x < y$
- Density: $\forall x, y : x < y \rightarrow \exists z : (x < z \land z < y)$

Higher-order postulates:

- Dedekind Continuity
- Homogeneity: Time is structural similar throughout
- Isotropy: There is no formally preferred direction of time

The Minkovski space M is an example of a temporal structure that is based on causal precedence between a point and its successor (cf. [1], p.5):

$T = (T, <)$ with $T = R \times R$ and $(x,y) < (u,v)$ if $x < u$ and $y < v$

This is a branching temporal structure, which still allows "confluence":

$\forall x,y : \exists z (x < z \land y < z)$ and $\forall x,y : \exists z (z < x \land z < y)$

R and Q have the same *first-order* theory, which differs from that of M or Z. R and Q have different *second-order* theories: the former is Dedekind continuous.

R, Q, and Z have the same *universal* first-order theory, which differs from that of M.

Russell and Walker ([42], Ch. 4) have shown that it is possible to construct a temporal continuum that is isomorphic with $\Re$ (reals) without the introduction of metric concepts. This means that there is still freedom in choosing a metric for the temporal continuum. As an example, Whitrow ([42], p.218) shows how a metric different from the standard metric can be defined. However, one further condition points to the standard metric as the most convenient, and is stated in the homogeneity principle: physical laws are formulated so as to be independent of the particular times of occurrence of the events to which they apply. Therefore, only the differences of the times of events and not the times themselves are regarded as significant.

The analysis presented by Whitrow is used by Joseph and Goswami (in [16]) to examine the relation between the actions of a computation and the nature of time associated with the execution. One of their main conclusions is that the real-number representation of time is needed when the relative starting times of different parallel components in a program are comparable to the execution times of commands, or when the execution times of commands have a continuous dispersion. Thus, the specification of time may need to assume that it takes values from the whole real domain. However, the limitations of measurement and the lack of effect of minor variations in time may make it possible for simpler domains to be used.

Technological Development

To conclude this introduction to the concept of time, we provide here a short account of the measurement of time as it has improved over the last two centuries in connection with technological development. In what follows, we make use of [2], [7], and [42].

Time services were founded for navigation purposes. In Britain, the correct time has been provided by the Royal Observatory in Greenwich since 1833. The introduction of telegraphy made possible the synchronization of clocks over long distances. To avoid the inconvenience of numerous differences in local times, an international system of 24 time zones was adopted in 1884. The mean solar time of the Greenwich meridian is denoted as Universal Time (UT). The time in any standard zone differs from UT by an integral number of hours; minutes and seconds remain the same.

Precise time and frequency have been broadcast by radio since the beginning of this century. Transmissions of time signals began as an aid to navigation (Boston in 1904, Germany in 1907: Radio Norddeich Deutschland); they are now widely used for many scientific and technical purposes. It soon became evident that time signals from different stations deviated from each other. Therefore, an international institute, Bureau International de l'Heure (BIH), was founded to coordinate the results of astronomical observatories all over the world so as to arrive at the exact time.

The fundamental unit of time interval in the metric system was, by usage but not by formal adoption, the second of mean solar time: 1/86,400 of the mean solar day. Because of unpredictable small variations in the earth's rate of rotation, astronomers decided in 1955 to introduce a more accurate unit of time based on the period of the earth's revolution around the sun. The second of ephemeris time (ET) was defined as 1/31,566,925.9747 of the tropical year for 1900, January 1 at 0 hours ET (tropical year is the period between two successive moments at which the sun crosses the celestial equator in the same direction). This definition was ratified by the General Conference of Weights and Measures in 1960. The initial values of UT (denoted as UT0) that were obtained at various observatories differ slightly because of polar motion. A correction was added for each observatory in order to convert UT0 into UT1. An empirical correction to take account of annual changes in the speed of rotation is then added to convert UT1 to UT2.

It was felt desirable to have a more fundamental unit than any that could be derived from astronomical observations. Such a unit is given by the radiation frequency of a particular atomic spectral line. Optical lines are not suitable because, although we can measure their wavelengths, we have no means of measuring their frequencies directly. The discovery of spectral lines in the new radio frequency band, however, led in 1957 to a new method of time measurement of remarkable accuracy (one part in 10^{10}): the Cesium atomic clock or frequency standard. This led in 1961 to a new standard of time, the international atomic time (TAI), that is consistent with the UT2-time of 1 January 1958, 0 hours. In this way, a fundamental time scale fully independent of and more accurate than any astronomical determination of time was obtained. As a result, in 1967 a new definition of the second was introduced in the International System of Units (SI); it was defined as the duration of 9,192,631,770 periods of the radiation corresponding to the transition between the two hyperfine levels of the ground state of the Cesium-133 atom. In recent years the accuracy of the Cesium atomic clock has been further improved to one part in 10^{13}, corresponding to a clock error of only one second in 300,000 years.

Coordinated Universal Time (UTC) is obtained from an atomic clock that is adjusted in epoch so as to remain close to UT2; the frequency of a clock on UTC is constant, that of atomic time. Beginning January 1, 1972, the practice was adopted of keeping UTC within about 0.5 second of UT2 by resetting a clock on UTC by exactly one second when necessary. Since 1972, some 15 seconds have been added in this way.

The United States Naval Observatory provides a service for synchronizing clocks around the world to about one microsecond through the use of portable atomic clocks. Artificial satellites of the Earth also have been used in experiments to synchronize widely separated clocks to about 0.1 microsecond.

In Germany, the institute responsible for broadcasting the exact time is the Physikalisch-Technische Bundesanstalt (PTB). It uses the DCF77-transmitter to provide for UTC as well as TAI signals. These transmissions are used not only for navigation purposes; they also provide a time standard for scientific institutes (astronomical and meteorological), public services (radio, television and railway), and a growing number of civil applications.

Conclusion

This brief introduction has shown that the concept of time is less well developed than the concept of space. Philosophical ideas concerning time are still inconclusive, while time admits of different mathematical structures. Improvements in the technology of measurement have resulted in time standards of very high accuracy.

REQUIREMENTS FOR REAL-TIME SYSTEMS

In the literature, the indication "real-time" is being used in connection with a number of different systems that are used in nuclear power stations, computer-controlled chemical plants, flight-control software for airplanes, and so forth. The adjective "real" is meant as a contrast to simulated time. Although some confusion exists as to where the dividing line between real-time systems and other computer systems should be drawn ([26],[40]), a concise, unambiguous definition of "real-time operation" is available in [9]. There it is defined as *"an operation of a computer system, in which the programs for the processing of incoming data are constantly operational so that the processing results are available within a given time interval. Depending on the application, the data may appear at random or at predetermined times."*

To some extent, the confusion mentioned above depends on the distinction between a problem-oriented and a technology-oriented description of a system. As an example, on-line interactive systems are often referred to as real-time systems. In this case, the problem formulation cannot be considered "real-time", but a typical implementation strategy is a system that has many characteristics in common with real-time systems. Without a clear dividing line between what belongs to a system and what belongs to the environment of a system, it is impossible to distinguish between the essentials of a problem and the technology with which it is solved.

If we look at the application areas for which real-time systems are required, real-time data processing and computer control of industrial processes, a common feature is the tight coupling to be maintained between computer-based processes and physical processes. This results in the following requirements for real-time systems:

1. timeliness
2. simultaneity
3. predictability
4. dependability

Timeliness

Apart from requirements for order in time, two requirements for the timing of system behavior can be distinguished that are specific to real-time systems:

Relative time: Some actions have to occur within a given interval of time relative to the occurrence of an event.

Absolute time: System behavior is specified for given points in time, with accuracy relative to a chosen granularity.

The occurrence of drift, which is permissible when only relative timing constraints are specified, cannot be allowed when absolute time is required.

Hard and soft real-time conditions: In connection with the timeliness requirement, an important consideration is the costs associated with a failure to meet specified timing constraints. A distinction can then be made between soft and hard real-time environments. In both cases, a due time or deadline can be associated with particular responses that the system has to perform, and hence with certain tasks. If the costs associated with completing a task after the deadline is a gradually increasing, continuous function of time, the deadline is called *soft*. A discontinuous, fast increase of this function characterizes a *hard* deadline. In the first case, if the due time is exceeded the result can still be used, but its value for the system as a whole is less (for instance, position information in a navigation system). In the second case, violation of a deadline may disrupt a control process or result in a catastrophe.

In the literature (see, for instance, [40]), the distinction between hard and soft real-time is often characterized in terms of typical reaction times; for the former, these were in the range of milliseconds, and for the latter in the range of seconds. However, this characterization is not essential for the distinction between hard and soft real-time conditions.

Yet another use of the terms hard and soft real-time can be found in [18]. In this article, a real-time system is defined as an interactional system that maintains an ongoing relationship with an asynchronous environment, i.e., an environment that progresses irrespectively of the real-time system, in an uncooperative manner. A real-time system is fully responsible for the proper synchronization of its operation with respect to its environment. In case a real-time system fails to synchronize properly, the result will be erroneous. If the system is able to restore proper synchronization transparently, i.e., without external assistance and without loss of data or control, the system is called *soft* real-time. If transparent recovery is not possible, the system is called *hard* real-time ([18],p 160).

Thus, no reference is made here to the costs of failure nor to the typical reaction times. The possibility of recovery after synchronization failure is the distinguishing criterion. This is consistent with the foregoing definition, since the distinction between transparent recovery and the inability to recover from synchronization failure can be described in terms of the shape of a cost function.

Simultaneity

The system may receive inputs from two or more independent sources, and must respond to each of them within given time limits. To satisfy this requirement, some form of parallel processing is necessary. A number of different implementations are possible, two major alternatives being parallel processing on more processors, and quasi-parallel processing (multiprogramming) on a single processor.

Predictability

The actual behavior of a system must be within limits that can be derived from the specification. This means that reactions undertaken by the system must be fully predictable with respect to the input from the environment. Even under worst-case conditions, the system has to obey strict timing requirements; thus, in case of overload, the system has to degrade its performance in a predictable manner. This means that, in general, nondeterministic choice is undesirable. But performance-enhancing features like virtual memories or caches also introduce some degree of unpredictability (cf. Chapter 4.4), so they are to be avoided.

Dependability

Dependability is a term that indicates the general requirement of trustworthiness. It can be broken down into several more specific requirements: correctness, robustness, and permanent readiness.

Correctness is the requirement that the functional behavior of a system satisfies the specification. This is a general requirement for all systems, but for real-time systems a strong relation exists between correctness and performance; producing the wrong result and producing the right result at the wrong time may be equally incorrect for a real-time program.

Robustness is the requirement that the system remains in a predictable state, even if the environment does not correspond to the specification (e.g., inputs are not within a given range), or when a part of the system (e.g., a peripheral device) fails. "Fail safe" behavior and "graceful degradation" of performance are desirable characteristics. This requirement is particularly important in the light of the aforementioned costs of failure.

To achieve robust operation, the failure of some external process to meet a specified behavior has to be detected as early as possible. An example of this is when an external process fails to provide a signal within a predefined amount of time; a "timeout" scheme is then used to invoke an error handler.

Another aspect of robustness is fault tolerance. In order to guarantee acceptable performance in the presence of software or hardware faults, some form of redundancy can be used. In effect, this amounts to multiple implementations of part of the system, with a comparison unit to check for consistency.

Readiness means that nonterminating processes provide for an ongoing interaction with the environment. Termination of these processes, e.g., as a result of failure, cannot be tolerated.

Conclusion

General requirements for real-time systems are highly involved with several aspects of time. Therefore, a specification language is necessary in which not only temporal order but also relative and absolute time relations can be expressed. Furthermore, for design and implementation of real-time systems, development methods are needed that incorporate these different notions of time.

THE PRIMITIVENESS OF TIME CONCEPTS IN COMPUTER SCIENCE

Current work on real-time systems is based on assumptions and praktices that are common to computer science as a whole. The concept of time in the field of computer science is rather primitive, as will be demonstrated for the following areas:

1. Computational models
2. Typing mechanism and program execution time
3. Formal methods for specification and verification
4. Synchronization concepts

Computational Models

The computational model that dominated computer science prior to the 1970's is based on the von Neumann architecture. In this model, the control mechanism is organized sequentially. Instructions are executed one by one on elementary data items. Temporal succession (the before/after relation) is the only relation that is available on the machine-language level, and it is sufficient to specify a program without explicitly referring to time. Thus, in the context of one-processor systems, the introduction of time as a parameter was not required.

Other models that have been applied in language design are mathematical functions and logical assertions. They have given rise to functional and logic programming, respectively. Programming languages in these categories are designed to abstract from procedural aspects of program execution, such as the order in which primitive statements are executed. Therefore, these higher-level programming languages have even fewer capabilities to exert control over time of execution than the lower-level, imperative languages.

This trend toward abstracting time away is also evident when parallelism is involved. In concurrent programming, the programmer has no direct control over the speed of the process execution. In particular, the user is not responsible for changing the state of a process from "ready" to "running", since this task is left to the underlying implementation. In a recent survey of work on parallel processing [22], no special attention is devoted to the concept of time. The distinction between sequential and nonsequential order, and also the speed of execution, are considered to be the main categories of the temporal aspect of concurrency.

Typing Mechanism and Program Execution Times

With respect to the support specification and programming languages provide to enforce correctness, Wupper and Vytopil ([41], p. 111) distinguish three generations of language constructs:

1. First generation constructs give no support at all; they merely allow the achievement, somehow, of any necessary computation.
2. Second-generation constructs allow or enforce automatic inclusion of run-time checks, synchronization, garbage-collection, stack administration, and so forth. They help to detect design errors and thereby to avoid error-prone programming. Second-generation constructs are always accompanied by powerful run-time systems, i.e., collections of precompiled algorithms for anticipated problems.
3. Third-generation constructs syntactically enforce consistency in order to make run-time checks unnecessary. They syntactically restrict the expressive power of a language in such a way that it is difficult to use error-prone structures but easy to use approved ones. Examples are static typing, module structure, and polymorphism. Construction errors are likely to be detected as syntax errors, rather than to give rise to inconsistent systems.

With respect to pure functionality, third generation constructs are predominant in several languages. But as far as timing and reliability requirements are concerned, there is no language support whatsoever to enforce consistency:

> "Timing and reliability are regarded as second-class citizens in the world of system development and verification."([41], p. 113)

The typing mechanism has been successfully applied to constrain data domains with respect to their structure. However, available data abstractions currently represent only the spatial aspect (memory space); temporal aspects of data have been neglected.

It is commonly thought that verification of temporal and of reliability aspects of systems is not possible without referring to concrete hardware, as opposed to functional correctness. However Wupper and Vytopil show that this idea is untenable, by developing a specification language for hard real-time systems that treats both functionality and timing in the same formal framework [41].

Formal Methods for Specification and Verification

The success of the predicate-transformation method in proving correctness of sequential programs has stimulated interest in formal methods to incorporate time into the specification of parallel processes. Existing formalisms that are applied in the specification of real-time systems may be grouped into four broad categories (see [32]):

1. Temporal logics
2. Process algebras
3. Automata methods
4. Assertional methods

Reports of the ongoing development of these methods can be found in Bakker et al. [1], Joseph [14], and Randell [30].

In general, these formal methods have no adequate capability to express time. Only various orderings of discrete events can be expressed and verified. The underlying assumption is that speed of execution and, hence, timing of programs is irrelevant to correctness. Thus, the category of time is left implicit in specification; the only aspect of time considered is sequential order. The absolute or even relative timing (in terms of durations) is not taken into consideration. The behavioral properties of programs that are analyzed with the help of these methods can be grouped in the following three categories ([22], Ch.4):

Safety properties or invariance properties: partial correctness, mutual exclusion, and absence of deadlock;

Liveness properties: total correctness and termination, accessibility, absence of livelock and starvation, and responsiveness;

Fairness properties: providing equitable or impartial treatment of all processes so that no process is neglected, that is, "any process which may be scheduled infinitely often will be executed infinitely often".

Thus, available specification methods are designed to analyze and verify qualitative timing properties. However, for real-time requirements, the purely qualitative specification and analysis are inadequate.

Recently attempts have been made to extend existing specification methods to allow expression of quantitative timing properties:

- Global Clock Temporal Logic and Quantized Temporal Logic (Pnueli and Harel [29]),
- Real-Time Temporal Logic (Ostroff [28]),
- Metric Temporal Logic (Koymans [21]),
- Timed Communicating Sequential Processes (Reed and Roscoe [33]),
- Real-Time Logic (Jahanian and Mok [17] and Mok [26]).

In general, these proposals have not shown much progress in reducing the complexity that is involved in the specification of parallel processes. Small examples have been used to demonstrate these methods, but no specification of a reasonably sized real-time system has been reported so far.

Synchronization Concepts

When there is a possibility of interaction between computational processes (tasks), it is necessary to synchronize their mutual behavior. To achieve this, the following synchronization concepts have been introduced in current high-level programming languages:

1. Busy waiting;
2. Semaphores;
3. Conditional critical regions;
4. Monitors;
5. Path expressions;
6. Rendezvous.

With respect to time, synchronization concepts are inferior as compared to synchronization practices used by people. This can be illustrated by comparing two representative examples, semaphore and rendezvous, with the synchronization practices of everyday life (cf. Chapter 3.2).

Semaphore: A process executing a P-operation may have to wait until some other process executes the V-operation belonging to the same semaphore. The waiting period is not bounded by some limit. Processes may be waiting for indefinite lenghts of time, depending on the set of active processes and the times they spend in the P-V sections. However, at a traffic light (which in Spanish shares its name with a semaphore), the waiting period is bounded by some maximum period. Thus, the solution used in coordinating traffic at an intersection limits the maximum waiting time to yield an appropriate regulation of the traffic flow. However, no such timing control is to be found in the corresponding synchronization construct in (real-time) high-level languages.

Rendezvous: A rendezvous in real life is an arrangement in which the persons involved, a predetermined time, and a location are essential. However, the rendezvous concept as implemented by the select-construct in Ada drops two requirements, in that an arbitrary partner may be chosen by nondeterministic selection, and the timing of the meeting is not predetermined. Thus, the rendezvous concept as currently implemented in Ada is really a misnomer, because essential characteristics, time in particular, are lost.

We conclude that constructs currently available for synchronization operations leave timing unspecified. Therefore, they are primitive in comparison with the methods people use to coordinate their mutual activities. While people use time to enhance the predictability of their interaction, the application of synchronization constructs introduces a factor of unpredictability into the system.

Suppression of Time

In the foregoing discussion it was demonstrated that the concept of time in computer science is in several respects inferior to everyday notions of time. In this sense, the concept of time in computer science is primitive. Moreover, this lack of a well-developed concept of time is maintained by an attitude of active avoidance regarding the introduction of any notion of time in requirements analysis or in specification formalisms. In this section, we want to look at the arguments and motives for doing so. In a pair of articles, Turski has most vehemently argued against the use of time within specifications of real-time systems ([37],[38]). From these articles, it is clear that he does not criticize any single aspect of the use of time in real-time systems. He argues against all unnecessary introduction of timing considerations, whether in problem specification, design analysis, verification, or programming. Since we believe that his position is not uncommon in the computer science community, we shall summarize his argument, in short statements, followed by our reply.

1. Time is an avoidable source of error when it is used as an independent variable in equations that approximate the behavior of natural variables (linearization, forecast time evolution).

   Reply: No adequate feedback can be given without some assumptions being made about the behavior of a system in the near future, because there is a time lag between measurement and remedial (or compensating) action. Turski's solution of using only "current measurements" cannot be realized. He seems to entertain the simplifying assumption, also underlying some real-time languages, that the execution of some primitive statements does not take time. Since some forecasting is inevitable, the parameter time enters in the specification of system behavior. Thus, assumptions about timing have to be verified, at least at some level of implementation.

2. "Time-out" unnecessarily introduces time considerations into program design.

   Reply: Time-out can be avoided in many cases, as Turski rightly points out. But in other cases, the appeal to efficiency simply does not make sense, because there is no sensible alternative to busy waiting. In such cases, the use of time-out is perfectly legitimate.

3. Temporal logics may have some useful contribution in specifying deterministic processes, but they are inadequate in dealing with indeterminism or cooperating distributed processes. Real-time cannot be incorporated into a formal system. Various temporal logics merely concern various orderings of discrete events. Thus, we cannot prove any timing properties of a program in the way we can prove its formally expressible properties ("We cannot prove that 'eventually' is before Friday noon"). By complexity analysis, we may estimate the running time of a program. But an estimate depending on experiment and/or trust is not a proof. So, formal methods do not capture real-time, and estimation methods do not prove timing properties.

   Reply: Temporal logics are limited in their applicability - by design. There are indeed problems of decidability (first order theories not being decidable) and complexity (the exponential time complexity of a semiautomatic proof system limits its usefulness for the specification of a real-time system).

4. Real-time properties are about programs implemented on a computer. As such, they are essentially experimental and depend on the properties of the machinery - certainly there is no degree of portability.

   Reply: Implementations of real-time are no longer machine dependent if use is made of time signals broadcast by official time service institutes (e.g., DCF77).

5. The scope of catastrophic errors of judgment in time-sensitive software is so large and fundamentally irreducible that any inclusion of time-related features should simply be avoided.

   Reply: Time-related concerns do not necessarily lead to an increased possibility of errors in software. On the contrary: without explicit use of time, there are often hidden assumptions about timing involved that may cause errors.

A DIFFERENT APPROACH: THE PRAGMATIC VALUE OF COMMON-SENSE CONCEPTS OF TIME

"Das Problem der Zeit erscheint häufig als ein Problem der Physiker und der Metaphysiker. Dadurch hat man beim Nachdenken über die Zeit den Grund unter den Füßen verloren."

N. Elias ([10], p. XV)

(Translation: "The problem of time often appears as a problem of physicists and metaphysicists. Therefore, in thinking about time, one has lost the ground under one's feet .")

In the foregoing, it was suggested that common-sense knowledge about time could be more fruitfully applied to problems encountered in the design of real-time systems. Therefore, we shall now investigate the assumptions and preconceptions that are inherent in our common-sense notions of time. In doing so, the historical development and social context of our concept of time will be considered. Some ideas of Norbert Elias (in [10]) have been helpful in this exploration.

Historical Factors in the Development of Concepts of Time

The idea that people have always experienced time in a manner similar to the currently prevailing conception is in contradiction to available information about past as well as current cultures. The correction of the Newtonian time concept by Einstein is a current example of the cultural character of concepts of time. Since Einstein, we have known that time is dependent on a frame of reference, i.e.,time presupposes an observer who attributes time values to the events he perceives. If we look beyond physics, we see that time necessarily has a social significance. Its units of duration as well as its absolute values reflect a cultural organization of relatively high complexity for which time has become indispensable.

From concrete to abstract concepts of time: In an early phase of civilization, the timing of actions and social events was dictated by natural conditions. We may assume that the earliest experience of time was highly involved with concrete events, e.g., the movements of heavenly bodies such as sun, moon, and stars. Where instruments to measure time were lacking, time could not be abstracted from the events or activities that served to demarcate it. Because of the absence of a standard unit of time, periods of time could not be compared beyond a certain duration. In early civilizations that lacked an integrated numbering system, it was equally impossible to give a clear account of the past. People would use stories about major events as milestones in their experience of the past. The concept of time in primitive cultures was thus embedded in the representation of concrete things and phenomena; there was not yet a general, distinct concept of time that synthesized all these implicit notions of time. This earlier concept may be called a *concrete* concept of time. A more *abstract* notion of time could only develop when civilized life freed itself from tight bonds with the natural environment.

From passive to active timing of activity: Like many other achievements, the skill of timing and of planning in time developed in connection with the increase of specific social demands. In the first phases of civilization, when hunting was the dominant means of existence, the need for active timing of events was at a minimum, and so were the means for it. People lived in small groups and did not need strict timing arrangements in their mutual relations. Basically, their needs and the natural conditions dictated the timing of their behavior. This may be called *passive* timing.

When agriculture developed, it was important to sow and to reap when natural conditions were optimal. This lead to more explicit and elaborate concepts of time, as witnessed by archaeological research of great agricultural civilizations.

In big, urbanized state-societies, especially in those in which specialization of social functions has progressed to a great extent and in which long chains of interdependence between society members exist, the practice of timing, as well as a well-developed sense of time has become indispensable. Therefore, each individual in these societies has to learn at an early age to regulate his behavior with respect to a social timetable.

This requires the discipline to postpone the expression of impulses, to endure a delay of gratification, and to hurry up in anticipation of being "too late". Together, these abilities are necessary in order to conform to the social requirement of being "on time". Here *active* timing is involved: active regulation and control of individual impulses is a precondition for social life.

Conclusion: The concept of time is more than just a physical dimension in which the course of events can be represented. "Time", as it is used in everyday life, is a concept on a high level of generalization and abstraction that presupposes a social reservoir of knowledge about regularity, time order, and time measuring. As such, it adds to the physical world a unifying framework that fulfills a vital function for the organization of society by providing the means for the following:

Orientation: By being sufficiently abstract, concepts of time provide a frame of reference to describe the historical aspect of the world, in the way that spatial frames of reference describe the geographical aspect of the world . Time is also the dimension in which future developments can be anticipated, planned, or predicted, thereby allowing an active orientation to the present.

Regulation: Each individual learns at an early age to adjust his behavior to timing requirements imposed by the social order. This time discipline is a precondition for his successful integration into the society in which he lives.

Coordination: The objective order given by clock time has grown into a social institution that is vital to the organization of society as a whole. It provides a powerful means to coordinate the behavior of many people.

IMPLICATIONS FOR SPECIFICATION OF REAL-TIME SYSTEMS

In the foregoing, we have argued that our concept of time is characterized by a functionality that is conducive to the organization of society. We shall examine the aforementioned instrumental values more closely in order to look for their possible application in the design and specification of real-time systems. In doing so, we make use of analogies that exist between a social organization and a computer system.

Time as a Framework for General Orientation

At present, real-time systems are characterized by a concept of time that we have here called concrete. The timing of actions is left implicit in the state-processing approach, and is determined by concrete events in the event-processing approach. In both cases, references to abstract time (of a clock) are absent, or at best indirect (via timer interrupt). With the growing complexity of real-time systems, the use of concrete time will be increasingly insufficient in satisfying timing requirements. Since the environment of the system proceeds according to its own laws, a system that interacts with this environment loses important information if its interaction history contains only the temporal order of events. Therefore, a real-time system should have access to a reliable source of time signals that is sufficiently accurate with respect to its purpose.

One simple solution to this problem in a distributed real-time system is the direct use of radio-transmitted time signals, made possible by providing each component with a radio receiver set. Hence, the construction of a common time reference of specified accuracy is technologically within reach, and can be used in specification of timing requirements. The current practice of using quartz clocks for local time measurements, with periodic resynchronization with an external time signal, has become obsolete. A clock-synchronization algorithm for distributed systems, as published in [19], is not necessary anymore, since radio-transmitted time signals (e.g., DCF77 or the satellite-based General Positioning System (GPS)) can be used directly by each component of the system. With GPS, the accuracy of synchronizing a computer's internal clock with TAI can be as high as 100 nsec.

If time is added as a parameter to the interface between a system and its environment, the information of a state transition or of an event can be supplemented by its time of occurrence. In this way, messages between different parts of a system can get a time stamp when sent and when received. The idea to assign a validity time to messages is used in the Real-Time Operating System of MARS [8]. Communication among tasks and components is realized by the exchange of state messages with a validity time. As soon as the validity time expires, the message is discarded by the operating system. This illustrates that the introduction of time enables a more active orientation toward the environment, thereby enhancing predictability.

Regulation

People regulate their behavior with respect to timing constraints deriving from their social interaction. If we draw an analogy between individuals as members of society and tasks (processes) as component parts of a real-time system, we should like to provide the system with a discipline of time, i.e., a flexible response strategy with respect to its timing requirements. In anticipation of a deadline at which some task must be fulfilled, it should be possible to choose from different program segments the one that maintains optimum performance. The decision as to which alternative is chosen should be local to the program and should depend on

1) the time remaining before the deadline expires, and
2) for each alternative, the expected amount of time needed to complete it.

Current implementation methods do not support the kind of regulation and time discipline required in this proposal. Process scheduling in real-time systems has, for reasons of simplicity and speed, almost invariably used one or more of three algorithms: fixed priority, First-in-First-out (FIFO), or round robin. With these algorithms, the choice as to which process is the next to be scheduled is independent both of remaining processing times and of deadlines. Jensen et al. [13] propose a time-driven scheduling model for real-time operating systems in which the effectiveness of most currently used process schedulers in real-time systems can be measured. With this model, they intend to produce a scheduler that will explicitly schedule real-time processes in such a way that their execution times maximize their collective value to the system. A value function takes the place of the conventional deadline. For each process, a function gives for each time the value to the system for completing that process. In addition, a facility for user-defined scheduling policies is proposed.

Coordination

In a real-time system with concurrent processes that have to synchronize among each other to make use of shared resources, a complex problem of coordination exists that in its full generality cannot be solved by scheduling theory (see [26]). Therefore, an appeal is made to "good software engineering methods" (in [17]) to simplify the task of design verification by restricting design complexity. When we see how problems of coordination in society are solved by making use of the generally available frame of reference that is provided by clock time, we have a collection of general solutions for synchronization problems based on time. It may be noted that some methods currently used in coordination of interdependent tasks already correspond with more primitive practices also encountered in everyday life, For example, queue (e.g., in a shop) and priority rule (in traffic, or emergency cases in hospital). However, in most "human resource allocation situations", time is used as a more sophisticated means to coordinate transactions, e.g., timetables to indicate when service is given, railway guides, reservation of places, and "Just In Time" policies between producer and consumer. These examples show how time is used to enhance predictability in the interaction between servers and customers. In addition, a time-based interaction enhances efficiency by reducing waiting times.

TIME SLOT BASED SYNCHRONIZATION

To apply the foregoing ideas to the specification and design of real-time systems, a synchronization method will be outlined that makes explicit use of time. This method illustrates how an active concept of time enables an active orientation toward the future.

Assume a set of tasks that share a number of resources. To avoid conflicts between tasks, a process that may conveniently be called the administrator takes care of a proper synchronization of access operations. The administrator maintains a list of time slots for each resource for which mutual access must be guaranteed. Time slots are intervals of time at which the pertaining resource is still free for any access. The characteristics of these intervals depend on the capabilities of the corresponding resource.

When a task wants to make use of shared resources, it has to deliver a request as soon as its program logic allows it to do so. The request specifies the future points in time at which access is desired. By comparing a new request with the pertaining list, the administrator can decide whether time slots can be assigned to the task. Synchronization conflicts arise when time slots corresponding to different requests overlap. These conflicts can be resolved by evaluating the relative urgencies of the tasks:

- If the requesting task has a lower urgency than the task already occupying the time slot, the request is turned down and the administrator proposes other suitable time slots.
- If the requesting task has a higher urgency, the time slot is withdrawn from the original occupant and assigned to the more urgent task. An exception signal is sent to the task originally holding the time slot.

An exception handler in each task selects and reserves another time slot among those offered by the administrator. In addition, the exeption handler can reevaluate the task's control algorithm by appropriately updating time parameters. Requests from tasks with high urgencies may cause requests from lower-priority tasks to be canceled if there are no suitable time slots left. However, this will be noticed before the actual synchronization conflict takes place, so that remedial action can be taken.

From the outline given here, it is obvious which operating system functions need to be implemented in order to support time-slot-based resource-access synchronization. The procedure outlined so far shows a number of advantages over the the conventional approach:

- Enhanced predictability. The procedure eliminates unpredictable waiting periods due to delay statements or synchronization operations.
- Greater problem orientation. The problem is stated in terms of time, and therefore reflects the user's way of thinking.

- Enhanced dependability. The procedure provides checking and early conflict resolution.
- Tools to establish alternatives and graceful system degradation.

SUMMARY

The concept of time and its role in specification and design of real-time systems has been investigated. After reviewing some major approaches toward time, we have shown that current specification methods are inadequate with respect to the requirements for real-time systems. A different approach is developed, motivated by the observation that in everyday life, time is used to achieve predictable synchronization of activities. Common-sense concepts of time provide a means for orientation, regulation, and coordination, and this practical significance can guide the informal part of the development of real-time systems. In particular, a resource-access synchronization method is outlined to demonstrate how time can be used to satisfy the predictability requirement of real-time systems.

REFERENCES

[1] J. W. de Bakker, W. P. de Roever, G. Rozenberg (Eds.), "Linear time, branching time and partial order in logics and models for concurrency", *School/Workshop*, Noordwijkerhout, The Netherlands, May 30-June 3, 1988, LNCS 354, Berlin-Heidelberg-New York: Springer-Verlag 1988.

[2] G. Becker, "Die Sekunde", *PTB-Mitteilungen*, vol. 85 (Jan. 1975) p.14-28, Physikalisch-Technische Bundesanstalt, Braunschweig, Germany.

[3] J. A. Bergstra, "Terminologie van algebraische specificaties", Deventer: Kluwer 1987.

[4] C. Bron, "Controlling Discrete Real-Time Systems with Parallel Processes", in [30].

[5] B. Cohen, W. T. Harwood, M. I. Jackson, "The specification of complex systems", Addison-Wesley Publishers Ltd. 1986.

[6] V. Cingel, N. Fristacký, "A temporal logic based model of event-driven nets", *Research report*, Department of Computer Science and Engineering, Faculty of Electrical Engineering, Slovak Technical University, Mlynska dolina, 812 19 Bratislava, Czechoslovakia, 1990.

[7] E. Dekker "De juiste tijd. Hoe klokken in de afgelopen eeuwen gelijk werden gezet", *Intermediair*, Vol. 26, No. 51, 1990, pp. 35-41, 1990.

[8] A. Damm, J. Reisiger, W. Schwabl and H. Kopetz, "The real-time operating system of MARS", *Operating Systems Review*, Vol. 23, pp. 142-157, 1989.

[9] DIN 44300: Informationsverarbeitung, October 1985.

[10] N. Elias, "Über die Zeit", Suhrkamp: Frankfurt am Main, 1984.

[11] V. H. Haase, "Real-time behavior of programs", *IEEE Transactions on Software Engineering*, Vol. SE-7, 1981.

[12] W. A. Halang, "Resource access synchronisation based on the parameter 'time': a suggestion for research" University of Groningen. Groningen, 1990.

[13] E. D. Jensen, C. D. Locke, H. Tokuda, "A time-driven scheduling model for real-time operating systems", *Proceedings of the 6th Real-Time System Symposium*, San Diego, pp. 112-122, 1985.

[14] M. Joseph (Ed.), "Formal Techniques in Real-Time and Fault-Tolerant Systems", *Lecture Notes in Computer Science*, Vol. 331, Berlin-Heidelberg-New York: Springer-Verlag 1988.

[15] M. Joseph and A. Goswami, "Formal description of realtime systems: a review", *Information and software technology*, Vol. 31 No. 2, 1989.

[16] M. Joseph and A. Goswami, "Time and computation" in [30].

[17] F. Jahanian and A. K. Mok, "Safety Analysis of Timing Properties in Real-Time Systems", *IEEE Transactions on Software Engineering*, Vol. SE-12, No. 9, 1986.

[18] R. Koymans, R. Kuiper, E. Zijlstra, "Paradigms for real-time systems", in [14], pp. 159-174.

[19] H. Kopetz and W. Ochsenreiter, "Clock Synchronization in Distributed Real-Time Systems", *IEEE Transactions on Computers*, Vol. C-36, No. 8, 1987.

[20] H. Kopetz, "Real-Time Computing - Basic Concepts", "Design of a Real-Time Computing System", and "Clock Synchronization", in [30].

[21] R. Koymans, "Specifying Real-Time Properties with Metric Temporal Logic", *Research Report*, Philips Research Laboratories, Eindhoven, Oct., 1989.

[22] E. V. Krishnamurthy, "Parallel processing: principles and practice". Addison-Wesley Publishers Ltd. 1989.

[23] L. Lamport, R. Shostak, M.Pease, "The Byzantine Generals Problem", *ACM Transactions on Programming Languages and Systems*, Vol. 4, No. 3, pp. 382-401, 1982.

[24] L. Lamport, "What good is temporal logic?", *Proc. IFIP,* Information Processing 83, R.E.A. Mason (Ed.), Amsterdam, The Netherlands: North-Holland, pp. 657-668, 1983.

[25] A. Moitra and M. Joseph, "Implementing Real-Time Systems by Transformation"in [30].

[26] A. K. Mok, "Real-Time Logic, Programming and Scheduling", in [30].

[27] W. Newton-Smith, "The structure of time", Routledge & Kegan Paul Ltd, London, 1980.

[28] J. S. Ostroff, "Mechanizing the verification of real-time discrete systems", *Microprocessing and Microprogramming 27,* pp. 649-656, 1989.

[29] A. Pnueli and E. Harel, "Applications of temporal logic to the specification of Real-Time systems", in [14], pp. 84-98.

[30] B. Randell, (Ed.) "Real-Time Systems", *Proceedings of the Joint Univ. of Newcastle Upon Tyne/International Computers Limited Seminar,* Newcastle, Sept., 1989.

[31] W. Reisig, "Towards a temporal logic for true concurrency, Part 1: Linear time propositional logic", *Arbeitspapiere der GMD (Gesellschaft für Mathematik und Datenverarbeitung mbH)* Nr. 277, Nov., 1987.

[32] W. P. de Roever, (Ed.) "Formal Methods and Tools for the Development of Distributed and Real-Time Systems", *Computing Science Notes,* Eindhoven University of Technology, 1990.

[33] G. M. Reed and A. W. Roscoe, "A timed model for communicating sequential processes", *Theoretical Computer Science,* 58, pp. 249-261, North Holland 1988.

[34] B. Russell,"On order in time", *Proceedings of the Cambridge Philosophia Society*, Vol. 32, pp. 216-228, 1936.

[35] A. C. Shaw, "Reasoning about time in higher-level language software", *IEEE Transactions on Software Engineering*, Vol. 15, No. 7, July, 1989.

[36] W. M. Turski and T. S. W. Maibaum, "The Specification of Computer Programs", Addison-Wesley 1987.

[37] W. M. Turski, "Time considered irrelevant for real-time systems", *BIT*, Vol. 28 , pp. 473-486, 1988.

[38] W. M. Turski, "Timing Considerations Will Damage Your Programs", "How To Cope With Many Processors In No Time At All" in [30].

[39] A. G. Walker, "Durées et instants", *Revue Scientifique*, Vol. 85 , pp. 131-134, 1947.

[40] P. T. Ward and S. J. Mellor, "Structured development for real-time systems", Vol. 1: Introduction and tools, Prentice-Hall 1985.

[41] H. Wupper, J. Vytopil, "A specification Language for Reliable Real-Time Systems", in [14], pp. 111-127.

[42] G. J. Whitrow, "The Natural Philosophy of Time", Oxford: Clarendon Press 1980.

[43] S. J. Young, "Real-time languages: design and development", Chichester, U.K.: Ellis Horwood Ltd. 1982.

3

LANGUAGE-INDEPENDENT SCHEDULABILITY ANALYSIS OF REAL-TIME PROGRAMS

Alexander D. Stoyenko
Department of Computer and
Information Science
New Jersey Institute of Technology
Newark, New Jersey 07102 U.S.A.

INTRODUCTION

The use of embedded real-time computing systems to control industrial, medical, scientific, consumer, environmental and other processes is rapidly growing. A failure of an embedded computing system to properly control its real-time process may lead to major losses, possibly including the loss of human life. A real-time application thus demands from its embedded computing system not only significant computation and control processing, but also, even more importantly, a guarantee of predictable, reliable and timely operation.

Due to the hazards its failure may cause, a real-time system must not be put to use without a thorough a priori assessment of its adherence to inherent, time-constrained, functional requirements of the process the system is intended to control. The focus of the assessment must be on predictability of system behaviour. This assessment, once completed, will serve as a basis for design and implementation evaluation, feedback, requirements correlation and, ultimately, official safety licensing of the system.

We believe that real-time computing systems with predictable behaviour can indeed be realized. The requirement of predictable system behaviour, given time-constrained, functional specifications of the environment, can be embodied into programming language, operating system, hardware and other components of real-time systems. In this Chapter it is demonstrated how the resulting real-time computing system is made subject to an a priori predictable behaviour assessment. We refer to this assessment as *schedulability analysis* - a term introduced by the author [23,49,48].

Currently, real-time systems programmers perform schedulability analysis in a manual, non-systematic, error-prone way. Attempts to systematize schedulability analysis have so far resulted in techniques which can adequately model and analyze only simple real-time systems. When applied to realistic real-time systems, these techniques yield overly pessimistic worst-case time bounds. This failure to produce tight bounds can largely be attributed to the lack of detail exploited. That is, practically all information about the language, its implementation, program organisation, and hardware configuration is not used or is reduced to a small number of parameters. Our schedulability analysis is based on a new technique referred to as *frame superimposition,* that utilizes knowledge of implementation- and hardware-dependent information, and provides tight worst-case time bounds and other schedulability information.

The rest of the chapter proceeds as follows. In the following section we state and defend the assumptions we make about predictable, verifiable real-time software and hardware. In the next section we present a complete set of new language-independent schedulability analysis techniques for real-time programs, that exist in an environment that satisfies the assumptions of the previous section. We explain how real-time programs are mapped to real-time program segment trees, how the trees are consequently mapped to an algebraically-expressed model, and how the model is then solved using frame superimposition. In next section a prototype implementation used to evaluate this work is described. The implementation consists of Real-Time Euclid [23], a language specifically designed with sufficient built-in schedulability analysis provisions, a compiler, a schedulability analyzer, and a two-layer run-time kernel, and a distributed microprocessor system. We explain how Real-Time Euclid programs map to language-independent real-time segment trees. In the following section we report on our evaluation of this work. Using the implementation described before, realistic Real-Time Euclid programs have been analyzed for guaranteed schedulability (with both our method and a state-of-the-art method described in the literature), and then run, with the actual guaranteed response times observed. Our schedulability analysis techniques have resulted in accurate predictions of guaranteed response times, while the other method has failed. The final section summarizes the main contributions reported here, reports briefly on some of our related, current work, and concludes with some goals for future research.

ASSUMPTIONS ABOUT PREDICTABLE, VERIFIABLE REAL-TIME SOFTWARE AND HARDWARE

A real-time application imposes critical timing constraints on the software and hardware that control it. The very nature of the application thus requires that the software and hardware adhere predictably to these constraints. The fundamental assumption we make is that only software and hardware that can be checked (that is, analyzed for schedulability) for adherence to the constraints *before the application is run* fit the requirements of predictable real-time applications.

Language Assumptions

We assume that a real-time language makes sufficient provisions for schedulability analysis. Thus, every program can be analyzed at compile time to determine whether or not it will guarantee to meet all time deadlines during execution.

We assume that the language allows concurrent, real-time processes. Each process is associated with a *frame* (a minimal period). The frame is usually dictated by the external environment, and corresponds to the minimal frequency of the occurrence of the physical task controlled by the process. The process can be activated periodically, by a signal (from another process or an external activity, this is denoted as an atEvent activation), or at a specific compile-time-specified time (this is denoted as an atTime activation). Once activated, a process must complete its task before the end of the current frame. Furthermore, once activated, a process cannot be reactivated until the end of the current frame. To enable atEvent activations, in between activations a process is assumed to be blocked attempting to open a special, busy activation bracket (see below). As expected, the activation signal closes that bracket.

We assume that the language has primitives for inter-process synchronization. At the implementation level, most if not all primitives of synchronization used in higher-level languages are supported essentially the same way. A kernel call blocks until a desired, shared resource is free, then claims the resource and returns. All consequent attempts to claim the same resource will block until the process that has the resource executes another kernel call that releases the resource. Some primitives of synchronization (critical sections, MUTEX BEGIN/END, LOCK/UNLOCK, semaphores and so on) map directly to this model.

Others, such as the rendezvous or monitors-waits-signals in the sense of [17] break down into groups of these claim-resource/release-resource brackets. We assume that the primitives used in the language are implemented using such possibly nested claim-resource/release-resource brackets, which we refer to simply as *brackets*. To the kernel calls that claim and release resources we refer to as *opening* and *closing* brackets.

We assume that the language has no constructs that can take arbitrarily long to execute. Thus, a number of familiar conventional constructs, such as, for instance, a general while-loop, are banned from the language. Indeed, an iterative computation that has a deadline must be expressible in terms of a constant-count loop. Otherwise, the computation may miss its deadline and thus fail. If, on the other hand, unbounded iteration is called for in the real-time application, then it must be the case that the iterative computation has no inherent deadline associated with it. Then, a proper way to implement the computation is to make every single iteration (or a bounded number of iterations) a process activation.

The only loops allowed are constant-count loops. We prefer constant-count loops over time-bounded loops, since it makes little semantic sense to time out somewhere in the middle of a loop. Recursion and dynamic variables are either disallowed or their use is restricted to allow compile-time-known bounds on their use (both time- and storage-wise). Wait- and device-condition variables time out if no signal is received during a specified time noLongerThan delay. Furthermore, all waits and signals, and device-access statements go through the kernel, where they can be modelled as open- and close-bracket operations.

System Software and Hardware Assumptions

The very nature of schedulability analysis requires predictable system software and hardware behavior. It must be known how much time each machine instruction takes to execute. The hardware must not introduce unpredictable delays into program execution. Hardware faults and recovery from such faults are not accounted for in this paper. Hierarchical memories can lead to unpredictable variations in process execution timing. Thus, caching, paging and swapping must either be disallowed or restricted. Cycle stealing slows down the processors in an unpredictable way. Consequently, DMA operations in a real-time computer system are a source of difficulty when determining process execution timing. Knowing the maximum flow rates from DMA device controllers allows calculation of a worst-case bound on process slow-down. However, it is difficult to use this bound in other than a very pessimistic way, unless the relative amount and relative positioning of DMA activity is known. Modeling such activity has not been incorporated into the schedulability analysis techniques described here. The actual prototype multiprocessor used to experimentally evaluate the schedulability analyzer is described later. There is no DMA cycle-stealing activity in this system. Neither paging nor swapping of processes is present. There are no memory caches in the system.

Much like their language counterparts, the system software and hardware assumptions certainly differ significantly from the traditional assumptions made in non-real-time areas of computing. However, these assumptions are not as unrealistic as they may seem. Not only can realistic systems closely fitting these assumptions be assembled from existing components, a number of sources indicate that entire such system software and hardware systems used in time-critical real-time applications can, should be and are being designed this way [14,15,6,24,21,39,54,55,38].

SCHEDULABILITY ANALYSIS DETAILS

A real-time program is analyzed for schedulability in two stages. The schedulability analyzer program consequently consists of two parts: a partially language-dependent front end and a language-independent back end. The front end is incorporated into the code emitter, and its task is to extract, on the basis of program structure and the code being generated (thus, the front-end may be as language-dependent as the code emitter) timing information and calling information from each compilation unit, and to build language-independent program trees. By "compilation unit", we mean anything that is separately compilable; that is, a subprogram, a process or a module. The front end of the analyzer does not estimate inter-process contention. However, it does compute the amount of time individual statements and subprogram and process bodies take to execute in the absence of calls and contention. These times, serving as lower bounds on response times, are reported back to the programmer.

The back end of the schedulability analyzer is actually a separate, language-independent program. Its task is to correlate all information gathered in recorded in program trees by the front end, and predict guaranteed response times for the entire real-time application. To achieve this task, this part of the analyzer maps the program trees onto an instance of a real-time model, and then computes the response time guarantees.

The particular real-time model we use satisfies the assumptions of the previous Section and consists of a commonly-found distributed configuration where each node has a processor, local memory and devices. Processors communicate via messages or a mailbox memory. The response time guarantees are represented as a set of constrained sums of time delays. The delays in each sum are worst-case bounds of various types of execution times and contention delays. Each sum represents a guaranteed response time of a process.

The results generated by the schedulability analyzer tell the programmer whether or not the timing constraints expressed in the real-time program are guaranteed to be met. If so, the programmer is finished with the timing constraint verification process. Otherwise, the results help the programmer to determine how to alter the program to ensure that the timing constraints are guaranteed.

Front End of the Schedulability Analyzer

Despite the major programming effort required to implement it, the front end of the schedulability analyzer is a relatively straightforward tool to understand. Hence, we only present a high-level description of this part of the schedulability analyzer.

Segment Trees: As statements are translated into assembly instructions in the code emitter, the front end of the schedulability analyzer builds a tree of basic blocks, which we call *segments*, for each subprogram or process.

The various types of segments are as follows.

1. A simple segment corresponds to a section of straight-line code. The segment contains the amount of time it takes to execute the section. As each assembly instruction is generated in the straight-line section, the front end of the schedulability analyzer adds its execution time to the segment time. Instruction execution time is computed, in a table-driven fashion, as a function of the opcode(s), the operands and addressing modes, following the procedure described in the appropriate hardware manual such as [41]. All times are recorded in processor clock cycles instead of in absolute time for some implementation of the architecture.

2. An internal-call segment corresponds to an internal subprogram call. The segment contains the name of the subprogram. Each such segment is eventually resolved by substituting the amount of time it takes for the called subprogram to execute.

3. An external-call segment is similar to an internal-call segment, except that the former corresponds to an external subprogram call.

4. A kernel-call segment is similar to an internal-call segment, except that the former corresponds to a kernel subprogram call. If the call attempts to open or close a bracket, the name of the bracket (which may correspond to a semaphore name, a critical section label, a monitor number, a condition variable name and so on), and a pointer to the bracket record are recorded in the segment record. Furthermore, if the call is an open-bracket, then its noLongerThan time is recorded.

5. A communications segment is similar to a kernel-call segment, except that the former corresponds to a call that triggers inter-processor communications.

6. A selector-segment corresponds to a selector (an if- or a case-) statement. The segment consists of n subtrees, where n is the number of clauses in the selector, and each subtree corresponds to a clause. A selector-segment is eventually resolved[1], only the subtree which takes the longest to execute is retained.

An iterator (a loop) does not by itself generate any new segments. If the body of the iterator is straight-line code, then the iterator just contributes to a simple segment an amount of time equal to the product

---

[1]By "resolution" we mean the purging of subtrees until the segment (sub)tree becomes a segment list.

of a single iteration time and the maximum number of iterations. If the body of the iterator contains a call-, or a selector-segment, then the iterator is unwound into a segment chain. That is, the body of the iterator is repeated the number of iterations times.

Logic analysis can certainly improve upon the schedulability analyzer performance. For example, such analysis can be used to detect and reduce if-statements with vacuous conditions, such as the one on Figure 1. However, logic analysis falls outside the scope of the work discussed here, and remains future open for future work. Thus, the schedulability analyzer does not do any analysis of the program logic.

```
x := 5
if x > 7 then
  < statements >
end if
```

Figure 1: An if-statement with a vacuous condition.

A reader familiar with execution graph software analysis will notice that the front end of the schedulability analyzer builds segment trees similar to execution graphs. However, while execution graph program analysis typically computes *expected*, or *averaged* response times [42], our analysis is *worst-case*. To demonstrate how segment trees are built, let us consider the example of Figures 2, 3 and 4.

```
H₁    x := y + z
H₂    y := 2 * y
H₃    B (x,y)
H₄    if y > x  then
H₅        z := z + 5
H₆    else
H₇        open-bracket  (b)
H₈    end if
H₉    z := x * y
```

Figure 2: A block of high-level language statements.

A_{01}	‹Add y and z and assign to x›
A_{02}	‹Shift y left one bit›
A_{03}	‹Push x,y on stack; Jump to B›
A_{04}	‹Restore stack›
A_{05}	‹Compare y and x›
A_{06}	‹If greater, jump to A_{09}›
A_{07}	‹Add 5 to z›
A_{08}	‹Jump to A_{11}›
A_{09}	‹Push b on stack; Jump to Kernel.OpenBracket›
A_{10}	‹Restore stack›
A_{11}	‹Multiply x and y and assign to z›

Figure 3: The corresponding block of assembly statements.

High-level statements H_1 and H_2 translate into assembly sections A_{01} and A_{02} respectively. H_3, a call to a procedure B, generates sections A_{03} and A_{04}. The procedure execution will take place right after A_{03}, a section comprising a push on stack of the variables x and y and a jump to B, and before A_{04}, a stack restore section. The if-test of H_4 results in a compare A_{05} and a conditional jump A_{06}. The assignment H_5 maps to A_{07}. The else H_6 becomes the unconditional jump A_{08}. The open-bracket H_7 (used for inter-process communication) generates A_{09}, an assembly section encompassing a push on stack of the bracket variable b and a jump to Kernel. OpenBracket, the kernel routine supporting the open-bracket primitive, and A_{10}, a stack restore section. The last assignment statement, H_9 generates section A_{11}.

The segment tree corresponding to the high-level code block is formed as follows. A_{01}, A_{02} and A_{03} form a contiguous section of straight-line code. Thus, they comprise a single simple segment S_1. The amount of time A_{01}, A_{02} and A_{03} take to execute is recorded in the segment. The call to B forms S_2, an internal-call segment.[2] A_{04} and A_{05} form a simple segment S_3. Now we must form two branches. If the conditional jump A_{06} fails, then the sections A_{07}, A_{08} and A_{11} will be executed. The amount of time it takes to execute A_{06} if the test fails is thus combined with the times of A_{07}, A_{08} and A_{11} to form a simple segment S_4. If the test succeeds, then the sections A_{09}, A_{10} and A_{11} will be executed. The right branch is therefore formed as a list of three segments: a simple segment

---

[2]or an external-call segment. It does not really matter for the purpose of the example.

S_5 encompassing A_{06} (if it succeeds) and A_{09}, a kernel-call segment S_6 corresponding to the open-bracket, and S_7, a simple segment comprising A_{10} and A_{11}. For simplicity, in this example we assumed that open-bracket is the native language construct used for inter-process communication, and we omitted a discussion of how the bracket variable name and other supplementary information are stored. We will address language-specific operations when we discuss how our language-independent schedulability analysis is done on Real-Time Euclid.

Condition, Brackets, Subprogram and Process Records: Apart from building segment trees, the front end of the schedulability analyzer also records information on the organization of brackets as well as calling information.

For every bracket it records the names of the objects associated with the bracket (such as semaphore or condition variable names), and also, which subprograms and processes do access the bracket and how (open or close). For every subprogram, it is recorded who calls it, whom it calls, and what brackets it opens or closes. Finally, for each process, its frame and activation information are recorded as well as whom the process calls, and what brackets it opens or closes.

Segment trees contain sufficient information to proceed with the rest of the schedulability analysis. However, the presence of the lists of bracket, subprogram and process records substantially speeds up the many searches and updates undertaken by the back end of the analyzer.

Front End Statistics: The front end of the schedulability analyzer provides the following statement-level timing information. The statistics for each statement involved consists of its line number coupled with the worst amount of time, in clock cycles, the statement may take to execute. The information is produced for the following statements.

1. Each language-level expression or non-iterative statement (such as assignments, asserts, binds, returns and exits).
2. Each selector (if- or case-) statement consisting of straight-line code, on the basis of which it is possible to determine the longest (timewise) clause.
3. Each iterator-statement consisting of a body of straight-line code.

The programmer can use this statement-level information to determine which statements and groups of statements run longer than he or she expects. The information can also be used at a later stage, when the rest of schedulability information, such as guaranteed response times, becomes available. Then the programmer can decide which process parts have to be re-written to increase their speed.

Back End of the Schedulability Analyzer

The back end of the schedulability analyzer derives the parameters of the schedulability analyzable real-time model, presented shortly, from the front end segment trees, and solves the resulting model.

Resolving Segment Trees: The back end of the schedulability analyzer starts off by resolving segment trees built by the front end. All segment trees and bracket, subprogram and process records are concatenated into one file, including the segment trees and records pertinent to predefined kernel and I/O subprograms. All non-kernel calls are recursively resolved by substituting the corresponding subprogram segment tree in place of each call-segment. As many selector-segments as possible are resolved. A selector-segment cannot be resolved when two or more of its subtrees result in a delay segment (see the next Section).

Converting Process Trees: After all non-kernel calls are resolved, only process[3] segment trees are left. These trees are converted to different segment trees, where each possible segment is one of the following.

1. An interruptible segment corresponds to an interruptible section of code, that is a section of code executed with interrupts on. The segment contains the amount of time it takes to execute the section.
2. A non-interruptible segment corresponds to a non-interruptible section of code, which we assume is found only in the language run-time kernel. The segment contains the amount of time it takes to execute the section.

---

[3]As opposed to both process and subprogram segment trees.

3. A specified-delay segment corresponds to a delay specified in a wait or a device-accessing statement in the program.

4. A queue-delay segment corresponds to a delay while waiting to open a bracket.

5. A communication-delay segment corresponds to a delay while waiting for inter-processor communication to take place.

6. A selector-segment corresponds to an unresolved selector-segment generated by the front end part of the schedulability analyzer. A selector-segment is considered to be unresolved when it has at least two clause subtrees such that it is impossible to determine (until the analysis described in the next Section takes place) which subtree will take longer to execute.

We now demonstrate how front-end segment trees are converted continuing with the example of Figures 2, 3 and 4. Figure 5 has the corresponding converted subtree, now a part of a process tree. The simple segment S_1 maps to an interruptible segment N_1. The call to B is resolved and N_2, the tree of B, is substituted. We now form two branches. Since one branch of the subtree involves a kernel call, both branches are kept. The segment S_3 is combined with the first interruptible segment of each branch. Thus, S_3 and S_4 together form N_3, an interruptible segment corresponding to the failed jump case, and S_3 also becomes a part of the first interruptible segment of the right branch. The way *Kernel.OpenBracket* operates is as follows: interrupts are turned off, queues are updated, interrupts are restored. The interruptible segment N_4 thus comprises S_3, S_5 and the interrupt disabling code of the signal. The queue-updating code of *Kernel.OpenBracket* forms a non-interruptible segment N_5. The return from *Kernel.OpenBracket* is combined with $S7$ to form an interruptible segment N_6. Each leaf of the tree of B is extended with the two branches we just formed.

Once the segment tree conversion is complete, the parameters of the model we are about to describe are ready to be derived.

53

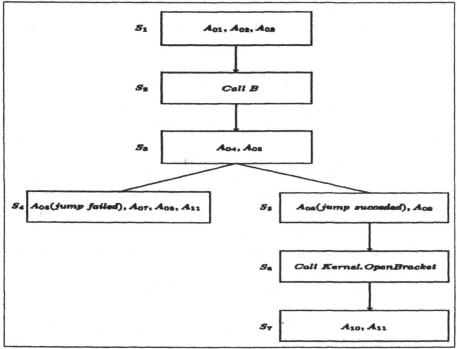

Figure 4: The corresponding tree of segments

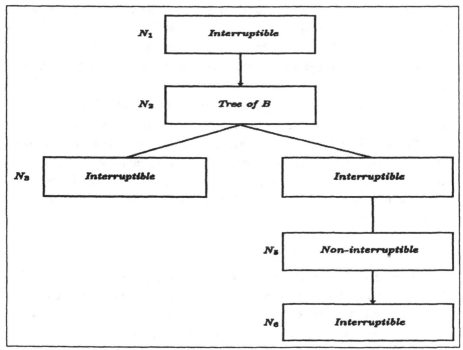

Figure 5: the corresponding converted subtree of segments

A Real-Time Model: The model we use assumes a distributed configuration where each node has a single processor, local memory and devices. Processors communicate a medium, such as a mailbox memory, which supports inter-processor messages.

Processes do not migrate across nodes. All intra-processor communication takes place via brackets. Inter-processor communication takes place in the following way (the mailbox protocol). To send a message to a process on another processor, a process deposits it, asynchronously and in a time-bounded way, into the medium of communication. To receive a message, a process retrieves one (or none, if none are available) asynchronously and in a time-bounded way from the medium. Only one process can access the medium of communications at a time. Thus, queueing delays are possible.

The mailbox protocol just described is not the only possible protocol for real-time communication. Any protocol will do, as long as no part of it takes an arbitrarily long amount of time. The mailbox protocol is used in the prototype evaluation system described later.

Each node has a copy of the language kernel. Each (hardware or software) resource is protected by a bracket. Ready processes are scheduled preemptive-earliest-deadline-first (that is, at every clock tick, the process with the earliest deadline is scheduled), while bracket and communication queueing is first-come-first-served.

The choice of the scheduling disciplines is summarized in the following argument. No process scheduling is optimal in the presence of shared resources [31]. Earliest-deadline-first ready process scheduling performs, in the sense of meeting deadlines, as well as other disciplines do [43],[47]. Earliest-deadline-first is the simplest deadline-driven scheduling discipline to implement. First-come-first-served is the simplest bracket and communication queue scheduling discipline to implement. Both earliest-deadline-first and first-come-first-served disallow indefinite queue waiting. Thus, they are good choices for our model.

A High-Level Model Description: For each process, we compute the longest amount of time it may take to execute. This time, referred to as the *guaranteed response time*, is computed as a sum of components. Each component corresponds to either the longest amount of time a particular class of process requests takes to execute, or to the longest amount of time the process is delayed due to a particular class of delays. If the process's frame is specified, then its guaranteed response time must not exceed the frame, and the process will not be activated more often than once per frame.

It is conceivable that the frequency of the occurrence of the physical tasks may fluctuate slightly with time. Moreover, fault-tolerance considerations may demand that the software be prepared to control some of the physical tasks more often than their normal physical characteristics dictate. Finally, the physical task occurrence frequencies may simply be unavailable early in the software development. Thus, it is useful to know what the minimum guaranteed achievable frame is for each process. To derive these frames, an initial set of frames is guessed, and the model is solved iteratively, substituting tighter and tighter frames until a guaranteed response time exceeds its corresponding current frame.

The following system of equations describes the high-level of our model, for each of the processors. The composition of each equation is somewhat similar to that in [25]. However, the terms QD_i, CD_i, and SD_i are new, and, as we are about to show, a radically different approach is used to derive these terms, and to solve the equations.

$$GT_i = I_i + N_i + QD_i + ID_i + ND_i + CD_i + SD_i \leq F_i \qquad (1)$$

for $i \in \{1,...,\#P\}$, where

$\#P$ is the number of processes on this processor,

I_i is the total CPU requirement for executing interruptible sections of process P_i,

N_i is the total CPU requirement for executing non-interruptible sections of P_i,

QD_i is the worst case amount of time P_i spends in bracket queues,

ID_i is the worst case amount of time P_i spends waiting due to other processes executing their interruptible parts,

ND_i is the worst case amount of time P_i spends waiting due to other processes executing their non-interruptible parts,

CD_i is the worst case amount of time P_i spends waiting due to interprocessor communication delays,

SD_i is the worst case amount of time P_i spends delayed in wait and device-accessing statements,

GT_i is the guaranteed response time of P_i,

F_i is the specified frame of P_i.

I_i and N_i are easily derived by adding up interruptible and non-interruptible segments of the process P_i respectively. To derive good upper bounds on QD_i, ID_i, ND_i, SD_i and CD_i, more work than simply adding up various process segment times is necessary.

Relationship to Earlier Models: In [48] we assess the state-of-the-art in real-time modeling. With the exception of one technique [27], all surveyed methods [31,43,47,25,26,30,32,35,36,44,53,59] deal with the difficult problems of resource queueing and various forms of contention by means of solutions which do not take into account segments' relative positions and inter-segment distances on the time line when estimating delays. These solutions are typically closed-form or polynomial, and thus work very quickly. However, they result in overly pessimistic worst-case queue delays, and ultimately in overly pessimistic guaranteed response times.

For example, a well-known solution [25,26] computes ND_i as:

$$ND_i = \sum_{\substack{j=1 \\ j \neq i}}^{\#P} N_j \left(\left\lceil \frac{F_i}{F_j} \right\rceil + 1 \right)$$

To put it simply, this method assumes that every non-interruptible segment of every process is delayed by every non-interruptible segment of every other process every time: an assumption that is never true.

Frame Superimposition: A technique defined in this paper as *frame superimposition* is used to derive all process delays. Since ours is a worst-case analysis, the purpose of the superimposition is to maximize the amount of various types of contention each process can possibly incur. Frame superimposition is a form of run-time simulation. It starts off by fixing a single P_i's frame start at time t_0. The frames of other processes are then positioned in various ways along the time line relative to time t_0. The algorithm shifts frames exhaustively, for every time unit, for every process, for every combination of frames possible. For every possible combination of frame positions, the amount of contention is recorded temporarily for each type of contention. The total contention, that is, the sum of all types of contention, is also recorded temporarily for each possible combination of frame positions. Throughout the course of the superimposition, the maximum amount of contention for each type of contention as well the maximum total contention are updated. These maximum amounts are substituted for the contention bounds at the end of the superimposition.

Frame superimposition uses relative positions of segments and distances between segments on the time line. Frame superimposition knows the correspondence between processor rates and clock cycles. Thus, it may "execute" only a part of a segment or more than one segment in a single time unit of time.

Frame superimposition can be refined in the following way to make use of the process activation information. The simulation starts when the first process is activated, and proceeds activating processes when necessary. To predict the time of an activation of a periodic process is trivial: the timing information is explicitly specified in the process definition. To predict the time of an atEvent process activation is more tricky. If the relevant activation bracket is closed via an interrupt, then all possible process activations, provided any consecutive two are at least a frame apart, must be tried. If the activation bracket is closed by another process, then the intervals defined by the earliest possible broadcast time and the latest possible broadcast time must be computed. All possible activation times within these intervals are then tried. Again, any two consecutive activations must be at least a frame apart. The simulation ends after

$2\ LCM_{i=1}^{N}\ (Fi) + \max_{i=1}^{N}\ (Fi) + \max_{i=1}^{N}\ (IAT_i)$ time unit steps (extension of [28],[29]), where

$$IAT_i = \begin{cases} atTime & \text{that is, the initial activation time of Pi if } atTime \text{ is defined} \\ 0 & \text{otherwise} \end{cases}$$

These many steps are taken to insure that all possible process mix combinations are tried.

Having described frame superimposition informally, we now present a more precise, formal description.

Let T_i be the segment tree of P_i for $1 \le i \le N$.

Let $L_{ik} = \langle S_{ik}^{(1)}, S_{ik}^{(2)}, ..., S_{ik}^{(n_{ik})} \rangle$ be a segment list of P_i, such that

L_{ik} is a subtree of T_i,

$S_{ik}^{(j)}$ is a segment of P_i, for $1 \le j \le n_{ik}$,

$S_{ik}^{(1)}$ is the root of T_i,

$S_{ik}^{(n_{ik})}$ is a leaf of T_i, and $1 \le k \le \Lambda(T_i)$, where $\Lambda(T_i)$ is the number of such unique lists L_{ik} within T_i.

Let $AL_i = \{L_{ik} \text{ such that } 1 \le k \le \Lambda(Ti)\}$ for $1 \le i \le N$.

The following events are undertaken once for each unique n-tuple $(L_{1k1}, L_{2k2}, ..., L_{nkn})$ in $\prod_{i=1}^{n} AL_i$.

Let $FSL = 2\ LCM_{i=1}^{N}\{F_i\} + max_{i=1}^{N}F_i + max_{i=1}^{N}\{IAT_i\}$ be the length , that is, the number of real-time units, of the time interval the frame superimposition will be run for.

Let $RATI_i = (REAT_i,\ RLAT_i)$ be the relative activation time interval of P_i, where $REAT_i$ and $RLAT_i$ are, respectively, the earliest and the latest relative activation times of P_i.

If P_i is activated periodically, then $REAT_i = 0$ and $RLAT_i = F_i$.

If P_i is activated by an external interrupt, then $REAT_i=0$ and $RLAT_i= FSL$.

If P_i is activated by an activation bracket close from another process, then we do not need to compute its $RATI_i$, for the reasons that will become clear shortly.

Having determined a $RATI_i$ for each periodic or externally-activated process, we simulate process executions for each possible combination of process activation sequences:

PAS_i, an activation sequence of P_i, is a sequence of activation times of $P_i(t_i^{(1)}, t_i^{(2)}, ..., t_i^{(k_{xi})})$, where $t_i^{(1)} = IAT_i$,

$FSL - F_i \le t_i^{(k_{xi})} \le FSL$, and

$F_i + REAT_i \le t_i^{(j+1)} - t_i^{(j)} \le RLAT_i$, for $1 \le j \le k_{xi} - 1$.

Therefore, a periodic process is activated exactly once per its frame, and an externally-activated process is activated no more often than once per its frame.

Having determined the only possible activation sequence for each periodic process and a possible activation sequence for each externally-activated process, we now have a totally deterministic schedule to simulate. This we do as follows:

- Set to *0* all global parameters (such as ND_i and I_i) to be maximized.
- Initialize bracket and active queues to empty lists, inactive queue to a list of all processes.
- For $T \leftarrow 0$ to *FSL* do

   - Activate every process *Pi* to be activated at time *T.*
   - Set to 0 all P_i's working parameters (such as shadows of *NDi* and *Ii*) to be maximized.
   - Deactivate every process *Pi* to be deactivated at time *T.* If a working parameter of *Pi* exceeds its global counterpart, set the counterpart to the working parameter (such as $GTi \leftarrow \max_{i=1}^N \{GT_i, localGT_i\}$).
   - Execute current segments of all active processes for one time unit or until they get blocked or terminate. Update local parameters, queues and lists accordingly.
   - If an activation bracket is closed, mark the activated process(es) for activation at the next time tick.

- end For

   In addition to maximizing parameters during each activation sequence simulation, the overall maxima observed across all possible activation sequence simulations are also recorded. The maxima are then reported to the programmer as overall guaranteed bounds.

   We now demonstrate how **frame superimposition** works by means of a simple example.

   Consider a system of three processes: P_1, P_2 and P_3. Their frames are equal, $F_1 = F_2 = F_3 = 3$. P_1 is periodic, P_2 is externally-activated and P_3 is activated by P_1. T_1, P_1's segment tree, consists of two segments: the first activates P_3 and takes negligible time to execute, the second is an interruptible segment that takes a single time unit to execute. The segment trees of P_2 and P_3 are identical, and each consists of a single interruptible segment of a single time unit size. $IAT_1 = 0$. The activation sequence of P_1 is $PAS_1 = (0,3,6)$, and an activation sequence of P_2 is one of the twenty-eight possible ones.

PAS2 $\in$ { (8),(7),(6),(0,8),(1,8),(2,8),(3,8),(4,8),(5,8),(0,7),(1,7),(2,7),
(3,7),(4,7),(0,6),(1,6),(2,6),(3,6),(0,3,8),(0,4,8),(0,5,8),(1,4,8),
(1,5,8),(2,5,8), (0,3,7),(0,4,7),(1,4,7),(0,3,6) }

The simulation computes GT_i and ID_i, $1 \le i \le 3$ for each of these twenty-eight possibilities.

For instance, if $PAS_2 = (8)$, then the simulation proceeds as follows:

T=0: P_1 is activated. P_1 closes the activation bracket of P_3. P_1 runs 1 time unit. P_1 terminates.

T=1: P_1 is deactivated. ID_1 and GT_1 are set to 0 and 1 respectively. P_3 is activated. P_3 runs 1 time unit. P_3 terminates.

T=2: P_3 is deactivated. ID_3 and GT_3 are set to 0 and 1 respectively.

T=3: P_1 is activated. P_1 closes the activation bracket of P_3. P_1 runs 1 time unit. P_1 terminates.

T=4: P_1 is deactivated. ID_1 and GT_1 are set to 0 and 1 respectively. P_3 is activated. P_3 runs 1 time unit. P_3 terminates.

T=5: P_3 is deactivated. ID_3 and GT_3 are set to 0 and 1 respectively.

T=6: P_1 is activated. P_1 closes the activation bracket of P_3. P_1 runs 1 time unit. P_1 terminates.

T=7: P_1 is deactivated. ID_1 and GT_1 are set to 0 and 1 respectively. P_3 is activated. P_3 runs 1 time unit. P_3 terminates.

T=8: P_3 is deactivated. ID_3 and GT_3 are set to 0 and 1 respectively. P_2 is activated. P_2 runs 1 time unit. P_2 terminates.

T=9: P_2 is deactivated. ID_2 and GT_2 are set to 0 and 1 respectively.

After this particular simulation, we thus get

$ID_1 = ID_2 = ID_3 = 0$ and

$GT_1 = GT_2 = GT_3 = 1$.

After all twenty-eight possible simulations are run, we get

$ID_1 = ID_2 = ID_3 = 2$ and

$GT_1 = GT_2 = GT_3 = 3$.

Thus, our simple system of processes is guaranteed to meet its deadlines under all circumstances.

The frame superimposition algorithm is clearly exponential. In fact, finding the optimal worst-case bound for resource contention in the presence of deadlines is NP-complete, given that even the most basic deadline scheduling problems are [13],[57]. However, so is compilation in general. Moreover, this part of the analysis operates on segment trees, and there are many instructions, even statements reduced to an aggregate segment. Finally, better delay bounds are derived by our algorithm, than by all but possibly one other algorithm.

This algorithm is the method of Leinbaugh and Yamini [27]. The work on this method took place in parallel with our work on frame superimposition. The [27] paper does not define their method algorithmically, and thus it is hard to tell exactly how it would compare to our method. However, because frame superimposition uses both position and timing segment information, while the Leinbaugh and Yamini method uses only position segment information, frame superimposition will not result in bounds more pessimistic than bounds derived by their method.

The timing complexity of the Leinbaugh and Yamini method is also uncertain. However, it does seem that to derive good worst-case contention bounds the reduction part of the technique has to consider all possible sets of mutually incompatible blockages, as well as the possible intersection subsets of these sets. Then, we believe that the Leinbaugh and Yamini [27] technique runs in exponential time, just as frame superimposition does.

To appreciate the power of frame-superimposition, consider the simple two process system of Figure 6. F_1 equals F_2 Moreover, the segment trees of P_1 and P_2 are equivalent. There are no resources in the system. Each process consists of two non-interruptible segments. All segments take the same time to execute, say x.

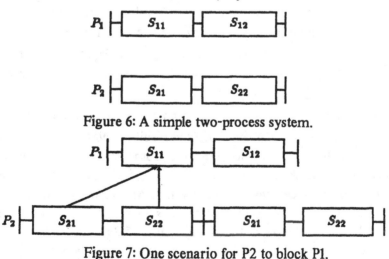

Figure 6: A simple two-process system.

Figure 7: One scenario for P2 to block P1.

Figure 7 demonstrates one possible way of positioning two frames of P_2 against a frame of P_1 so that ND_1 is maximized. Both segments of the first frame of P_2 block the first segment of P_1. A set of any three of the eight possible ways to block P_1 (see Figure 8) contains two incompatible blockages because (1) a segment of P_2 can block at most one segment of P_1, and (2) no segment of P_1 can be blocked by two segments from different activations of P_2. The latter kind of blockage cannot occur for the following reason. Since processes are scheduled earliest-deadline-first and $F_1 = F_2$, therefore once the first activation of P_2 completes, the new deadline of P_2 is passed the deadline of P_1, and consequently P_1 is given control. Thus actual contention for P_1 cannot exceed 2x. On the other hand, the contention can certainly be 2x (for example, Fig. 7 has one such scenario). Thus, in our simple example frame superimposition derives the best bound possible, namely $ND_1 = 2x$ (and similarly, $ND_2 = 2x$).

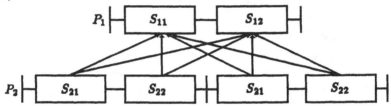

Figure 8: All possible ways P_2 can block P_1.

Note that if the aforementioned Leinbaugh and Yamini [25,26] solution were to be used in our example, we would have to assume that every non-interruptible segment of P_1 is delayed by every non-interruptible segment of P_2. We would have then gotten $ND_1 = ND_2 = 4x$, an overly pessimistic result.

Observe that if $F_1 = nF_2$, then frame superimposition can result in $ND_1 = nN_2$, and the Leinbaugh method results in $ND_1 = (n+1)N_2$. Thus, if n is large, then we seem to gain little from using the superimposition instead of a quick solution. However, practical real-time systems are typically partitioned into process families. All processes within a family have the same frames, and perform subtasks of a common larger (family) task. Moreover, the Leinbaugh [25,26] and other methods discussed in [48] do not take into account such important information as process starting times, atEvent process activations and bracket access order. All of these are taken into account by frame superimposition.

Thus frame superimposition is an effective technique for generating good worst-case delay bounds. We will see later that frame superimposition consistently outperforms the Leinbaugh [25,26] method.

Delays: An open-bracket operation starts a bracket segment. Each bracket segment itself is interruptible, but is necessarily in between a queue-delay segment (waiting to get into the bracket) and a non-interruptible segment (updating bracket queues and state variables upon leaving the bracket). QD_i's are computed by frame superimposition.

SD_i's are computed as follows. Around each close-bracket operations we build intervals defined by the earliest and the latest possible close-bracket times. The relevant SD_i's are then set to the minimum of the maximum amount of time before a close-bracket occurrence, and the delay specified in the   noLongerThan clause of the open-bracket operation.

ND_i's and CD_i's are also computed by frame superimposition. In fact, all delays are computed at once and *the sum* of QD_i, SD_i, ND_i and CD_i is maximized. We maximize the sum because the sum is a part of GT_i, and GT_i is the entity whose upper bound we are ultimately interested in.

Frame superimposition spans across all processors, because while QD_i, SD_i and ND_i are caused by *intra*-processor contention, CD_i is the result of *inter*-processor contention.

Interruptible Slowdowns: ID_i is the delay caused by other processes executing their interruptible segments while P_i is ready.

If P_i is in a bracket, then, it cannot be preempted by another process trying to use the same bracket. We thus compute ID_i as

$$ID_i = \sum_{k=0}^{\#B} ID_{ik} \qquad (2)$$

where ID_{ik} is the kth bracket's contribution to ID_i. Brackets 1,2,.., $\#B$ are real brackets, while bracket 0 corresponds to the non-bracket interruptible segments.

ID_{ik} can be computed in two ways: by frame superimposition, or as
$$ID_{ik} = s_{ik} \, WID_{ik} \qquad (3)$$
for $k \in \{0,...,\#B\}$, where WID_{ik} is the worst-case interruptible interference for the kth bracket, computed as

$$WID_{ik} = \begin{cases} k \neq 0 & \sum_{\substack{j=1 \\ j \neq i}}^{\#P} \sum_{\substack{t=0 \\ t \neq k}}^{\#B} I_{jt} \\ k = 0 & \sum_{\substack{j=1 \\ j \neq i}}^{\#P} \sum_{t=0}^{\#B} I_{jt} \end{cases} \qquad (4)$$

and s_{ik} is the slow-down rate for the kth bracket, which is further constrained by equations (5) and (6)

$$0 \le s_{ik} \le 1 \tag{5}$$

$$\sum_{i=1}^{\#P}(1 - s_{ik}) \le 1 \tag{6}$$

Overall Solution Algorithm: If we are only interested in verifying a given set of frames, then we only have to solve the set of systems of equations (1) through (6) once. Recall that the sum of QD_i, ID_i, ND_i, SD_i and CD_i is maximized, and not the individual parameters. It is possible for different combinations of values of individual parameters, across all processes, to add up to the same largest sum for each process. Naturally, the individual sums do not have to be maximized for all processes in the same pass of frame superimposition. In principle, it is possible to construct a set of processes for which there is an exponential (in the sum of the number of unresolved selector-segments) number of possible parameter combinations, though this was certainly not the case with the realistic programs we used to evaluate this methodology.

There can be different combinations of individual parameters resulting in the same largest sum. However, these largest sums are unique for their respective processes (frame superimposition is, after all, deterministic, and there can be but one maximum over any finite set of numbers). Therefore, since the other parameters (I_i and N_i) are computed exactly once and thus have unique values, a single set of guaranteed response times GT_i always results. A solution to (1) may or may not exist, depending on whether every derived GT_i is less than or equal to its F_i. However, if a solution does exist, then it is unique, in the sense that any other response time, not in the set of GT_i's, is either overly pessimistic (too large) or not guaranteed (too small).

If we want to determine the tightest set of frames we can guarantee, then we solve the following fixed-point iteration.

$$GT_i^{(x+1)} \le F_i^{(x)} \tag{7}$$

where

x is the iteration count, $F_i^{(0)}$, is given or guessed, $GT_i^{(x)}$ is obtained by solving (1) through (6), and

$$GT_i^{(x)} \le F_i^{(x+1)} < F_i^{(x)}. \tag{8}$$

Our set of systems of equations is underdetermined. To make it determined, we can make a number of constructive assumptions.

In some systems, processes naturally group into families. Each family is responsible for performing synchronized, related tasks. Often, system engineers would define frame ratios for each process family. As a result, we would then add the following equation to our system.

$$\frac{F_i^{<x>}}{F_j^{<x>}} = \frac{F_i^{<0>}}{F_j^{<0>}} \qquad (9)$$

for $i, j \in (1, ..., \#P)$, for all x.

Unless there is a specific reason to do otherwise and if frame superimposition is not used to compute ID_i's, we can assume that for each process, slow-down rates are the same for all brackets:

$$s_{i0} = s_{i1} = , ..., = s_{i \#B} \qquad (10)$$

for $i \in (1, ..., \#P)$. That is, a process has the same slow-down rate throughout its execution.

To derive $F_i^{(x+1)}$, any operation on $F_i^{(x)}$ and $GT_i^{(x)}$ satisfying (8) will do.

Once we determine (1) through (10), we iterate until we cannot reduce F_i's anymore, or alternatively, until we get no more reduction in the sum of F_i's. Should an $F_i^{(x+1)}$ be chosen to be less than $F_i^{(x)} - 1$, and the resulting $GT_i^{(x+1)}$ exceed $F_i^{(x+1)}$, then the iteration should not be stopped, but should be tried again with $F_i^{(x+1)} < F_i^{(x+2)} < F_i^{(x)}$.

Observe that there are no numerical approximations involved in these equations, since all parameters are expressed in terms of natural integers (integral time values, typically in clock ticks or instruction cycles). Furthermore, the iteration is clearly convergent, since integers are not dense, the frames get smaller and the guaranteed response times get larger with every iteration (with the exception for the case when we overshoot, and in that case we iterate no more than the largest (and finite) difference between the last two frames of the same process).

Every iteration solves (1) through (6) once. Thus, given a set of $F_i^{(x)}$'s, there is either no solution (the frames are too tight) or a unique (in the sense of the guaranteed response times) solution of (1) through (6), as already indicated. However, one can choose different combinations of values of $F_i^{(x)}$'s for every iteration where there is still room for the frames to get tighter. Thus, the tightest set of frames is not unique. Let $GT_i^{(x)}$ stand for the amount of time it takes P_i to execute in the absence of any other processes. Given a set of $F_i^{(0)}$'s, then there can be no more than

$$\prod_{i=1}^{\#P}(F_i^{<0>} - GT_i^{<0>}) \tag{11}$$

different sets of tightest frames. Naturally, this number may be quite large. However, given a set of tightest frames $F_i^{'x'}$, for some appropriate x, it is unlikely, if not impossible, that the number of different sets of tightest frames may exceed

$$\prod_{i=1}^{\#P}(F_i^{<*>} - GT_i^{<*>}) \tag{12}$$

Having solved for guaranteed response times either a single time or iteratively, we then report the F_i, GT_i and s_i for each process, as well as timing information for every subprogram, call, bracket access, even every statement. The programmer will then study the results, and decide whether the program should be run as it exists, or whether it should be modified.

If the purpose of the analysis was to verify a given set of frames, then, in the case of a positive solution of our set of systems of equations, the programmer need not do any changes to the program to achieve guaranteed schedulability. Similarly, if the purpose is to guarantee as tight a set of frames as possible, and the resulting set of frames and response times is satisfactory, no changes to the program are needed.

If, however, the guaranteed response times or frames obtained are too large, the supplementary timing information will be of help to determine where the processes can be streamlined.

IMPLEMENTATION

In this Section we report on the prototype implementation of a real-time language and its schedulability analyzer.

The Language

A descendant of the Pascal [58], Euclid [18], and Concurrent Euclid [8] line of languages, and the first designed, implemented (along with a schedulability analyzer) and evaluated schedulability analyzable language, Real-Time Euclid has been presented in detail elsewhere [23]. A short description suffices here.

Real-Time Euclid (originally known as Real-Time Turing [46]) has been designed with a sufficient set of provisions for schedulability analysis and it fits the model presented earlier in the paper. Thus, every Real-Time Euclid program can be analyzed at compile time to determine whether or not it will guarantee to meet all time deadlines during execution. The language has processes. Each process is associated with a frame, and can be activated periodically, by a signal, or at a specific compile-time specified time. Once activated, a process must complete its task before the end of the current frame. After activation, a process cannot be reactivated until the end of the current frame.

Real-Time Euclid has no constructs that can take arbitrarily long to execute. The only loops allowed are constant-count loops. Recursion and dynamic variables are disallowed. Wait- and device-condition variables time out if no signal is received during a specified time delay. Process synchronization is achieved through monitors, waits, outside-monitor waits and broadcasts. Monitors can call other monitors.

The language has structured, time-bounded exception handlers. The handlers are allowed in any procedure, function or process. Exception propagation is single-thread, in the opposite direction to a chain of calls. There are three classes of exceptions, based on their severity.

Real-Time Euclid is modular, procedural, and strongly-typed. It is structured and small enough to be remembered in its entirety, thus facilitating programming-in-the-small. Modularity and separate compilation make Real-Time Euclid a suitable language for programming-in-the-large.

The Compiler

We have designed and built a Real-Time Euclid compiler [23,47,48,22,33]. Due to our limited resources and time, we had to omit exception handling and floating-point arithmetic from the compiler.

Based on a production Concurrent Euclid compiler, the compiler has four passes: scanner/parser, semantic analyzer, allocator and coder. The assembly code generated is for the NS16000/32000 microprocessor family.

The control-data flow rules for each pass of the compiler are defined in S/SL (Syntax/ Semantic Language) [7]. The rule drivers are written in Sequential Euclid (the sequential part of Concurrent Euclid).

The compiler is about 30K lines of Sequential Euclid code, including S/SL rule table generation code (2/3 of all code), and is thus quite small. It generates both straightforward code with run-time checking and optimized code. The optimized assembly code is as compact and fast as the standard C compiler under a 4.2 Berkeley Unix System.[4]

The Kernel

To support Real-Time Euclid programs at run-time, we have designed and built a two-layer Real-Time Euclid kernel. The upper layer, referred to as *the Time Manager*, keeps track of time in real-time units. The Time Manager maintains the status of each process, next activation times, frames and other timing information. A Concurrent Euclid module of about 1K lines of code, the Time Manager, also handles timer interrupts.

The lower kernel layer, referred to as *the Basic Kernel*, is responsible for initializing the hardware, transferring timer interrupts to the Time Manager, handling device interrupts, maintaining various process queues, and basic earliest-deadline-first scheduling. It is written in assembly language and consists of about 4K lines of code.

The Schedulability Analyzer

A prototype schedulability analyzer, as described before (see "Schedulability Analysis Details"), has been implemented.

The front end of the schedulability analyzer is embedded in the coder pass of the compiler. While assembly code is generated, this part of the schedulability analyzer extracts timing information, and builds lists and segment trees. The front end is written in Sequential Euclid, like the rest of the coder, and is about 4K lines long.

The back end of the schedulability analyzer acts along the lines of the corresponding theoretical section (see "Back End of Schedulability Analyser"). All delays, including ID_{ij}'s, are computed using frame superimposition. It is assumed that process frames (F_i's) are given. The analyzer verifies the suitability of the frames by solving the equations of the latter section once[5] for guaranteed response times. This part of the schedulability analyzer is written in Turing [19], and is about 5K lines long.

[4] Unix is a trademark of AT&T Bell laboratories.

[5] That is, the fixed-point iteration is performed for a single iteration only.

To express Real-Time Euclid inter-process synchronization using brackets, we do the following: Every monitor and condition variable maps to a unique bracket. `enter-monitor` and `exit-monitor` map to `open-bracket` and `close-bracket`, respectively. `Broadcasts` generate a sequence (as many as there are waiting processes) of (cascaded) `close-brackets`. A `wait` closes the bracket of the corresponding monitor and opens the bracket of the condition variable. To ensure that no process gets into a monitor in between the `close` and the `open`, they are done in a single non-interruptible segment, and the interrupts are turned back on right after the open. Similarly, a signal closes the bracket of the condition variable and opens the bracket of the corresponding monitor, indivisibly, in a similar non-interruptible segment.

To ensure that all resource synchronization is done via brackets, all software and hardware resource access is done through monitors. That is, we insist the programmer build a monitor around every resource. To ensure that resource queueing is first-come-first-served, priority condition variables are not used.

An important problem in real-time modeling is that of incorporating overhead into the models. We include overhead as follows. The amount of time it takes to process a timer interrupt, increment the time count, and return from the interrupt is bounded. This timer-interrupt bound is then used to reduce the processor rate for each processor. Process activation code is bounded and a chain of segments (interruptible segment, non-interruptible segment, interruptible segment) is generated to correspond to this code. A copy of the chain is then prefixed to each process segment tree. Similarly, overhead chains corresponding to deactivation, monitor access, waits, signals, broadcasts and timeouts are also inserted appropriately into the trees.

The Hardware

The hardware configuration used in the evaluation of this work is shown on Figure 9. This distributed configuration consists of three computer nodes and an independent megabyte mailbox memory board, interconnected by a Multibus. Each node has a NS32016 processor, a megabyte of local memory, and a terminal. The processors can communicate via the mailbox memory board.

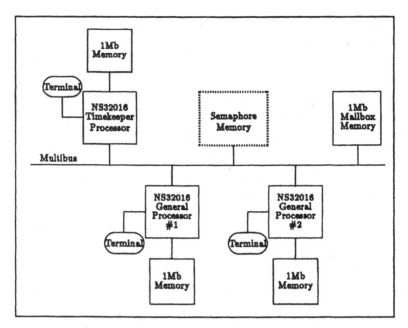

Figure 9: The system used in the evaluation

The Real-Time Euclid kernel supports the abstraction of mailbox communication. The mailbox is viewed as a Real-Time Euclid module, possibly containing monitors, but not processes or condition variables. The mailbox variables are mapped onto the independent mailbox memory board, while the code is replicated for each processor.

Mailbox memory access was designed to be serialized by the use of a semaphore memory component, shown by the dashed-line box in Figure 9. Hardware faults in this component prevented its use during the evaluation period. Therefore, serialization of mailbox memory accesses was achieved by the use of software techniques (Dekker's algorithm [10]) in conjunction with inhibiting interrupts on the processor attempting access. Because of the small number of processors, this solution was feasible, though inefficient. The software solution involved busy-waiting to support mutual exclusion in mailbox monitors. There was no way to suspend a process executing a *wait*. Therefore, *waits/ signals* in the mailbox monitors were not supported in the evaluation tests.

In distributed real-time systems, it is essential to keep sufficiently accurate track of time. That is, it is essential to prevent time values at different nodes from drifting too far apart. When interrupts are turned off it is possible to miss a hardware timer tick. Even if the hardware timer can always interrupt the running process, it may take different amounts of time on different nodes to serve the interrupt before resetting the timer. For these reasons, we keep the time variable in the mailbox, and limit access to it to two procedures: *IncrementCurrentTime* and *GetCurrentTime*. The two subprograms are kept outside of all mailbox monitors regardless of the application. Thus, bus contention (bounded by a constant) is the only contention in accessing the time variable.

One of the nodes is dedicated to just keeping the time, and its only process calls *IncrementCurrentTime* every real-time unit. This node never misses a tick, and never turns its interrupts off. The other two nodes are used as general purpose nodes. Their Time Managers call *GetCurrentTime* at every local timer interrupt to restore the time. Thus, a local time variable cannot differ from the correct time by more than the number of ticks that fit in the longest local non-interruptible code section at any given time, and is always restored to the correct current value at every timer interrupt occurring after a non-interruptible section.

EVALUATION

The evaluation of our schedulability analysis techniques has taken place as follows. Simulated software applications have been designed and implemented using the Real-Time Euclid implementation as described in the former section. The applications have been analyzed with the schedulability analyzer, and then run on the actual system. The guaranteed response times predicted by the analyzer have been contrasted with those actually observed, and all times as well as their corresponding deviation (actual from predicted) margins have been tabulated

Applications

Two advanced undergraduate students have designed and implemented two programming projects [56],[34]. A third project, undertaken by two other students, have aimed at a considerably more ambitious application, and unfortunately have failed to complete in time to be included in this evaluation. Each successful project involved producing a medium size (between 3K and 5K lines of code) distributed Real-Time Euclid program, targeted at the hardware described in last Section.

Each project has served as a beta-test of Real-Time Euclid and its schedulability analyzer. Thus, the students have been exposed to the language through conventional teaching techniques, and then the students have been asked to provide their own simulated process control applications and implement them in Real-Time Euclid. The students have been allowed to interact with each other as well as with those better familiar with the language for as long as only the functionality of the language had been discussed and not ways of using the language to construct their programs. Thus, the programs resulting in the projects have not been ``cooked" in any sense, to fit the desired results.

We now present these projects. The tabulated evaluation results are presented below.

A Simulated Power Station: For the purposes of this experiment, a simplified model of a power station is used [56]. In this model, the only factors needing monitoring are the boiler temperature, the boiler pressure, and turbine rotation speed. All these factors are related to the burn rate of the fuel heating the boiler and the current load on the generators. The control system for this power station is loosely based on the control system for the Drax coal fired power station in the United Kingdom [1]. In the Drax station, each boiler/turbine generating unit is supervised by seven ``data centers" which monitor the unit and report to a central CPU. The central CPU is responsible for tracking faults as reported by the data centers. Each data center consists of a PDP-11/23 while the central CPU consists of a PDP-11/44.

In the simulation, only five data centers are used. Each data center collects information from five sensors monitoring the boiler temperature, the steam pressure, and the turbine speed. Each data center is designed to cope with sensor failures and the central CPU is designed to cope with data center failures. A data center is responsible for validating the sensor input by comparing the readings with readings from other sensors. The central CPU validates recommendations of the data centers by comparing the recommendations from the data center with one another. If a sensor fails, it is deactivated. If a data center fails, it is deactivated. If enough sensors or data centers fail, the central CPU shuts the boiler/turbine and all data centers down.

The simulation software consists of the *DataCenter* module, the *CentralCPU* module, the *Failure* module, the *Simulator* module, the *MailBox* module, and the *Console* module. The *DataCenter* module simulates the five data centers. The *CentralCPU* module simulates the central CPU. The *Failure* module simulates sensor and data center failures by deactivating communication units in the *MailBox* module. The *Simulator* module simulates the boiler and turbine and sets the values for the sensors to be read by the *DataCenter* module. The *MailBox* module provides communication units that serve two purposes. First, the communication units allow the various modules to communicate with each other regardless of the physical CPU the modules are located on. Second, the existence of the units enables the *Failure* Module to deactivate the units to simulate sensor or data center failures. The *Console* module provides a means for the *DataCenter* and *CentralCPU* modules to communicate with the human operator by printing output on a console.

Console Module

The *Console* module provides a means for the power station control program to communicate with the human operators of the power station. It exports two routines: *DisplayString* which displays a string using the *PutString* procedure, and *DisplayInt* which displays an integer using the *PutInt* procedure. In a real system, this module would be replaced by a module whose routines would print output on a special printer or monitor.

Mailbox Module

The *MailBox* module provides facilities that allow the other modules to communicate with each other via communication units. A communication unit is a 32-bit memory location accessible from any physical CPU. The value of a communication unit can be set with the procedure *WriteComm* or read with the procedure *ReadComm*. In addition, communication units can be deactivated by the procedures *KillComm* or *LooneyComm* and reactivated with the routine *RepairComm*. *KillComm* deactivates a communication unit by setting a flag which instructs *ReadComm* not to return a value for that communication unit. *LooneyComm* deactivates a communication unit by instructing *ReadComm* to return a random number rather than the value of the communications unit. *RepairComm* repairs a communications unit by clearing the flags set by *KillComm* and *LooneyComm*.

Mutually exclusive access to the communication units is assured by placing all communication units in a single monitor. This solution is not optimal since separate communication units have no relation to one another and thus there is no reason why reading or writing to a communications unit should prevent another process from reading or writing to a different communication unit. However, due to the large number of communication units (twenty five), and the fact that all accesses to the communication units consist of only a few machine instructions, the advantages of allowing higher concurrency by placing each communication unit in its own monitor are far outweighed by the awkwardness of managing a large number of monitors that are carbon copies of each other. Also, the solution to the readers and writers problem cannot be used since hardware faults (see last Section) prevent the use of conditions in the memory shared by the CPU's on the NS32000 multiprocessor computer used by the current implementation of the Real-Time Euclid system.

DataCenter Module

The *DataCenter* module contains five periodic processes, and one procedure. Each process simulates a data center by calling the procedure *DataSim*. The *DataSim* procedure first polls all sensors for the boiler temperature, the steam pressure and the turbine speed. If a sensor read fails to return a value, a message is printed on the console to inform the operator of the sensor failure. Once all the sensors have been read, the values for each type of sensor are averaged and all the sensor results that are outside a tolerance range from the average are discarded. If the readings for a sensor are discarded, a message is printed on the console informing the operator that a sensor is returning an inconsistent value and the error count for the sensor is incremented.

If the error count for a sensor exceeds a certain value (five), then the sensor is deactivated until repaired. If fewer than three sensors are still active and are returning consistent values, the data center informs the central CPU that it wishes to shut down the power station. Once all the sensors have been read and validated, the data center computes a recommended fuel burn rate for the boiler and informs the central CPU of this recommendation.

Central CPU Module

The *CentralCPU* Module contains one periodic process which simulates the central CPU. The process polls each data center to obtain a shutdown or a fuel burn rate recommendation. If the majority of active data centers recommend a system shutdown, all data centers are deactivated, a message is printed on the console, and the power station is shut down. If a data center recommends a system shutdown, but the majority of the data centers do not recommend a system shutdown, the error count for the data center recommending system shutdown is incremented. The burn rate recommendation for all active data centers is averaged and any recommendations outside a tolerance zone of the average are discarded. If the recommendation for a data center is discarded, the error count for that data center is incremented. When the error count for any data center reaches a critical level (five), that data center is deactivated until repaired. If fewer than three data centers are active and returning consistent values, the central CPU deactivates all data centers, prints a message on the console, and shuts the power station down.

Failure Module

The *Failure* module contains a single periodic process. The process deactivates communication units with *LooneyComm* and repairs communication units with *RepairComm* at random intervals. The *Failure* module schedules the deactivations so that there is one failure every 60 real-time units and schedules repairs so that a deactivated communication unit is repaired after 180 real time units.

Simulator Module

The *Simulator* Module contains a periodic process which reads the fuel burn rate from the *CentralCPU* and sets all the sensors to the appropriate values. Since the status of a communications unit is handled when the communications unit is read rather than when the communications unit is written to, the *Simulator* module need not be concerned with the current status of the sensors as it sets them.

Suitability of Real-Time Euclid

Real-Time Euclid provides many features that simplify the implementation of the power station control system. Identical periodic frames are used for the data center processes, the central cpu process, and the simulator process, since they are generating and manipulating the same data. The *Failure* module process has a longer periodic frame to reduce the frequency of system failures and repairs. Time-bounded waits are used to insure that read access on sensors and the data from the data centers is obtained within a deadline or emergency measures are taken if the deadline is not met.

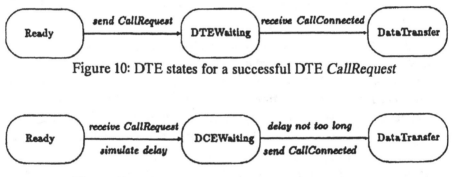

Figure 10: DTE states for a successful DTE *CallRequest*

Figure 11: DCE states for a successful DTE CallRequest

A Simulated Packet-Level Handshaking in X.25: The purpose of this project is to reproduce the sequence of packet type exchanges that occur between DTE and DCE during the call set-up and disconnect phases of the X.25 protocol [34]. The brand of X.25 modeled is DATAPAC's Standard Network Access Protocol (SNAP) [37],[9]. The Real-Time Euclid programming language is employed successfully to implement the simulation. The multiple-microprocessor run-time environment allows dedicated CPUs to be assigned to both DTE and DCE simulation modules. The Mailbox module construct of Real-Time Euclid facilitates easy inter-processor communication for simulating the inbound and outbound packet channels between the DTE and DCE.

In general, hosts (DTEs) connected to a packet network can activate many virtual circuits and carry on many conversations at once, each delineated by its logical channel number. In this simulation, it is assumed that the DTE only sets up and maintains one connection at any time. Hence only the packet types of the handshaking between DTE and DCE are simulated. Other aspects, such as logical channels, virtual addresses, and packet formats, are not incorporated in the simulation.

When the DCE receives a request packet from the DTE, it must send the packet out onto the sub-net to the remote DCE and receive a positive response before acknowledging the DCE request. In this simulation, if the delay is too long the DCE assumes a network error has occurred and sends a *NetworkError* packet (not specifically contained in SNAP) to the DTE. Both DTE and DCE assume the transaction has been canceled and change their states appropriately.

Figures 10, 11, 12, and 13 illustrate the DTE/DCE state transitions. The states inside the ovals correspond to the states of either the DTE or DCE or both. The edges represent actions taken. Referring to Figures 10 and 11, assume that both DTE and DCE are in the *Ready* state. To make a connection with a remote host, the DTE sends a *CallRequest* packet to the DCE and goes into the *DTEWaiting* state. Upon receiving the *CallRequest* packet from the DTE, the DCE goes into the *DCEwaiting* state and simulates a remote DCE connection delay. Assuming the delay is not too long, the DCE goes into the *DataTransfer* state and sends a *CallConnected* packet to the DTE. Upon receiving the *CallConnected* packet, the DTE also goes into the *DataTransfer* state.

As previously described, the DCE may return a *NetworkError* packet in response to a request by the DTE. Figures 12 and 13 illustrate the state diagram for a failing DTE *CallRequest*.

Figure 12: DTE states for a failed DTE *CallRequest*

Figure 13: DCE states for a failed DTE *CallRequest*

[9] specifies actions to be taken by the DCE when unexpected packets are received while in various states. The same is not provided for the DTE. For the purposes of the simulation it is assumed that the DTE acts in a manner symmetrically equivalent (in relation to the state diagrams) to the way the DCE acts.

The implementation is comprised of three modules: *DTE*, *DCE*, and *Mailbox*. The simulation of a packet transaction consists of passing integers corresponding to specific packet types between the *DTE* and *DCE* modules via the *Mailbox* module. The *DTE* and *DCE* modules keep track of their own state and co-ordinate the initiation of packet transactions based on requests entered at a terminal connected to their respective processor.

DTE Module

This module contains three processes. The *PollKybd* and *DTEreceiver* processes poll for input to the module. The *DTEsender* process is awakened to send output to the *DCE*. The *DTE* module also contains the Finite State Machine (*FSM*) monitor. This monitor holds the *DTE* present and previous (to the latest transaction) state and decides what is to be done when inputs are received. The previous state is kept so that state transitions can be printed as part of the trace of a transaction. Once the *DTEsender* process has awakened and entered the monitor, if a packet is to be sent, the packet type is determined and relayed to the

DTEsender process. The module receives input from two sources: the terminal and the *InboundChannel* mailbox monitor. The terminal is used by a human operator, and the *InboundChannel* monitor holds packets sent to the *DTE* from the *DCE*.

The *PollKybd* process periodically prompts the terminal for input commands. Once a recognizable command is given, the process enters the *FSM* monitor to see if the command can be executed. For example, if a request for a data packet transmission is made and the *FSM* is not in the *DataTransfer* state, then the command is not executed and a message is printed on the *DTE* terminal. If the command can be executed then the *packetToGo* variable is set to the corresponding packet. The *DTEsender* process is then awakened via a broadcast statement, and a confirmation is printed on the terminal.

The *DTEreceiver* process periodically polls the *InboundChannel* mailbox monitor to see if a packet has been placed there by the *DCE*. If a packet is obtained, the process calls the *FSM* monitor procedure

ReceivePacket. This procedure decides whether a state change or a packet response or both should take place. The process learns of the *DTE* state before and after receiving the packet, and awakens the *DTEsender* process if a response is required. *DTEreceiver* then prints a trace of the transaction at the *DTE*'s terminal.

When awakened by a broadcast statement, the *DTEsender* process calls the *FSM* monitor procedure *PacketToSend.* The procedure determines what packet is to be sent, causes a state change (unless the packet is pure data), notes the old and new state of the *FSM,* and prints a trace of the transaction on the *DTE* terminal. The process then enters the *OutboundChannel* mailbox monitor, places the packet there, and sets the *packetInChannel* flag.

DCE Module

This module contains symmetrically equivalent structures and three of its processes execute in a similar manner to those of the *DTE* module. However, the *DCE FSM* expects to see and sends packet types different from those of the *DTE*. The *DCEreceiver* process polls the *OutboundChannel* mailbox monitor for packets sent by the *DTE*. The *DCEsender* process places packets destined for the *DTE* in the *InboundChannel* mailbox monitor.

The *DCE* also contains a fourth process, *NetworkDelay.* This process simulates the delay of a packet request from the *DTE* traversing the sub-net to the destination *DCE* and back. *NetworkDelay* is awakened whenever the *DCEreceiver* process enters the *DCE FSM* monitor with a request packet from the *DTE*. A delay value is determined from a pseudo-random number generator, and the process is delayed for that long. The process then decides whether the delay time was too long and signals the *DCEreceiver* process waiting for the delay to pass. The *DCEreceiver* process is made aware whether the delay was too long. If it was too long, the *packetToGo* variable is set to the *NetworkError* packet and the *DCE* state goes back to its state prior to receiving the packet from the *DTE*. Upon receiving this error packet from the *DCE*, the *DTE* resets its state to the state it was in prior to sending the request packet. Thus, the complete transaction is effectively nullified in the case of a network error.

Mailbox Module

The *Mailbox* module accommodates packet deliveries between *DTE* and *DCE* modules. It contains two monitors corresponding to packet flow in each direction: *InboundChannel* and *OutboundChannel*. There is no buffering of packets beyond one in either direction. Both monitors provide procedure entries to set and to read the packets in transit.

Suitability of Real-Time Euclid

The periodic process construct is a convenient way of implementing a polling process, with the polling frequency being the process frame length. By knowing how often processes poll in real-time, one has a definite upper bound on the frequency with which requests for action can occur. Thus CPU time can be allocated without being concerned by activities executed during request service time, if a good estimate of the maximum duration of a poll is known.

The **broadcast** statement is used in the program whenever a sending process, such as *DTEsender*, needs to pass a packet between processors. This is a useful construct, allowing a designer to have a process started up from any point within the code. This construct accommodates situations where the frequency of execution of a process varies as is the case in this simulation.

Summary of Schedulability Analyzer Evaluation: As reported in the former section (Applications), Real-Time Euclid was found to be a suitable language to use in real-time application writing. We now present the results of the evaluation of Real-Time Euclid's schedulability analyzer, based on the simulated real-time applications of that section. The static decompositions of the programs are tabulated in Tables 1 and 2. Table 3 summarizes the process content of the programs.

In our evaluation, we varied process resource requirements (by inserting idle for-loops into various parts of the programs) while keeping the frames constant. We gathered sets of predicted and actual response times. We measured load in terms of ratios of actual process response times to their corresponding frames. The evaluation took place in the context of four different loads (and thus four different sets of resource requirements): light load of 7% to 20%, medium load of 13% to 50%, heavy load of 35% to 80%, and overload (some processes missed their deadlines) of 42% to 120%.

Program #	Computation	Control	Synchronization	Communication
Power Plant	53%	19%	15%	13%
DTE/DCE	54%	15%	18%	13%

Table 1: Static Program Decomposition By Function

Program #	Process	Monitor Subprogram	Non-monitor Subprogram	Mailbox
Power Plant	24%	20%	33%	23%
DTE/DCE	26%	16%	28%	30%

Table 2: Static Program Decomposition By Construct

Program #	process frames	periodic
Power Plant (cpu1)	15, 15, 15, 15, 15	Yes, Yes, Yes, Yes, Yes
Power Plant (cpu2)	60, 15, 15	Yes, Yes, Yes
DTE/DCE (cpu1)	10, 2, 14	Yes, No, Yes
DTE/DCE (cpu2)	10, 4, 14, 3	Yes, No, Yes, No

Table 3: Evaluation Program Process

The predictions were made by our method and the Leinbaugh and Yamini [25],[26] method. We used this method for comparison with our method, because of all methods surveyed in [48] it is the one second best suitable for the analysis of the real-time systems like ours. The best existing method is probably the Leinbaugh and Yamini [27] technique. Unfortunately, because no algorithmic description of it is available, it is not clear how to implement it. Moreover, one of the aims of this evaluation is to demonstrate the superiority of frame superimposition over existing closed-form and polynomial techniques. The Leinbaugh and Yamini [27] method is not such a technique.

Tables 4 through 11 presents the results of evaluation.

As expected, both algorithms predict less accurate bounds in the case of the DTE/DCE program than they do in the case of the Power Plant program. This is due to the fact that while the Power Plant program contains periodic processes only, half of the DTE/DCE processes are aperiodic (atEvent). The times of periodic process activations are easily and accurately predicted by both algorithms. However, only ranges of the times of aperiodic process activations (as

Program #	maxRespTimes	predictedRespTimes	difference (%)
Power Plant (cpu1)	all 3	all 3.10	all 3.33
Power Plant (cpu2)	4, 2, 2	4.11, 2.07, 2.04	2.75, 3.50, 2.00
DTE/DCE (cpu1)	1, 0, 1	1.04, 0.23, 1.05	4.00, n/a, 5.00
DTE/DCE (cpu2)	1, 0, 1, 0	1.02, 0.31, 1.04, 0.45	2.00, n/a, 4.00, n/a
average (%)	n/a	n/a	3.33

Table 4: Performance of Our Method Under Light Load

Program #	maxRespTimes	predictedRespTimes	difference (%)
Power Plant (cpu1)	all 3	all 3.69	all 23.00
Power Plant (cpu2)	4, 2, 2	5.10, 2.51, 2.49	27.50, 25.50, 24.50
DTE/DCE (cpu1)	1, 0, 1	1.48, 0.93, 1.42	48.00, n/a, 42.00
DTE/DCE (cpu2)	1, 0, 1, 0	1.31, 1.03, 1.30, 1.20	31.00, n/a, 30.00, n/a
average (%)	n/a	n/a	28.63

Table 5: Performance of Leinbaugh Method Under Light Load

Program #	maxRespTimes	predictedRespTimes	difference (%)
Power Plant (cpu1)	all 6	all 6.37	all 6.17
Power Plant (cpu2)	8, 4, 4	8.43, 4.24, 4.20	5.38, 6.00, 5.00
DTE/DCE (cpu1)	2, 1, 2	2.12, 1.08, 2.17	6.00, 8.00, 8.50
DTE/DCE (cpu2)	2, 1, 2, 1	2.09, 1.10, 2.17, 1.08	4.5, 10.00, 8.50, 8.00
average (%)	n/a	n/a	6.71

Table 6: Performance of Our Method Under Medium Load

Program #	maxRespTimes	predictedRespTimes	difference (%)
Power Plant (cpu1)	all 6	all 8.25	all 37.50
Power Plant (cpu2)	8, 4, 4	11.34, 6.31, 6.28	41.75, 57.75, 57.00
DTE/DCE (cpu1)	2, 1, 2	3.16, 1.58, 3.72	58.00, 58.00, 86.00
DTE/DCE (cpu2)	2, 1, 2, 1	2.85, 1.61, 3.12, 1.58	42.50, 61.00, 56.00, 97.50
average (%)	n/a	n/a	50.90

Table 7: Performance of Leinbaugh Method Under Medium Load

Program #	maxRespTimes	predictedRespTimes	difference (%)
Power Plant (cpu1)	all 12	all 13.39	all 11.58
Power Plant (cpu2)	21, 12, 12	22.94, 13.83, 12.96	9.24, 15.25, 8.00
DTE/DCE (cpu1)	7, 1, 10	8.27, 1.26, 11.38	18.14, 26.00, 13.80
DTE/DCE (cpu2)	7, 3, 11, 2	8.12, 3.41, 12.63, 2.40	16.00, 13.67, 14.82, 20.00
average (%)	n/a	n/a	14.19

Table 8: Performance of Our Method Under Heavy Load

Program #	maxRespTimes	predictedRespTimes	difference (%)
Power Plant (cpu1)	all 12	all 17.34	all 44.50
Power Plant (cpu2)	21, 12, 12	34.23, 21.76, 21.68	63.00, 81.33, 80.67
DTE/DCE (cpu1)	7, 1, 10	13.45, 1.97, 22.77	92.14, 97.00, 127.70
DTE/DCE (cpu2)	7, 3, 11, 2	13.57, 5.86, 22.34, 3.95	93.86, 95.33, 103.09, 97.50
average (%)	n/a	n/a	76.94

Table 9: Performance of Leinbaugh Method Under Heavy Load

defined by **broadcast** statements) can be determined by our algorithm, and the Leinbaugh algorithm has no means of doing even that (and thus this latter algorithm assumes that aperiodic process activations can occur at any time at all).

Program #	maxRespTimes	predictedRespTimes	difference (%)
Power Plant (cpu1)	all 16	all 21.65	all 35.31
Power Plant (cpu2)	25, 16, 16	31.23, 21.09, 20.60	24.92, 31.81, 28.75
DTE/DCE (cpu1)	9, 3, 12	12.87, 4.14, 15.21	43.00, 38.00, 26.75
DTE/DCE (cpu2)	9, 5, 12, 4	12.97, 6.95, 15.98, 5.09	44.11, 39.00, 33.17, 27.25
average (%)	n/a	n/a	34.22

Table 10: Performance of Our Method Under Overload

Program #	maxRespTimes	predictedRespTimes	difference (%)
Power Plant (cpu1)	all 16	all 37.14	all 132.13
Power Plant (cpu2)	25, 16, 16	53.35, 35.11, 34.04	113.40, 119.44, 112.25
DTE/DCE (cpu1)	9, 3, 12	21.86, 7.83, 35.13	142.89, 161.00, 192.75
DTE/DCE (cpu2)	9, 5, 12, 4	22.14, 14.18, 37.64, 10.12	146.00, 183.60, 213.67, 153.00
average (%)	n/a	n/a	146.61

Table 11: Performance of Leinbaugh Method Under Overload

As we mentioned previously, our frame superimposition algorithm is exponential in nature. Thus an important question is how much time our schedulability analysis contributes to the compilation-and-analysis phase of real-time software development. While we do not attempt to answer this question in general, our experience with the two applications described in this Section has been that the analysis increases the pure compilation time by about 60-100%. For example, one of our programs took 5 minutes to just compile and 10 minutes to compile and be analysed. Thus we feel that the analysis does not take unacceptably long to run -- though this is just a speculation for the case of very large programs with complex resource contention patterns.

Despite the relative difficulty our algorithm has with aperiodic processes, it performs well on the absolute scale. The method gives very accurate (within 2% to 5%) predictions in the light load case, good (within 4.5% to 10%) predictions in the medium load case, reasonably accurate (within 8% to 26%) predictions in the heavy load case, and only marginally inaccurate (within 24.92\% to 44.11%) predictions in the overload case. The performance of the Leinbaugh method is in sharp contrast with the performance of our method. Under every load, the Leinbaugh method predicts guaranteed response times that differ quite considerably from the actual response times. In fact, the Leinbaugh method even falsely predicts missed deadlines in the case of heavy load (see, for example, Table 9 in which the predicted response times for the Power Plant (cpu1) processes are all 17.34 › 15 (their frames) › 12 (their observed maximum response times)).[6]

CONCLUDING REMARKS

We have presented a set of language-independent schedulability analysis techniques. Utilizing knowledge of implementation- and hardware-dependent information in a table-driven fashion, these techniques provided good worst-case time bounds and other schedulability information. The techniques employ frame superimposition to derive interprocess synchronization, non-interruptible, specified-delay and communication segment contention, and use slow-down rate analysis to bound processor contention in the presense of preemption.

To demonstrate the effectiveness of these techniques, a prototype schedulability analyzer has been developed. Using a prototype implementation of the compiler, run-time kernel and analyzer, two model real-time application programs written in Real-Time-Euclid (a new language specifically designed with schedulability analysis provisions built-in) have been developed and analyzed. The programs were run on a realistic multiple-microprocessor system. The worst-case time bounds predicted by the analyzer were compared with the actual times, and found to differ only marginally. When applied to the same two programs, a state-of-the-art method resulted in considerably more pessimistic predictions.

While frame superimposition seems to generate good bounds on mailbox communication delays, it is not clear whether the same would be true in the case of a more sophisticated inter-processor communication system. This is an important research direction in schedulability analysis techniques.

We believe that applying schedulability analysis to Real-Time Euclid programs has been a success. However, much of world's existing real-time software is written in assembly languages, Fortran [45], PEARL [11],[12] and Ada [20]: languages that do not make sufficient provisions

[6]Since our method also results in somewhat pessimistic predictions, it is possible to find a load for any system of processes under which no actual deadline will be missed while the predictions will indicate a missed deadline. However, since our method appears to be quite accurate, the cicumstances under which such a false alarm would occur should be quite rare.

for schedulability analysis. Thus, a very important question is whether and how we can apply schedulability analysis techniques developed here to these languages. At first, the problem seems unsolvable. Indeed our own survey demonstrated that hardly any existing languages make provisions for schedulability analysis. Does this mean that we are out of luck and should abandon all hope? We feel that the answer is ``no''. While Ada and Fortran may not be real-time, the external environments we are concerned about are. These real-time environments impose critical timing constraints on their embedded software. Thus the programmer is forced to write this software in a schedulability analyzable way, regardless of the programming language used. Thus, real-time software, if written correctly, is schedulability analyzable.

Our original question now reduces to two simpler ones:
1. How do we recognize schedulability analyzability of software written in a non-analyzable (in general) language?
2. How do we actually do the analysis?

There are no easy answers, though some work has already taken place. Baker's work [2],[3],[4],[5] on Ada is a good example. As a first step, Baker introduced constraints into Ada at the package level. The constraints made Ada task scheduling more deterministic and thus easier to analyse for guaranteed schedulability. Baker intends to explore the ways in which real-time Ada software can be reduced systematically to be provably schedulability analyzable. Another example of such work is due to Sha and Goodenough [40]. Sha and Goodenough's project aims to cope with the priority inversion problem (due to Ada's native synchronization) by modifying the Ada run-time scheduler, to both minimize the time a higher-priority task is blocked by a lower-priority task, and to prevent related deadlocks from occurring.

While it seems that to change the definition of Ada is quite hard, we are currently attempting to influence PEARL'9X, which is to become the new standard of PEARL. So far enough initial interest has been generated to invite the author as Keynote Speaker to the PEARL'90 annual workshop, and present his views on how PEARL can benefit from the language-independent schedulability analysis described here and language design ideas introduced in Real-Time Euclid [51]. An initial design for a schedulability analyzer for PEARL'9X has been presented at Prozeßrechensysteme'91 [52].

While frame superimposition seems to generate good bounds on mailbox communication delays, it is not clear whether the same would be true in the case of a more sophisticated inter-processor communication mechanism. A related question has to do with the amount of time schedulability analysis would take in the case of large, complex communicating systems. Will the analysis still increase the overall compilation time by a small factor, or will the exponential nature of the analysis make it infeasible? Research is thus quite open in the area of real-time systems with complex communication mechanisms.

Most real-time software is written in a redundant, fault-tolerant way. An important question is: How do we determine guaranteed schedulability in the presence of exceptions? On the one hand, it is overly pessimistic to assume that every exception happens every time. On the other hand, what if all possible exceptions will in fact occur at least once at the same time? One possible solution to this problem is to make the programmer write software in a way that is data-flow balanced in time. That is, the amount of time it takes to execute normal, exception-handling-free code is not substantially different from the amount of time it takes to execute the exception-handling code that replaces the normal code. It is not clear whether or not all real-time software can be written this way. The question of real-time exception handling in the context of schedulability analysis is thus quite open as well.

While we have failed to find, in the existing body of scientific literature, an accurate polynomial-time schedulability analysis algorithm, it is conceivable that one could be developed. We are beginning an investigation into semantics-preserving real-time program transformations and incremental polynomial-time analysis heuristics that may offer some hope.

DISCLAIMERS AND ACKNOWLEDGEMENTS

This Chapter is based on some of the research undertaken by the author during his Ph.D. studies at the University of Toronto. This and related work have been reported in a number of scientific articles, some already printed [53,64,1] and some currently under review, the author's Ph.D. thesis [63], and in a book co-authored with Wolfgang Halang of Groningen University of the Netherlands [4]. This research has been supported in part by NSERC (Canada) Grants A-5192 and A-8647, and an NSERC Postgraduate Scholarship.

The author would like to thank a great many people who or whose work have had direct influence on both the author's way of thinking and on his continuing development as a scientist. Among these, special thanks go to Carl Hamacher and Ric C. Holt whose official records at the University of Toronto simply state that they co-advised the author during his study, but who executed their duties as Masters better and beyond any expectations that the Apprentice who was the author had, or thought possible for any apprentice, any place or any time. The author is in much debt to Wolfgang Halang, whose pioneering work in real-time languages, systems and architectures has already had and is certain to continue having a lasting, major impact on the very focus of the field of predictable real-time systems, and whose influence on the author's professional development has been profound. The author is very grateful to his many colleagues in the area of real-time computing, among them most notably, Flaviu Cristian, Susan Davidson, Marc Donner, Farnam Jahanian, David Jameson, Kevin Jeffay, Hector Garcia-Molina, Jerry Gechter, Jack Goldberg, Gary Herman, Kane Kim, Insup Lee, John Lehoczky, Kwei-Jay Lin, Jane Liu, Doug Locke, Fabrizio Lombardi, Nancy Lynch, Al Mok, Swami Natarajan, Jonathan Ostroff, Fred Schneider, Lui Sha, Alan Shaw, Kang Shin, Jack Stankovic, Raj Rajkumar, Krithi Ramamritham, Andre Van Tilborg, Hide Tokuda, Dick Volz, Horst Wedde, who have commented on and criticized author's work, actively participated in great many debates on predictable real-time systems, and whose numerous achievements have served as constant inspiration to the author. The Real-Time Euclid project which the author ran and which produced the results reported here simply would not have happened nor succeeded without the project team members: Gene Kligerman, Chris Ngan, Scott Thurlow, Gerry Parnis, Greg Nymich and Victor Anderson. The author's wife Lana and parents David and Irene endured through much and always provided the author with their love and support, during those intensive and wonderful years of study.

THE BIBLIOGRAPHY

[1] "Advances in Power Station Construction", *Generating Development and Construction Division, Central Electricity Generating Board,* Barnwood, Gloucester, UK, Pergamon Press, 1986, pp 313-328.

[2] Theodore P. Baker, "A Corset for Ada", *TR 86-09-05,* Computer Science Department, University of Washington, 1986.

[3] Theodore P. Baker, "A Lace for Ada's Corset", *TR 86-09-06,* Computer Science Department, University of Washington, 1986.

[4] Theodore P. Baker, "Implementing Timing Guarantees in Ada", *Proceedings of the Fourth IEEE Workshop on Real-Time Operating Systems,* Cambridge, Massachusetts, July 1987, pp. 129-133.

[5] Theodore P. Baker, "Improving Timing Predictability of Software", *Working Paper,* Department of Computer Science, Florida State University, August 1987.

[6] G. Chroust, "Orthogonal Extensions in Microprogrammed Multiprocessor Systems - A Chance for Increased Firmware Usage", *EUROMICRO Journal,* 6,2,104-110, 1980.

[7] James R. Cordy, Ric C. Holt, David B. Wortman, "S/SL Syntax/ Semantic Language Introduction and Specification", *Technical Report CSRG-118,* Computer Systems Research Group, University of Toronto, 1980.

[8] James R. Cordy, Ric C. Holt, "Specification of Concurrent Euclid", *Technical Report CSRG-133,* Computer Systems Research Group, University of Toronto, August 1981.

[9] "DATAPAC Standard Network Access Protocol Specifications", Computer Communications Group, TCTS, Ottawa, 1976.

[10] Edsgar W. Dijkstra, "Cooperating Sequential Processes", *Technical Report EWD-123,* Technological University, Eindhoven, the Netherlands, 1965.

[11] DIN 44300: Informationsverarbeitung, Nr. 161 (Realzeitbetrieb), March 1972.

[12] DIN 66253: Programmiersprache Pearl, Teil 1 Basic Pearl, Vornorm, July 1981; Teil 2 Full Pearl, Norm, October 1982.

[13] M. R. Garey, D. S. Johnson, "Complexity Results for Multiprocessor Scheduling under Resource Constraints", *SIAM Journal on Computing,* Vol. 4, No. 4, December 1975, pp. 397-411.

[14] Wolfgang A. Halang, "On Methods for Direct Memory Access Without Cycle Stealing". *Microprocessing and Microprogramming*, 17, 5, May 1986.

[15] Wolfgang A. Halang, "Implications on Suitable Multiprocessor Structures and Virtual Storage Management when Applying a Feasible Scheduling Algorithm", *Hard Real-Time Environments, Software - Practice and Experience*, 16(8), 761-769, 1986.

[16] Wolfgang A. Halang, Alexander D. Stoyenko, "Constructing Predictable Real-Time Systems", to be printed by *Kluwer Academic Publishers*, Dordrecht-Hingham, 1991.

[17] C. A. R. Hoare, "Monitors: An Operating System Structuring Concept", *Communications of the ACM*, Vol. 17, No. 10, October 1974, pp. 549-557.

[18] R. C. Holt, D. B. Wortman, J. R. Cordy, D. R. Crowe, J. H. Griggs, "Euclid: A language for producing quality software," *Proceedings of the National Computer Conference*, Chicago, May 1981.

[19] Ric C. Holt, James R. Cordy, "The Turing Language Report", Technical Report CSRG-153, Computer Systems Research Group, University of Toronto, December 1983.

[20] Jean D. Ichbiah, "Reference Manual for the Ada Programming Language", U.S. Department of Defense, 1980.

[21] "KE-Handbuch", Periphere Computer Systeme GmbH, Munich, 1981.

[22] Eugene Kligerman, "A Programming Environment for Real-Time Systems", *M.Sc. Thesis*, Department of Computer Science, University of Toronto, 1987.

[23] Eugene Kligerman, Alexander D. Stoyenko, "Real-Time Euclid: A Language for Reliable Real-Time Systems," *IEEE Transactions on Software Engineering*, Vol. SE-12, No. 9, pp. 940-949, September 1986.

[24] R. Lauber, "Prozessautomatisierung I", Berlin-Heidelberg-New York, Springer-Verlag, 1976.

[25] Dennis W. Leinbaugh, "Guaranteed Response Times in a Hard-Real-Time Environment," *IEEE Transactions on Software Engineering*, Vol. SE-6, No. 1, January 1980, pp. 85-91.

[26] Dennis. W. Leinbaugh, Mohammed-Reza Yamini, "Guaranteed Response Times in a Distributed Hard-Real-Time Environment", *Proceedings of the IEEE 1982 Real-Time Systems Symposium*, December 1982, pp. 157-169.

[27] Dennis W. Leinbaugh, Mohammed-Reza Yamini, "Guaranteed Response Times in a Distributed Hard-Real-Time Environment," *IEEE Transactions on Software Engineering*, Vol. SE-12, No. 12, pp. 1139-1144, December 1986.

[28] John Y.-T. Leung, M. L. Merrill, "A Note of Preemptive Scheduling of Periodic, Real-Time Tasks", *Information Processing Letters*, Vol. 11, No. 3, pp. 115-118, November 1980.

[29] John Y.-T. Leung, J. Whitehead, "On the Complexity of Fixed-Priority Scheduling of Periodic, Real-Time Tasks", *Performance Evaluation*, Vol. 2, pp. 237-250, 1982.

[30] C. L. Liu, J. W. Layland, "Scheduling Algorithms for Multiprogramming in a Hard-Real-Time Environment", *JACM*, Vol. 20, No. 1, January 1973, pp. 46-61.

[31] Aloysius K. Mok, "The Design of Real-Time Programming Systems Based on Process Models", *Proceedings of the IEEE 1984 Real-Time Systems Symposium*, December 1984, pp. 5-17.

[32] Aloysius K. Mok, Michael L. Dertouzos, "Multiprocessor Scheduling in a Hard-Real-Time Environment", *Proceedings of the 7th Texas Conference on Computing Systems*, November 1978, pp. 5.1-5.12.

[33] Chris Ngan, "Implementing the Real-Time Euclid Compiler", *Student Project Report*, Department of Computer Science, University of Toronto, August 1986.

[34] Gerry Parnis, "Simulation of Packet Level Handshaking in X.25 Using the Real-Time Euclid Programming Language", *Student Project Report*, Department of Computer Science, University of Toronto, April 1987.

[35] Krithivasan Ramamritham, John A. Stankovic, "Dynamic Task Scheduling in Distributed Hard Real-Time Systems", *Proceedings of the IEEE 4th International Conference on Distributed Computing Systems*, May 1984, pp. 96-107.

[36] Krithivasan Ramamritham, John A. Stankovic, S. Cheng, "Evaluation of a Flexible Task Scheduling Algorithm for Distributed Hard Real-Time Systems", *IEEE Transactions on Computers*, Vol. C-34, No. 12, December 1985, pp. 1130-1143.

[37] A. M. Rybezynski and D. F Weir, "DATAPAC X.25 SERVICE CHARACTERISTICS", Computer Communications Group, TCTS, Ottawa, 1977.

[38] K. Schleisiek-Kern, "Private Communication", *DELTA t*, Hamburg, 1990.

[39] G. Schrott, "Ein Zuteilungsmodell für Multiprozessor-Echtzeitsysteme", *PhD Thesis*, Technical University, Munich 1986.

[40] L. Sha, J.B. Goodenough, "Real-Time Scheduling and Ada", *Technical Report CMU/SEI-89-TR-14, ESD-TR-89-22*, Software Engineering Institute, Carnegie-Mellon University, 1989.

[41] Series 32000 Instruction Set Reference Manual, National Semiconductor Corporation, Santa Clara, June 1984.

[42] Connie U. Smith, "Independent General Principles for Constructing Responsive Software Systems", *ACM Transactions on Computer Systems*, Vol. 4, No. 1, February 1986, pp. 1-31.

[43] Paul G. Sorenson, "A Methodology for Real-Time System Development", *Ph.D. Thesis*, Department of Computer Science, University of Toronto, 1974.

[44] Paul G. Sorenson, V. Carl Hamacher, "A Real-Time System Design Methodology", *INFOR*, Vol. 13, No. 1, February 1975, pp. 1-18.

[45] "Specifications for the IBM Mathematical FORmula TRANslating System, FORTRAN", IBM Corporation, New York, November 10, 1954.

[46] Alexander D. Stoyenko, "Turing goes real-time...", *Internal Programming Languages Report*, Department of Computer Science, University of Toronto, May 1984.

[47] Alexander D. Stoyenko, "Real-Time Systems: Scheduling and Structure", *M.Sc. Thesis*, Department of Computer Science, University of Toronto, 1984.

[48] Alexander D. Stoyenko, "A Real-Time Language with A Schedulability Analyzer", *Ph.D. Thesis*, Department of Computer Science, University of Toronto, 1987.

[49] Alexander D. Stoyenko, "A Schedulability Analyzer for Real-Time Euclid," *Proceedings of the IEEE 1987 Real-Time Systems Symposium*, pp. 218-225, December 1987.

[50] ***not quoted in text*** Alexander D. Stoyenko, "A Case for Schedulability Analyzable Real-Time Languages," *Proceedings of the IEEE 1987 Workshop on Real-Time Operating Systems*, April 1987, Cambridge, Massachusetts.

[51] Alexander D. Stoyenko, "Real-Time Euclid: Concepts Useful for the Further Development of PEARL," *Keynote Speech, Proceedings of PEARL'90*, November-December 1990, Boppard, Germany.

[52] Alexander D. Stoyenko, Wolfgang A. Halang, "Analysing PEARL Programs for Timely Executability and Schedulability" ("Analyse zeitgerechter Zuteilbarkeit und Ausf\"uhrbarkeit von PEARL-Programmen"), *Proceedings of the Fachtagung Prozeßrechensysteme'91 Conference*, February 1991, Berlin, Germany.

[53] T. J. Teixeira, "Static Priority Interrupt Scheduling", *Proceedings of the 7th Texas Conference on Computing Systems*, November 1978, pp. 5.13-5.18.

[54] T. Tempelmeier, "A Supplementary Processor for Operating System Functions", *1979 IFAC/IFIP Workshop on Real-Time Programming*, Smolenice, 18-20 June 1979.

[55] T. Tempelmeier, "Operating System Processors in Real-Time Systems - Performance Analysis and Measurement", *Computer Performance*, Vol. 5, No. 2, 121-127, June 1984.

[56] Scott A. Thurlow, "Simulation of a Real-Time Control System Using the Real-Time Euclid Programming Language", *Student Project Report*, Department of Computer Science, University of Toronto, April 1987.

[57] Jeffrey D. Ullman, "Polynomial complete scheduling problems", *Proceedings of the 4th Symposium on OS Principles*, 1973, pp. 96-101.

[58] Niklaus Wirth, "The Programming Language Pascal," *Acta Informatica 1*, pp. 35-63, 1971.

[59] Wei Zhao, Krithivasan Ramamritham, `Distributed Scheduling Using Bidding and Focused Addressing", *Proceedings of the IEEE 1985 Real-Time Systems Symposium*, December 1985, pp. 103-111.

4

WHICH THEORY MATCHES AUTOMATION ENGINEERING?

PETRI NETS AS A FORMAL BASIS

E. Schnieder

Institut für Regelungs- und Automatisierungstechnik
Technische Universität Braunschweig
D-3300 Braunschweig

INTRODUCTION

Automated systems are economically successful; however, many problems in this technical field still remain to be solved. This is surely due to the complexity of the problems of automation technology and furthermore due to the many disciplines involved, which finally leads to a great deal of heterogeneity. At present, no abstract theory exists that allows the integration of the many different aspects of this discipline. This chapter will analyze the situation in automation technology, refer to other developed and theoretically based related technical domains and specify requirements for a theory of automation technology. Furthermore, the chapter will investigate whether a net theory can meet these requirements and will discuss the realization of this theory.

WHAT IS MEANT BY AUTOMATION TECHNOLOGY?

Automation technology is understood to be the technical solution of an automated and defined process operation in technical systems. The technical solution comprises the technical design as well as the means to realize in a definitive way a specific automated device.

Even with this clear and simple definition, a divergence of different semantic aspects cannot be avoided: the large number of equally important aspects is therefore exemplary and significant. The following list, although incomplete, shows what is meant by *aspects of an automation:*

- Object-specific aspects (*object* referred to in automation):
  - manufacturing system, production system
  - supply and disposal system (e.g., gas, water)
  - energy supply and distribution (e.g., energy distribution companies, power stations and plants)
  - process engineering (e.g., chemistry, refineries, biotechnology)
  - traffic system (e.g., roads, water-, air-, rail-transport)

- Technical-device aspects (*hardware* of automation)
  - implement-oriented (e.g., programmable controllers, process control computer, microprocessor, ASIC, architecture)
  - communication-oriented
  - assembly technology

- Program-oriented aspects (*software* of automation)
  - universal language (procedure-oriented, declarative)
  - special language
  - programming language for embedded systems

- Design-oriented aspects (*design* of automation system)
  - requirements engineering
  - software engineering
  - experts- and know-how-based systems
    artificial intelligence (AI)

- Methodical-functional aspect (*function* of automation)
  - closed-loop-controlled (continuous or discrete)
  - open-loop-controlled (binary-, time- and event-controlled)
  - traffic (behavior, performance)
  - reliability

This list could be extended and continued. Furthermore, these different semantic aspects cannot be regarded in isolation: they are characteristic for a system in any combination. The interconnections of these elements complicate a consequent realization. Critical problems of automation technology are the divergence of specific aspects and the huge number of possible combinations of such aspects. The lack of an underlying theory is obvious, with the result that the high number of orientation alternatives finally leads to a loss of orientation; unfortunately, a local optimization is not always good for the optimum of the whole system.

WHAT HAPPENS WITHOUT A THEORY?

By focusing on such individual aspects, the lack of an organizing idea is often overlooked. Before developing a solution for these problems, the situation must be analyzed in more detail to be able to weigh the expected usefulness of a theoretical framework for automation engineering problems against the expected technical effort and the possible (intellectual) costs [1,2].

No automated system is realized as planned! Automated systems often grow in an unstructured way. When possible side effects cannot be avoided and complexity cannot be controlled, system behavior at large can run out of control. Thus it appears as if the developmental work of complex problems is not precisely foreseeable and cannot be calculated. Thus wasteful detours may have to be taken into account. System documentation presents a typical problem: even if the description corresponds to the realization of a system, not every following project can necessarily make use of already realized projects. The desire for "reusing" or "Software ICS" (Software Integrated Circuits) is, however, growing .

In conclusion, without a theoretical foundation, the costs for complex systems will increase while efficiency may decrease. Rationalization and increase of efficiency, which were former aims of large industrial projects, thus cannot be achieved. Often if such a realized system gets into an unforeseen state, certain effects can neither be interpreted nor reproduced which underlines the need for a theoretical foundation. Later corrections often make the system worse. Symptoms are corrected, and latent faults appear in other places and are still more problematic.

Although well-developed and highly specialized "tools" improve the productivity of developing automated systems as well as the "manual" quality of the product during the implementation phase, latent weak points can thus hardly be corrected.

Simulations, too, can only be applied in a certain range of application. Due to the inner complexity of larger systems, simulations and tests can only detect a limited, characteristic system behaviour because only states which are foreseeable can be simulated, excluding physically undesirable or generally "faulty" states.

WHAT IS A THEORY AND WHAT CAN IT ACHIEVE?

All the above mentioned problems demonstrate the urgent need to support design planning with a theoretical basis. Other related and established disciplines already make use of theories and are very successful. Examples are as follows:

Control engineering: Application of function theory to allocate the dimension to control loops (Nyquist criterion). Application of Laplace transform to investigate system behavior.

Theory of reliability: Application of the Laplace transform and probability analysis to calculate parameters and failure modes of complex and redundant systems.

Communication technology: Application of probability and code theories to calculate transmission failures.

Data processing: Application of Boolean algebra to design and minimize switching devices

This experience in other fields should be used as an orientation and model for work in automation engineering in order to establish an appropriate theoretical basis for this field as well.

Before answering the question of how to develop a theory for automation engineering, a definition of a *theory* must be given. Following the Encyclopaedia Britannica, a theory is

> "a systematic ideational structure of broad scope, conceived by the imagination of man, that encompasses a family of empirical (experimental) laws regarding regularities existing in objects and events, both observed and posited — a structure suggested by these laws and devised to explain them in a scientifically rational manner".

The words of Einstein that *"nothing is more practical than a good theory"* can also be considered as a definition in the form of a programmatic negative exclusion. In practice, however, the emergence of positive or useful results should describe the requirements of a theory.

A mathematical theory is understood as a set of laws that is closed or open in itself and that concerns objects or quantities based on axioms that are in themselves free from contradiction. For the application of a theory in practical situations, it must be shown that there is a clear equivalency between the set of laws concerning objects and the observable behavior of real quantities and objects in a wide range of applications.

More specifically, this means that

- A theory can model classes of real systems, i.e., the systems' structure and behavior, which more specifically means behavior of
  - its components
  - of its parts and the relations between its parts
  - its components and their relations

- A theory can be used to structure synthetic systems and to calculate future behavior. That means that a theory provides a planning basis for predictions.

HOW CAN A THEORY FOR AUTOMATION ENGINEERING BE DEVELOPED?

Before making the easier choice of an appropriate mathematical theory from existing abstract rule systems or the more difficult design planning of new self-standing appropriated control systems (Leibniz and Newton solved this problem by using infinitesimal calculus, which they independently invented in 1665), the requirements for a theory have to be specified.

Problem

It is of special significance that a theory either has to consider directly all the important aspects of automation engineering or, as an alternative, has to provide a completely new theoretical foundation based on a deeper systematic and structural view. In the latter case, the theory must be relatively general in expressive modeling in order to be able to meet the requirements. To restrict the means of descriptions, symbols, operators, relations, and symbols it is important to avoid complexity of solutions during the design phase. By an appropriate decomposition strategy, the necessary, successive refinement can be achieved. This strategy is similar to the state description of control systems in which complicated differential equations (of nth order) are replaced by a system of simple first order differential equations. Thus the problem becomes clearly structured, enabling a homogenous model.

Approach.

The proposition of introducing a theory is based on the apparently trivial idea to consciously consider the automation system as a single entity. The term *consciously consider* emphasizes the fact that the term *system* has to be precise and axiomatically based. This term has to comprise the special marks of a system concerning its structure and behavior as well as its design, which in the end is a developmental aspect [4,5].

This kind of system is marked by the following aspects [6]:

Structural principle (axiom 1)

The system consists of a set of parts interacting with each other and the system's peripherals. The parts of a system are defined by quantities. The value of a quantity of a system represents the system's state. To make a clear distinction between system and peripherals, the system needs to be independent. Resistance to peripheral influences is characteristic.

Principle of decomposition (axiom 2)

The system consists of a set of parts that can further be subdivided into subparts interacting with each other. The subparts also represent a certain complexity, i.e., general system aspects.

Causal principle (axiom 3)

The system consists of a set of parts. Their interactions with each other and the resultant changes are clearly determined. In a causal connection, later states can only be the consequence of former ones. Causality is understood as the logic of operations.

Temporal principle (axiom 4)

The system consists of a set of parts. Their structure or state depends on changes in time. Temporality is the temporal sequence of operations and changes.

Intuitively, it is possible to apply these abstract axioms to automation engineering and its detailed facilities when using the area of interpretation inherent in the axioms.

In the following, the verbal formulations of the systems' axioms will be presented with mathematical precision along with the theory.

Choice of a theory.

One of the main bases of the proposed theory is the special formulation of *causality* before *temporality* [7]. Both, the technical and controlling systems operate independently from each other. Their processes are partly coupled after the control loop is closed. The temporal axiom takes into account the relation between physical variables and time. The cause for automatic processing has to be seen in the causal link of individual and concurrent processes.

The diagrammatic view of the structure of an automation system, which is basic for an automation (e.g., cascade arrangement or state-controlling structure), as well as the graphic abstract representation of the causally conditioned relation of cause and effect, resembles the structure of nets. This knowledge is not new, but it has as yet not been considered from the point of view of dynamic system behaviour.

Therefore, a mathematical net-referring theory was sought that would lead to the net theory founded by C.A. Petri. Here, too, the motivation to develop a theory was the task *"to homogenously and exactly describe as many phenomena as possible appearing during information transmission and conversion"* as Petri wrote in his fundamental work [8]. Later we shall see, to what extent this aim was accomplished and whether net structures can really bear system models. At present, contributions concerning net theory comprise some 1000 publications [9].

Nets as system models

Technical systems are structurally analogous to nets. This observation is valid, e.g., for electrical networks with concentrated components as well as for discretization of distributed structures, modeling effectively with the grid network of a different microstructure, e.g., electric fields.

This is valid, too, for grid networks of thermal and mechanical systems of which there are many well-known examples. Nets symbolize behavior structures of land and air traffic and of gas, water, and electricity supplies. Net structures of linked control systems have already been mentioned. Nets are recognizable as models of data processing using computer structures and architecture. In such different fields as biochemical process systems or real-time data processing systems, modeling by nets is a helpful means for structure as well as for effective links.

After the feasability of net-oriented considerations has been demonstrated, it shall be asked whether the theory of Petri nets can match the axioms of automation systems and whether it will be able to homogenously describe entire automation systems of different designs and aspects.

CAN NET THEORY MATCH THE REQUIREMENTS OF AUTOMATION ENGINEERING?

This question shall be answered in four steps.

First it shall be shown that Petri nets as a mathematically based theory comprise clear and easily understandable representations of all system axioms, which is a great advantage.

Second, other more advanced types of Petri nets are presented, applying practical consideration of automation technology.

The third paragraph abstracts the bases of net theory considering automation engineering.

In the fourth paragraph, different concrete automated realizations prove the applicability of the net theory.

System axioms and net theory

In its simplest form - and the proof on this level will be sufficient - nets are mathematically and graphically described by the following structures, elements, and behaviors (cf. Fig. 1):

Structural principle

The system (net N) consists of a set of parts (place S, transition T) interacting with each other (flow relation F). The parts of the system (places) are described by quantities (markings). The values (markings) of quantities of a system represent its state (case, situation, state).

This can be mathematically represented as follows:

$$N = (S, T ; F, M,) \qquad S = (S_1, S_2,...) \qquad T = (T_1, T_2,...) \qquad (1)$$

$$F = (SxT \cup TxS) \qquad M = (M_1, M_2,...) \qquad M_i \in (0,1)$$

C represents the incidence matrix of flow relation. The flow relation can graphically be described by directed arcs between places and transitions or between transitions and places. In the net, the markings represent the system's actual state. The marking is symbolized by black tokens which cover the marked places (cf. causal principle). Hence, the net is described as a bipartite graph.

Principle of decomposition
Parts (positions, transitions) can be subdivided into nets.

$$N = (S, T; F, M,)$$

$$S \rightarrow N_S = (S_S, T_S; F_S, M_S) \qquad (2)$$

$$T \rightarrow N_T = (S_T, T_T; F_T, M_T)$$

Causal principle
The system consists of a set of parts. Changes (markings) in these parts are clearly determined. State changes in the net operate as follows: the tokens of the indicated places (S) move to places not yet marked by passing switched transitions. In this causal connection, later states can only be the consequence of former ones. That means that all places leading to a transition are marked and that all places leading away from a transition are free.

The transition operates as follows: the marked places transmit their tokens to the following unmarked places.

$$M(v) = M(v-1) + C\, T_F\, (v-1) \qquad (3)$$

The operational logic corresponds to the switching and marking sequence of the net (reachability graph).

Temporal principle
The temporal sequence of operations can be obtaine by the use of timed nets. In timed nets, temporally weighted, i.e., delayed arcs (Z_{pn}) can carry the temporal behaviour[12]. Here a marking remains on its place until

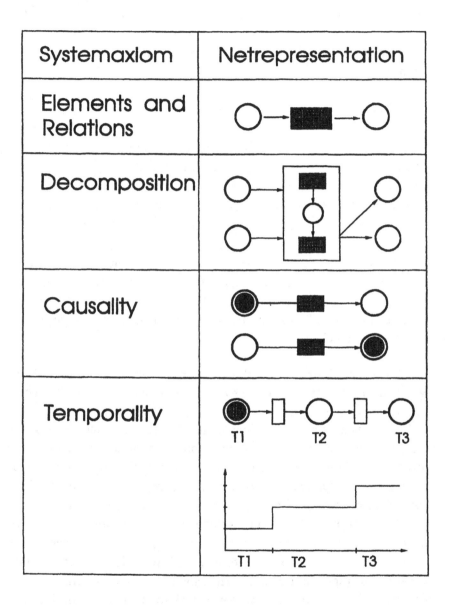

Figure 1: System axioms and Petri nets

the corresponding time has passed and the following transition switches.

$$N = (S, T; M(0), Z_{pre}) \qquad\qquad Z_{pre}((s, t)) = \begin{cases} 1, \omega \\ r, \omega \end{cases} \quad (4)$$

(Besides, there are other concepts to realize the temporal principle by Petri nets).

Aspects of net theory and its analytical reference

Application of nets in automation engineering requires practical mathematical modeling and especially interpretability of components, structures, and behaviors of and in automation systems. The analytical character has to be emphasized. On the other hand, net theoretical models can be synthesized and implemented in automation technology.

Table 1 opposes formal objects of net theory and their corresponding realizations in automation technology; following the principle of decomposition, these realizations represent different degrees of detail on different inspection levels. Thus, the structural similarities between net theory and automation technology are proved.

This section deals with the theoretical statements that are possible using net theory, i.e., the set of laws and their equivalences and references to automation technology.

Besides elementary definitions of its objects, the foundations of net theory are terminological definitions about its proprieties in specific arrangements or situations. The proprieties are described by terms such as *conflict, contact, confusion, restriction, safety, reachability, livelock, deadlock* and so on. *Safety* of a net, for example, means that at no place in the net is there more than one marking. In an automation system, this can mean that in one production cell there is only one workpiece. *Conflict* can mean that a token can alternatively leave a position, which demonstrates the impossibility of determination. In a multitasking system, this corresponds to an unsolved conflict as to whether a running task should be reset or finished when calling an operational system.

Each of the above-mentioned terms could be taken to interpret a characteristic situation or structure of an automated system, but such an investigation is beyond the scope of this chapter. Nevertheless, it is worth mentioning that specific net classes, i.e., elementary net constructions with specific structural proprieties, represent specific behaviors, too; thus, the so-called *free-choice nets* are always safe and restricted. By these basic or elementary structures, problems of complexity can be restricted, in a manner similar to what is achieved by the low-variant construction following Nassi-Shneiderman as compared to classic program flow diagrams.

Due to the causal dynamics of nets determined by switching, a behavior reachability graph (RG) can be determined theoretically by marking sequences corresponding to specific initial markings. Reachability graphs consider all junction alternatives due to conflicts. RGs can be interpreted, e.g., as a trace in a software system or as a trajectory in a state space for an object process-control system.

Recently, a solution for the treatment of timed nets by RGs has been attempted to solve [15]. Temporally determined deadlocks in particular can be detected.

A performance analysis of nets is the invariant calculus based in principle on a special solution following the calculus of marking sequences:

$$M(\upsilon) = M(\upsilon\text{-}1) + C \; T_F \, (\upsilon\text{-}1)$$

Specific markings - e.g., invariant relations - can be represented as follows:

$$I^T \, M(\upsilon) = I^T \, M(\upsilon\text{-}1) = \dots$$

The corresponding places invariants (invariant matrix) result from linear algebra:

$$I_S{}^T \, C = 0$$

This is also valid for specific switching sequences of transitions, leading to the so-called transition invariant

$$C \, I_T{}^T = 0.$$

Petrinet elements	Control-System				controlled system			
	development	control computer	software	reliability	production system	chemical engineering system	energy conversion	traffic/transport system
places anonymous definitive	development phase documents programs requirements	memory register screen plotter bus fault-state	memory variable semaphore program-state operand exception-state	fault-state	store buffer machine state tools worked-article	tank pipeline material state actuator/measure state, reactor state	heap power state pipes links	store buffer container pallet station siding track
transitions determined stochastic	specification construction simulation analysis programming put into operation	processor ADC DAC transmission	task procedure instruction	fail repair put into operation	assembling processing converting forming	reactor pump compressor mixer cracking	turbine alternator superheater pump transformator	fork joint point
relation weight of arc	logically temporal material	logically temporal material energy	logically temporal	logically temporal				
token anonymous individual	actual development state achieved results	data, parameter, value instruction interrupt system fault	data value exception-state	operation/fault-state	active tool machines phase	matter-state powerstate		venticles goods cars

Table 1

Invariants provide information about specifying dynamic structures of a system or about separate switching ranges. Thus a system can be appropriately modeled and analyzed.

Furthermore, certain exclusive conditions can be proved by invariant calculus, which is especially decisive for synchronization and coordination in multitasking and when applying primitives by operating systems.

Connecting the matrix notation of the net, linear algebra and graph theory provide further analytical methods to clearly analyze the given static structure and the net's behavior [16]. These three theories are especially convenient for modeling and analysis of behavior of automated systems. In this context the construction of complex systems by known parts and its top-down development with its improving accuracy should be mentioned. The net theory provides here precious advice concerning preservation and check of consistence [21] (it works on the consistency of dynamic constructs that are running) by its rules of embedding, decomposition, and design.

Finally, the simulation of the token game should be mentioned. These can be implemented independently from the application, thanks to the formal net representation. Thus, a large field of application, covering protocol specification as well as distributed flexible manufacturing systems, has been universally made available for investigation.

Other net types and their practical reference.

The development of the net theory during the last three decades made it possible to develop still more performant net types having the same basis but differing in their switching behavior, individualization of their tokens, and function. The majority of net types are supposed to be specially important in automation technology:

Colored and predicate-transition nets: Tokens are individualized (e.g., colored marks) and therefore individually different processes can run on the same nets [10]. In multitasking, for example, the state graphs of each task are identical but their states are different. In predicate-transition nets (PrT), mathematical predicates of places represent complex statements about the proprieties of their tokens. The originally binary-logical condition of place becomes a predicate-logical statement of the validity of a propriety [11]. Colored and PrT nets belong to the so-called *higher* nets. In fact, they are only a more compact representation.

Timed (determined) nets: In addition to the pure causal behaviour of a net, there will be a temporal component. The time (e.g., of a token remaining on one place) will be temporally determined or dependent on the switching behavior of a transition [12]. Thus, time-out conditions or watch-dog processes, for example, can be modeled.

Timed Stochastic nets: The switching of transitions behaves in a time-stochastic manner. The temporal firing delay now follows a probability distribution function, e.g., a Poisson distribution [13]. Thus, interrupt-controlled requirements of a process- controlling computer system, for example, can be modeled.

Fuzzy nets: In this relatively recent net type, the changeable switching behavior of a transition, for example, depends on the marking's previous position number [14]. Thus, Petri nets of this type allow the modeling of quasi continuous model behavior, offering an interesting perspective concerning the description of linear control systems. Connecting fuzzy and sharp PTr-nets allows a homogenous modeling of discrete and continuous systems.

By combining these extended net classes together with many other classes, a widespread descriptive area based on a fundamental theory is created.

All of the more recent net types have a more or less founded and developed theory. Their laws allow far-reaching conclusions and deep understandings about the corresponding modeled systems. Furthermore, treatment of nets will be more and more computer aided. For editing, simulation, and theoretical analysis, a great number of tools are in parts and already commercially available and well developed, albeit mostly incompatible [11].

Nets and automated systems.

This section will briefly sketch the correlation between nets and automated systems, while taking into account the different aspects discussed above.

Object-specific example 1:

The modeling of an experimental setup of local traffic using timed CE nets consisting of around 50 places and transitions is shown in Fig. 2 [13]. The control function was modeled at the same time, and the interaction of process and control was investigated.

Example 2:

The complex production of an industrial company and its logistical control were modeled and simulated using timed PTr nets (about 1000 places and transitions); see Fig. 3.

Technical device aspect, example 3:

To develop a single board microcomputer with ordinary ICs, a timed PTr-model was developed. Its functionality was proved by means of a reachability graph by taking into account time conditions [18]; see Fig.4.

Programming aspect, example 4:

Program systems and their components are often developed by using Petri nets, especially for automation systems with multitasking operations. Fig. 5 shows a net for implementation of protection functions in a process system of rails.

Design-oriented, example 5:

The practical process of developing an automation system can easily be presented as a net (cf. Fig. 6). Following the system design's method BASYSNET based on nets, innovative control systems were developed in the traffic field [20].

Project control, example 6:

Project planning and supervising of large development projects are often supported by net plans. Here the use of PTr nets is also possible (cf. Fig 7) thus enabling more developed planning declarations [19].

Functional-oriented aspect, example 7:

Specification of a control system can be modeled concretely and formally by using nets. Thus on an abstract level there is already a concept description that is mathematical, checkable, and automatically implementable in the future. **Figure 8** shows control functions specified in nets for a traffic control.

The above list can only be exemplary. Conclusions to be drawn from these examples are as follows:

1. Having recognized in the beginning the divergent aspects as a special problem, it can be noted that nets cannot only be applied as a means of description because processes in automation technology typically run in parallel.

2. Taking into account the many phases in the cycles of design, development, and operation of an automated system, nets can be used as a homogenous means of description. Inconsistencies during the developmental process when passing from one phase to another can be avoided.

3. The description of nets is clear, mathematical founded and proved.

4. For the description of nets, high-performance computer tools are already available on the market.

All in all, nets can be regarded as a theory of automation engineering that largely disallows harmful consequences of an engineering discipline without theory.

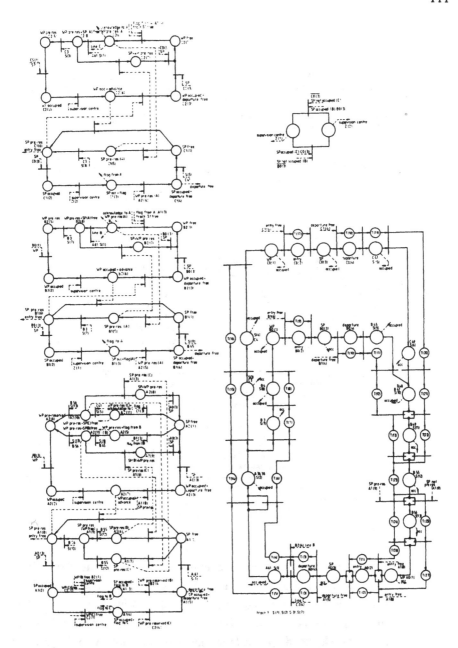

Figure 2: Timed Petri net for the process operation and control of an urban traffic system (see H-Bahn in [13]). Left: nets for the control functions for station areas A,B,C; right: net for the process operations of vehicles in the track-arrangement.

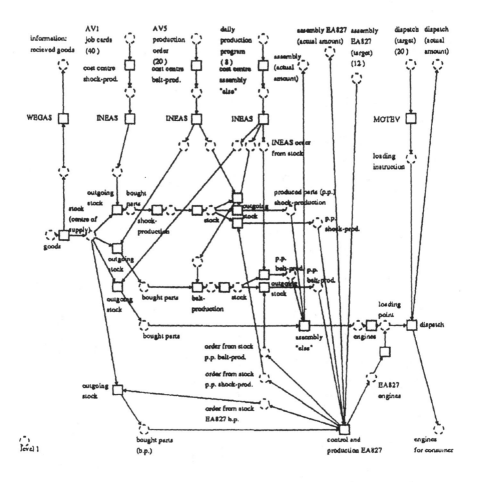

Figure 3: Petri net for logistic processes of a manufacturer. The dotted places and the transitions in the upper area contain the logistic control of the complex production net in the lower part. The production consists of four parallel lines, the schock production, the belt production, which join in the assembly transition, and a special control and production transition. All output tokens are distributed by the dispatch transition to the consumer.

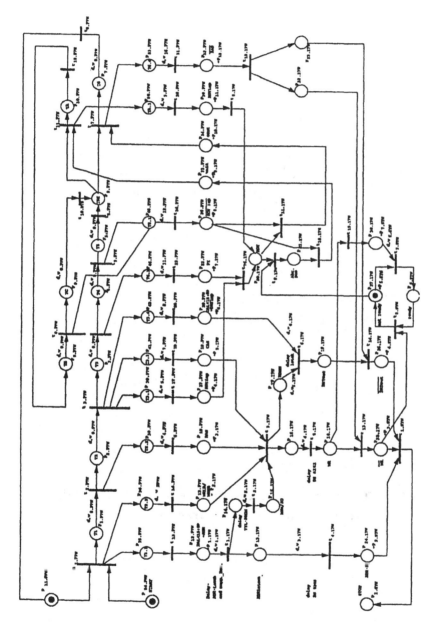

Fig. 4: Timed Petri net for the WRITE transaction in a micro-processor system (see [14])

114

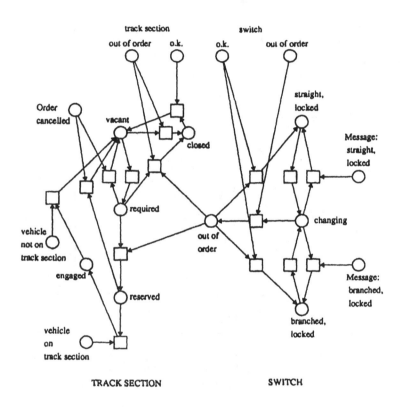

Figure 5: Petri net for protection functions of a traffic guidance system (Magnetbahn Transrapid)

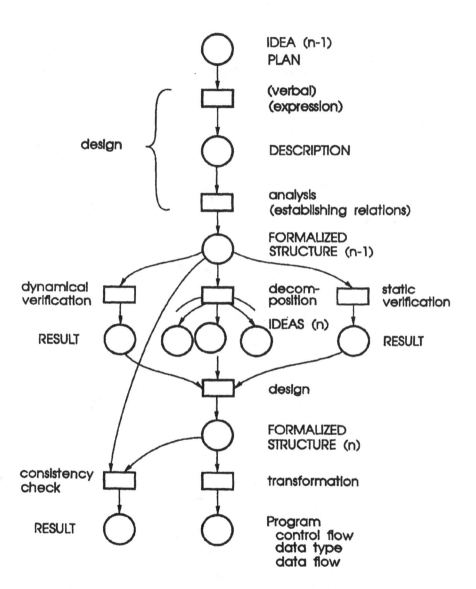

Figure 6: Petri net for design following BASYSNET (see[16])

116

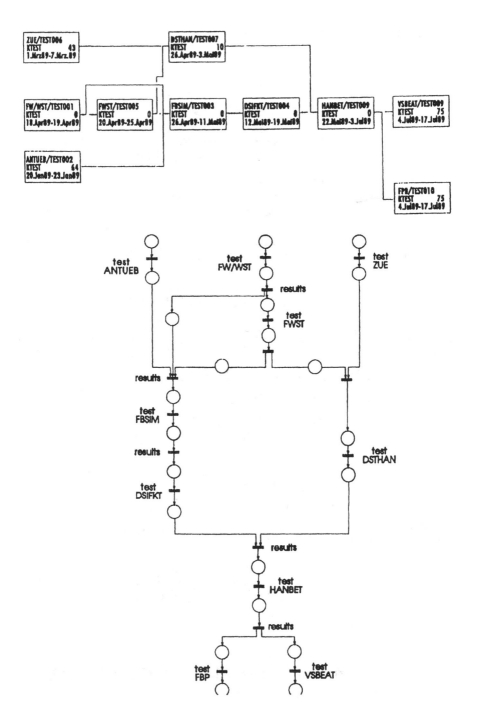

Figure 7: Net Plan of a prototype and corresponding Petri net

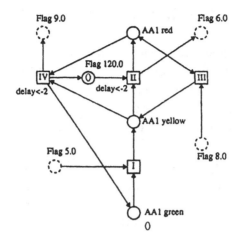

```
NETZWERK 1        0000           Traffic light AA1
0000      :U    M    5.0         If Flag 5.0 on
0001      :U    A    0.2         and AA1 green on
0002      :R    A    0.2         then reset AA1 green
0003      :R    M    5.0              reset Flag 5.0
0004      :S    A    0.1              set AA1 yellow
0005      :
0006      :SPA FB 247             starting yellowtime
0007 NAME :DELAY 2
0008      :
0009      :UN   T    9           If yellowtime over
000A      :U    A    0.1         and AA1 yellow on
000B      :U    M    120.0       and Flag 120.0 on
000C      :R    M    120.0       then reset Flag 120.0
000D      :R    A    0.1              reset AA1 yellow
000E      :S    A    0.0              set AA1 red
000F      :S    M    6.0              set Flag 6.0
0010      :
0011      :
0012      :
0013      :U    M    8.0         If Flag 8.0 on
0014      :U    A    0.0         and AA1 red on
0015      :R    M    8.0         then reset Flag 8.0
0016      :S    A    0.1              set AA1 yellow
0017      :
0018      :SPA FB 246             starting red/yellowtime
0019 NAME :DELAY 2
001A      :
001B      :UN   T    10          If red/yellowtime over
001C      :U    A    0.0         and AA1 red on
001D      :U    A    0.1         and AA1 yellow on
001E      :R    A    0.0         then reset AA1 red
001F      :R    A    0.1              reset AA1 yellow
0020      :S    A    0.2              set AA1 green
0021      :S    M    9.0              set Flag 9.0
0022      :S    M    120.0            set Flag 120.0
0023      :BE
```

Figure 8: Petri net for the control of a visual signalization device and corresponding program for a programmable control (programming language STEP 5)

HOW CAN A NET THEORY BE INTRODUCED IN PRACTICE?

It must be noted that the founders of net theory and their protagonists, with very few exceptions, have so far done very little to spread their theory into practice. This situation is slowly changing, and now the users themselves are making some efforts, probably because they gain the practical experience of using the net theory while the researchers only get the honor.

To spread net theory in the field of automation engineering, studies in the literature are useful: For automation technology, these especially include [6, 9, 12, 21, 22, 23] and others, Petri Net Newsletter published by Gesellschaft für Informatik(GI), and the excellent and actualized net bibliography [9].

Scientific aspects

Conferences taking place every year about Petri nets and their applications include reports of meetings of Lecture Notes on Computer Science published by Springer Verlag.

Educational aspects

The methodology of Petri nets at universities in Germany is, of course, particularly wide-spread. At the University of Hamburg, for example, studies of computer science are intensively supported by net theory [22]. At the Technical University of Braunschweig, the use of nets is taught to students of electrical engineering, computer science, and recently mechanical engineering after two years of their studies [6]. Furthermore, a lecture on advanced theory of automated systems is offered. These courses have been of advantage to our students when they start work outside universities.

We think the most important and successful method of transferring know-how into practice is *learning by doing*. Observations over the years and exchanges of information with colleagues who are active in this field in industry and research centers support this opinion. The active application of net theory motivated by the will to solve a new problem, shows after a short time of learning that already basic know-how of net theory is applied successfully and that people think in terms of nets. This feedback reinforces the willingness to learn and student's own ability to solve problems. People working with this theory commented that they can now solve their problems much better, whereas without this theory they could not have solved them at all. Some industrial projects thought to be unsolvable and inexplicable were later solved by problem analysis using nets.

CONCLUSIONS

All in all, the conclusion can be drawn that the application of net theory for the design of automation systems is already economically necessary. Thus, the intuitive, artistic, and manual process of making net theory a reality can be placed on a broad, scientifically based theoretical foundation. The effects of this can be especially noticed in quality, performance, and costs. If automated systems are more intensively treated with this means of design and its theory, new and interesting results can be expected.

All in all, the title question of this chapter "WHICH THEORY MATCHES AUTOMATION ENGINEERING?" can be answered precisely by the constructive suggestion of *Petri net theory*.

REFERENCES

[1] Steusloff, H.: Systemengineering für die industrielle Automation. Automatisierungstechnische Praxis 32, 129, 1990

[2] Zemanek,H.: Gedanken zum Systementwurf. in *Maier-Leibnitz, H. (Ed.):Zeugen des Wissens*, Mainz, Hase & Köhler, 1986

[3] Dieter Lorenz, Jürgen Schmid: Methoden für eine ganzheitliche Technikgestaltung. FhG-Berichte 2, 1988

[4] Rohpohl, G.: Eine Systemtheorie der Technik., München-Wien, Hanser Verlag, 1979

[5] Wunsch, G.: Eine Geschichte der Systemtheorie. München-Wien, Oldenburg Verlag

[6] Schnieder, E.: Prozessinformatik. Braunschweig, Vieweg, 1986

[7] Smith, E.: Kausalität und Temporalität bei der Modellbildung, *4. Symposium Simulationstechnik, Informatik-Fachberichte 150*, S. 127 - 134. Zürich, Springer 1987

[8] Petri, C.A.: Kommunikation MIT Automaten. *Schriften des rhein.-westfäl. Instituts für instrumentelle Mathematik an der Universität Bonn. Nr. 2*, Bonn 1962

[9] Bibliography of Petri nets. ISSN 0723-0508. Arbeitspapiere der GMD 315, 1988

[10] Jensen, K.: Coloured Petri Nets. *In Brauer, W., Reisig, W. und Rosenberg, G. (Eds.): Petrinets: Applications and relationship to other models of concurrency. LNCS 255*, S.248-299, Berlin-Heidelberg-New York, Springer, 1987

[11] Leszak, M., Eggert, H.: Petri-Netze-Methoden und Werkzeuge. Informatik Fachberichte; Berlin-Heidelberg-New York, Springer 1987

[12] König, R., Quäck, L.: Petri-Netze in der Steuerungs- und Digitaltechnik. München-Wien, Oldenburg Verlag, 1988

[13] Ajmone Marsan, M., Chiola, G.: Petri Nets with Deterministic and Exponentially Distributed Firing Times. *In Rozenberg, G. (Eds.): Advances in Petri Nets LNCS 266*, pp.132-145, Berlin-Heidelberg-New York, Springer, 1987

[14] Lipp, H. u.a.: Unscharfe Petri-Netze. Wiss. Schriftenreihe der TU Karl-Marx-Stadt 7, 1989

[15] Quäck, L.: Aspekte der Modellierung und Realisierung steuerungstechnischer Prozesse mit Petri-Netzen, Automatisierungstechnik 4, 116-120, 5, 158-164, 1990

[16] Abel, D.: Modellbildung und Analyse ereignis-orientierter Systeme mit Petri-Netzen. Fortschritt-Berichte VDI Nr. 142, VDI-Verlag, Düsseldorf, 1987

[17] Schnieder, E. u.a.: Analyse und Simulation von Verkehrssystemen mit Petri-Netzen am Beispiel der H-Bahn. Eisenbahntechnische Rundschau 30, 409-413, 1981

[18] Müller, T.: Nutzung von Petri-Netzen bei Entwurf und Verifikation eines SKR-softwarekompatiblen 16-Bit-Mikrorechnersystems. Dissertation (in prep) TU Dresden 1990

[19] Franz, V.: Planung und Steuerung komplexer Bauprozesse durch Simulation mit modifizierten höheren Petri-Netzen. Dissertation, Gesamthochschule Universität Kassel 1989

[20] Schnieder, E., Gückel, H.: Petri-Netze in der Automatisierungstechnik. Automatisierungstechnik 5, 173-181 and 6, 234-241, 1989

[21] Reisig, W.: Systementwurf mit Netzen. Berlin-Heidelberg-New York, Springer, 1985

[22] Jessen, E., Valk, R.: Rechensysteme - Grundlagen der Modellbildung. Berlin-Heidelberg-New York, Springer, 1986

[23] Rosenstengel, B.: Entwicklung eines Netz-Modells zur Erfassung einer petrochemischen Produktion. Bergisch Gladbach-Köln, Josef Eul Verlag, 1985

Further reading:

A. Kündig, R.E. Bührer and I. Dähler: "Embedded Systems", Berlin-Heidelberg-New York, Springer, 1986

X. Cao, Y. Ho: "Models of Discrete Event Dynamic Systmes", IEEE Control Systems Magazine, 6, 69-76, 1990

T. Stonier: "Information and the Internal Structure of the Univers", Berlin-Heidelberg-New York, Springer, 1990

T. Murata: "Petri Nets, properties, analysis and applications", Proc. IEEE, 77, 541-580, 1989

J. L. Peterson: "Petri Net Theory and the Modelling of Systems", Englewood Cliffs, NJ, Prentice-Hall, 1981

S. Chen, J. Ke and J. Chang: "Knowledge Representation Using Fuzzy Petri Nets", IEEE Trans. Knowledge and Data Eng. 2, 311-319, 1990.

5

REQUIREMENTS ENGINEERING FOR REAL-TIME AND EMBEDDED SYSTEMS

Peter Hruschka
Systemhaus GEI
Pascalstrasse 14
D5100 Aachen

> *NEW PROBLEMS REQUIRE NEW SOLUTION STRATEGIES*

INTRODUCTION

From batch systems to embedded real-time systems

Comparing software systems developed today with systems developed some years ago, one can often observe an enormous increase in complexity. Consider the size of the systems: in the fifties programmers were content to solve mathematical equations or formulas with programs containing some hundred lines of code. Today, new software projects often involve many 100,000 lines of code, addressing complete factory automation or complicated weapon systems.

Another kind of complexity is the embedding of software systems into the "real world". For some time, programs were written to perform their job whenever the operator started the program (*batch programs*). Later on, programs performed their jobs online, in a *transaction oriented* way, such as in banking or airline reservation systems. Today, a wide variety of systems require software to react immediately to events coming from the "real world": programs have to react automatically to various sensor inputs, to interrupts, to alarm conditions, etc.. Such systems are called *"embedded systems"*. A subset of these embedded systems that deals with high frequencies of external events or has to provide its reactions under severe timing constraints is called *"real-time systems"*.

To handle these complexities and produce good quality software within feasible schedules and budgets, new techniques addressing the specific issues arising in such systems are required. The following section describes how methods and tools have evolved to handle the problems of today's embedded real-time systems.

From an idea, via methods, to tools.

As soon as problems in system development are identified, a typical cycle of evolution takes place (Fig. 1): it begins with a good idea, from which a systematic method arises for handling the problem. At this stage, the method is normally only available to a selected few. But as soon as the method is stable, well tested and refined to overcome initial weaknesses, tools are then made available to automate whatever portions of the new method can be turned over to a computer (or a mechanical tool). At this time the idea becomes available to a wider group.

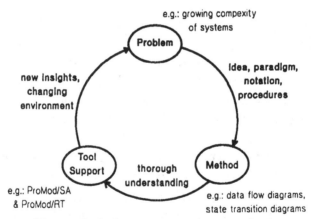

Figure 1: Software technology evolution

Where are we today?

For more than 20 years, experts from the academic worlds and industry have tried to overcome a problem known as the "software crisis": software projects are difficult to plan and control, they often run out of budgets and schedules, and finally - when delivered - often do not fulfil the user's requirements (cf. Chapter 2.3).

Methodologies and tools have been developed following the evolution cycle described above: in the late sixties, the software life cycle was described by W. Royce and B. Boehm, identifying major phases during the development and the activities to be performed in these phases. Various methods were subsequently developed. Around 1970, most research was in the area of structured programming: E.W. Dijkstra and N. Wirth introduced top down development, stepwise refinement and recommended restriction to 3 basic structured constructs: sequence, iteration and selection.

In the mid seventies, people concentrated on using these ideas for the design phase: D. Parnas introduced ideas for structuring large systems into "information hiding modules". M. Page-Jones, E. Yourdon, L. Constantine and G. Myers are just some of the researchers who then developed these ideas and are now the fathers of "Structured Design".

In the late seventies, T. DeMarco and D. Ross were among the first to refine these structuring ideas for the analysis phase [1]. "Structured Analysis" was the result. For some time, many developers of embedded real-time systems used Structured Analysis as best available method to capture their requirements. Although the method was adequate, they shared a feeling that it was not complete in the sense that it did not take into account the specific problems of this system category. Between 1983 and 1986 D. Hatley and I. Pirbhai adapted and extended the ideas of Structured Analysis to cope with the specific complexities arising from embedded real-time systems [2].

Tools supporting these structured methodologies always came a few years later: PASCAL compilers supporting structured programming showed up about 5 years after N. Wirth introduced the language and his prototype compiler. At about the same time, Caine & Gordon's PDL (the Program Design Language) supported Pseudocode for top down development and stepwise refinement. GEI's ProMod - originally supporting Structured Analysis, Design and Implementation - became available in 1982. This chapter will show how Hatley and Pirbhai's extensions for real-time analysis are supported in ProMod/RT.

MULTIPLE VIEWS TO CAPTURE REQUIREMENTS

As mentioned above, the new kinds of complexity require additional techniques to handle them. In requirements engineering, most experts today agree on using a number of different abstract models to capture different categories of requirements. The idea is that real world systems can be seen from different viewpoints. The three views considered to be most important portray the following three major aspects of a system and answer the following questions:

- What functions are performed by the system: the *"function view"*. What are the major functions? What subfunctions do they consist of? How are the functions combined? How are the functions to be achieved and what is the strategy of their business policy?
- What data does the system work on: the *"data view"*. What are the inputs? What are the outputs? How are they structured? How many of them have to be handled? What other characteristics do these data have?
- How does the system behave dynamically: the *"control view"*. Under which conditions are functions performed? What states or modes does the system have? What are the triggers or events that start certain activities or stop them? What events lead to changes in the state behavior of a system? Which data control the operation of the system?

Modeling functions and data

Considering the functions and data of a system is as old as data processing. Functions and data are not only relevant for embedded real time systems, but for many kinds of system. Therefore, the "classical" requirements techniques like Structured Analysis (DeMarco, 1978) can handle these kinds of models very well. Before going on to specific modeling techniques for real-time systems, let us shortly review some features that made "classical" Structured Analysis (SA) so popular among analysts.

Structured Analysis is based on graphics: Following the old saying "a picture is worth a thousand words", SA tries to capture a system in the form of a hierarchical set of diagrams. Each diagram portrays a limited portion of the system, its functions and the connections between the functions, the data flows. Therefore these diagrams are called *data flow diagrams*. Instead of the traditional approach of enumerating the functions and describing their aspects in isolation, with SA one tries to draw the "big picture" first. The diagrams show the complete decomposition of the system, how the various subsystems cooperate, what kind of information they exchange. Data flow diagrams have proven to be an excellent communication tool between clients and systems analysts, allowing errors to be identified very early in the development process.

SA documents are clearly structured: By arranging the complete requirements specification of a system in a hierarchy of diagrams, it is easy to distinguish between the important aspects of the system (usually modeled on the top levels) and detailed aspects (found in lower levels). Everything has its place in the models: information about the decomposition and the interfaces is found in the diagrams; more detailed information about the data of the system is written down in the data dictionary and the details of the functions (how the data are processed) is written in mini-specifications (one mini spec. for each function in the system).

SA permits early error checking: The method does not only provide guidelines for creating models, it also provides a set of rules that have to be obeyed when constructing such system models. These rules ensure the completeness of the system model (e.g., each function in a data flow diagram has to have at least one input and one output; each data flow has to be defined in the data dictionary; each function must be described in a mini specification, ...) and they ensure the consistency of the system model (e.g., what goes in and comes out on one level must also go in or come out on the next level of decomposition; the interfaces of a function in the diagram must also be consistently described in the corresponding mini specification; two identically named data flows in the diagram can only have one description in the data dictionary, ...).

These three arguments have proven to be very helpful for systems analysts. Any extension of the method should therefore strive to retain those advantages.

Modeling control

In the seventies, the typical answer to the questions "How will the system be controlled?" and "How does the system behave dynamically?" was: "We will deal with this problem *after* specifying the requirements, in the design". But people involved in the development of real-time systems want to address the third view - the control view - very early in the development process. The design phase seems to be too late. Let us consider some characteristics generally associated with real-time systems, in order to understand the need for modeling dynamic behavior and control early in the software life cycle.

Real-time systems usually consist of a set of functions (or processes, or tasks) that have to work in parallel, i.e., these functions mostly perform their jobs independently; they only they have to synchronize their activities with others in order to communicate and exchange data. After that they go on independently again.

Real-time systems are also characterized as being highly time dependent and time critical. Often data are only available for a certain limited time period and answers or reactions from the system must often be given within a very short time frame.

Real-time systems often work at the edge of technology. While hardware is not so important for batch systems or online transaction systems (they can perform their job whether the hardware is a little bit slower or faster), for embedded real-time systems the hardware selected and the interfaces chosen for a process environment are often crucial in the decision whether a system can be automated or not. Therefore these aspects have to be part of the analysis; characteristic performance data have to be known in time.

To cope with these specific characteristics of real-time systems, we do not only need to model function and data, as emphasized by Structured Analysis, we also have to have a formalism to model the control aspects of systems.

STRUCTURED ANALYSIS FOR REAL TIME SYSTEMS (SA/RT)

A formalism to model control

A widely known and used way of modeling control is the use of state transition diagrams. They allow the analyst to decompose the behavior of a system into a set of externally observable states and specify the conditions (or the events that have to occur) when the system changes from one state to another. State transition diagrams are always used whenever an action does not only depend on the input values, but on previous events that have brought the system into a certain state. So, signals or events are only valid if the system is in a specific state. Otherwise they may be ignored or may trigger a different operation, depending on how the model is constructed.

Alternative methods and notations are available to describe control of a system. If for instance there is no need for states (i.e., any output of the system can be described as logical combination of input conditions) then we may use decision tables or activation tables. These tables describe all the input signals that are relevant to the part of the system to be controlled. They also show the possible combinations of values these signals may have and specify output signals or process activations that can result from the input signals.

In simple cases a short paragraph in natural language may be sufficient to explain the dynamic behavior adequately.

Integrating the control model with the data and function model

Now that we have a formalism to model control, the question is: how does this formalism relate to Structured Analysis? How do we make sure that models developed with state transition diagrams fit together with data flow diagrams, mini specifications and the data dictionary?

It was Derek Hatley who answered this question in 1984. He introduced **control flow diagrams** in addition to the data flow diagrams. A data flow diagram shows the flow of data to any place in the system where these data are needed. In the context diagram, one can see the input data from and output data to the external world (outside the system under consideration). On lower levels, one can see how data are passed between the different parts of the system. The same is true for control signals. In the control context diagram, one can see the external stimuli coming into the system, e.g., from various hardware devices. One can also see signals created by the system going out to the external world. At lower levels, signals flow between parts of the system. So the diagrams are identical with respect to the functions modeled, the only difference is that data flow diagrams show the flow of data between these functions while control flow diagrams show the flow of control.

The separation between data and control is done for very practical reasons. Modeling control adds another dimension to the static models of Structured Analysis. If both aspects are shown in the same diagrams, they often become cluttered with information and simply unreadable, thus violating one of the basic rules of Structured Methodologies: keep models simple and readable, since they are mainly used for human communication. Only in simple systems (such as in the following case study) can the two views can be combined.

Another problem that frequently arises is that data and control are often difficult to distinguish. An easy way is to ask the question, "Will this 'thing' be used directly in a process or be transformed in some way? If the answer is "yes", it's probably *data*. Or you might ask: "Will this 'thing' primarily influence or direct other actions or processes?" If the answer to this question is "yes", then it is probably *control*. If the distinction can't be made or if the answer is "yes" to both questions, model the 'thing' as data where it acts as data and control where it acts as control.

In Structured Analysis, mini specifications are used to describe how data are processed (how the input data of one process are transformed to create the output of that process). The corresponding items in SA/RT are now called **control specifications**. Control specifications are not written for a single process, but for a complete diagram. They specify how the control signals in the diagram are processed, creating new control signals or activating processes in the diagram. The notation we use for writing such control specifications is state transition diagrams (or decision tables, or activation tables, or short natural language paragraphs) as described above.

The control model now has a clearly defined place in the overall model: the hierarchy of diagrams defines the overview. The diagrams can be drawn and interpreted from the two different viewpoints (data flow or control flow). The data flow diagrams are augmented by mini specifications to give more details about data processing; the control flow diagrams are augmented by control specifications to give more details about control processing. The component missing from the overall model is the dictionary. Of course, not only data are defined in the dictionary (as described above in the section on Structured Analysis) but also the control signals are entered into the dictionary (specifying their structure, frequencies, values, boundaries, etc.)

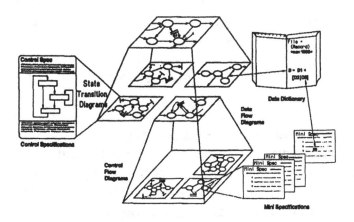

Figure 2: The SA/RT model: how the components fit together

One of the most important aspects of integrating control models with Structured Analysis in this way, is that all the checking and balancing rules can now be applied to the extended model. One rule ensures that any signal going into or out of a process appears again on the next level of decomposition (as input or output of some of the subprocesses). Of course, these signals may be decomposed on their way down through the hierarchy, but then the rule applies to the sum of the components of the signal. So we can ensure that no signal gets lost on the way from its origin to the place where it is finally needed and used.

Another rule checks signals in a control flow diagram with the corresponding control specification. In order to graphically distinguish signals flowing up and down in the hierarchy from signals flowing from a diagram to the control specification or vice versa, the latter are marked with a short bar at the beginning or at the end of the arrow. This short bar indicates the control specification. So arrows flowing into a bar are to be used in the control specification as input signals, and arrows flowing out of a bar are signals generated in the control specification.

Before summarizing the advantages of SA/RT two caveats should be discussed: As mentioned above, real-time systems are usually time critical systems and nearer to the edge of technology than other systems. Therefore, many analysts concentrate on the efficiency of the system right from the beginning of a real-time project. The major question is always: is the suggestion for a certain decomposition going to work fast enough? There is nothing wrong with thinking about efficiency, too! But there is something wrong with *only* thinking about efficiency. Even a real-time system needs a good structure, well defined tasks with clear interfaces and a well understood control behavior.

For systems operating under severe timing constraints, we require a thorough understanding of what goes on at what time. But before one can add behavioral information one has to have some structural information (to rephrase an old engineering principle: know what to control before you define how to control it). This does not mean that one has to start drawing data flow diagrams first and then add control specifications. It is often easier to draw a state transition diagram for a portion of the system first and then use the signals of this diagram to find a good decomposition. It only means that the dynamic behavior alone is not sufficient as a specification; the final model should specify both.

So the suggestion in SA/RT is: do not overemphasize control, model only the control portions that are inherent in the problem statement and ignore (for the moment) those that are the result of current technology or other implementation constraints. They should be dealt with in the design phase, when the SA/RT model is transformed into an implementation model taking all known constraints into account. The distinction between control aspects that should be considered in the analysis phase and those that should be ignored is more complicated than with typical batch or on-line transaction systems. If the analyst does not have sufficient experience to determine whether a proposed solution will work on given hardware or with given peripherals, some prototyping experiments may prove helpful.

The advantages of SA/RT are derived from the original methodology: again we use graphics (state transition diagrams and control flow diagrams) to improve communication and understanding. The new overall model shows a clear structure: one always knows where to find information. (For the extensions, the numbering and naming scheme of Structured Analysis is extended, too. The data flow diagram, the corresponding control flow diagram and its control specification all have the same name and the same number, thus indicating that all these partial models belong together.)

Early error detection in SA/RT models is possible because of the rules defined for the extensions. So we can again check the completeness and consistency of the model.

In addition to Structured Analysis, SA/RT provides powerful new modeling techniques which cope adequately with the problems of analyzing real-time systems.

EXCERPTS FROM A CASE STUDY

Excerpts from an industrial automation project will demonstrate some of the points discussed above. The goal of the project was to control the filling of paint cans so that on the one hand all government regulations about minimum filling amounts are obeyed and on the other hand overfilling is avoided. Figure 6 shows the schematics for this project.

The context diagram (Fig. 3) shows the interfaces between the computer controlled filling process and its environment: the hardware, in which the system is embedded and the operator. (Note: Due to the simplicity of this example we have combined the control flow diagram and the data flow diagram into a single diagram. In more realistic models a separation of the two is recommended. As supporting tool for the method, ProMod allows for both possibilities: one can combine the views on the screen and in the printed output, or separate the views at any given time.)

134

On the next level, there are four processes which can work independently. The two processes connected to the scales are triggered by the corresponding interrupts ("scales ready"), the operator interface is triggered
by "commands" and the major controlling process ("monitor filling") is
triggered whenever a new "net weight" arrives. Note the balancing between the two diagrams: All inputs and all outputs from the single node
in the context diagram show up again as inputs and outputs in this first
level of detail.

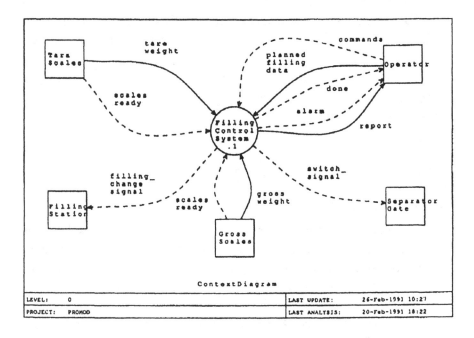

**Figure 3: The context diagram: An industrial process in its environ-
ment**

The secrets of how the process control algorithm works becomes clearer on studying the state transition diagram below (**Fig. 5**). You see the major states of the monitor process (drawn as rectangles) and the transitions between the states (drawn as lines). The labels on the transitions describe the conditions and events necessary to cause a state transition (above the separation line) and the signals that are generated when the transition takes place (below the separation line).

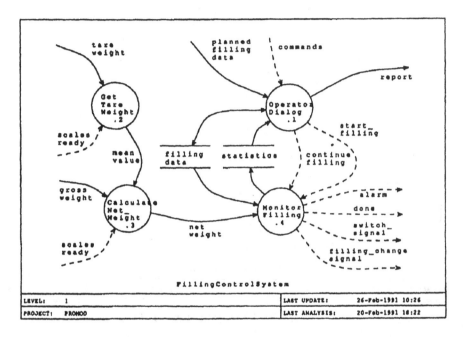

Figure 4: A first level decomposition

The flow diagram belonging to this state transition diagram (**Fig. 6**) is a refinement of process 4 in **Fig. 4**. It shows which of the subprocesses generate or receive the signals shown in the transitions in **Fig. 5**.

136

The four diagrams are only a short excerpt from the complete requirements specification for this case study. All the detailed definitions of the data and control signals in the data dictionary have been omitted. To complete the model, one also has to write mini specifications for all the functions shown in the diagrams. The excerpts should, however, be sufficient to give a flavor of the general procedure that systems analysts have to perform in order to come up with a complete requirements specification.

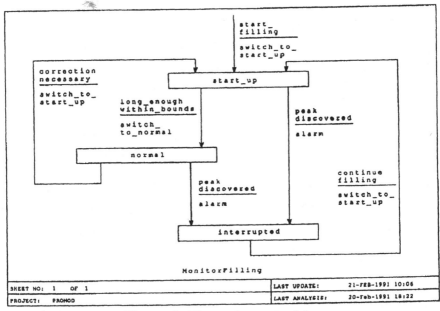

Figure 5: The main control loops

TOOL SUPPORT

The SA/RT method as described above is well supported by CASE tools. Let us discuss some tool features using GEI's CASE environment ProMod [3] as an example. Five major kinds of activities are supported:

- editing diagrams and textual specifications
- checking the consistency and completeness of the resulting model
- managing the central repository
- preparing structured documentation
- suggesting the next steps in the life cycle

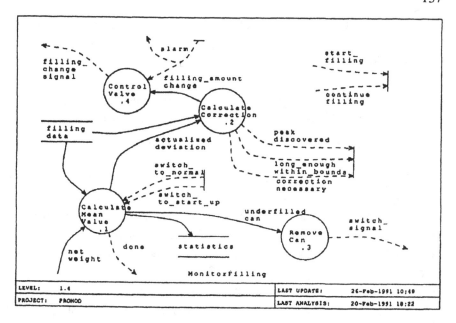

Figure 6: Part of the second level decomposition

Since the SA/RT model consists of many different components (as shown in **Fig. 2**), CASE tools should allow flexible editing of such components. ProMod offers a multi-window environment, where analysts can work on a control flow diagram in one window, add some definitions to the dictionary in another window, use a third window to draw state transistion diagrams, etc. This highly interactive way of specifying systems is shown in **Fig. 7**.

As soon as parts of the requirements model are edited, extensive checks can be preformed by the ProMod analyzers. The analyzers will give warnings whenever necessary details have not been entered. They will also give error messages whenever the analyzers discover contradictions or unbalanced situations in the model. These analyzers are highly effective help for system analysts, since they pinpoint trouble spots in the requirements that would lead to high costs later on for error correction and to missed deadlines because of rework.

The heart of today's CASE tools is the central repository, where all important components of the requirements model are kept and all the links and relationships between them are maintained. ProMod provides a highly efficient, multi-user repository which allows rapid changes between aspects that are of interest to the analysts and efficient updates to all dependent components whenever one component is changed.

The information in the repository and the results of the checks on the model are the basis for thorough reports. In addition to top-down ordered models and messages showing inconsistencies and incompleteness, a wide variety of tables, cross reference lists and statistics about the requirements specification are made available as basis for walkthroughs or reviews.

Whenever the system developer has collected enough details about the requirements, the ProMod analysis-design transformer can suggest an architecture for the planned system and the code generation tools in ProMod/SI can transform the system into Ada, C or PASCAL code. Thus the SA/RT specification is the basis for fast prototyping of real-time systems.

The ProMod tools are available on a wide variety of platforms including VAX/VMS, various UNIX platforms such as DEC RISC, SUN or HP9000 and IBM OS/2 systems. Handling the ProMod tools is especially easy, since the analysts will find a familiar "look and feel" on their platform, which they already know from other tools on the same hardware (e.g. MOTIF, Presentation Manager, Open Look, DECWindows).

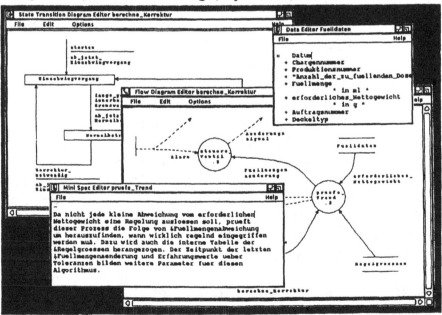

Figure 7: Multiple ProMod windows showing components of Sa/RT model

SUMMARY AND OUTLOOK

System analysts today have effective, industry proven methods and tools available for specifying RT&E systems requirements. Structured Analysis has proven to be a stable basis for specifying functional and data requirements; the real-time extensions developed by Hatley and Pirbhai take care of the dynamic aspects and the control aspects [4]. The thorough integration of these methods offers excellent guidance to analysts on their way from first ideas and sketches to solid, structured models. It also provides the basis for tools like ProMod/SA and ProMod/RT to do the necessary error checking in early stages of a project.

Nevertheless, research and development in the area of requirements engineering for real-time and embedded systems go on.

One of the near term goals is to further improve the communication with the end user and the feedback at this very early stage in a project. To do this, the SA/RT model is augmented with simulation and animation aspects. As soon as the model is complete, the user can visibly control the behavior of the system and "play" with various scenarios. The results of this animation are not only presented in terms of the formal SA/RT model but also in other graphical ways, where one can use predefined symbols for switches, gauges, indicators, etc. to build mock-ups of the user interface. The first tools supporting this animation (based on SA/RT) are already available.

At the strategic level, large European research projects have been set up. Two examples for such programmes, in which GEI is envolved, are the Eureka Software Factory (ESF) and COMPLEMENT. Within the ESF program a complete factory for the development of real-time and embedded systems is being produced. In a cooperative effort between Germany, France, Great Britain and Scandinavian countries, methods and tools for RT&E systems are integrated. Special emphasis is put on a communication driven archtitecture for the tools in this factory. Requirements tools, design tools, performance modeling tools, management tools, etc. will exchange information through a "software bus". This allows consistent development of RT&E systems in multi-site, multi-national projects without a centralized repository, using tools from different vendors from different countries in a uniform environment.

Another large effort has been started within the framework of the ESPRIT program: the COMPLEMENT project aims at improving the effektiveness of industry in the area of RT&E systems by improving the state of awareness of and capabilities in methods, tools and training. The project started with a Europe-wide assessment of industrial practice, methods and tools. The resulting information will be used to identify gaps in the areas of usage, methods and tools. These gaps will then be addressed by integrating appropriate techniques and when necessary by complementing them with additional techniques. Practice gaps will be addressed by a significant technology transfer program, including the development of tested training material, publishing results in form of books and establishing a training center.

Technology transfer seems to be the major challenge for the years ahead. Requirements methods and tools for RT&E systems are available. Now engineers must learn more about it so that CASE becomes common practice, like compilers and text editors.

REFERENCES

[1] DeMarco, "Structured Analysis and System Specification"8, *Yourdon Press*, Prentice Hall, New York, 1978

[2] Derek Hatley, "The Use of Structured Methods in the Development of Large", *Software-Based Avionic Systems, Proc. AIAA*, pp. 6-15, Baltimore, 1984

[3] Peter Hruschka, "ProMod at the age of 5", *LNCS 289, ESEC '87* (Nichols, Simpson Eds.), Springer, Hamburg, 1978

[4] Derek Hatley, Imtiaz Pirbhai, "Strategies for Real-Time System Specification", Dorset House, New York; in Europe: John Wiley, London, 1987

6

REAL-TIME PROGRAMMING LANGUAGES

Wolfgang A. Halang
Rijksuniversiteit te Groningen
Department of Computing Science
9700 AV Groningen/The Netherlands

Karlotto Mangold
ATM Computer GmbH
D-7750 Konstanz/Germany

INTRODUCTION

Historical Development of Real-Time Languages

Longer than in other areas of data processing, assembly languages were prevailing for the formulation of real-time applications. The reason for this were the high prices for processors and memories enforcing optimal programming with respect to execution time and storage utilization. However, also for developing real-time systems, the demand for high-level languages grew. The most natural approach was to augment existing and proven languages oriented toward scientific and technical applications by real-time features. This resulted in a large number of vendor-specific real-time dialects of FORTRAN and later also of BASIC and a few other languages.

The idea of creating new languages especially dedicated to the programming of real-time software was first realized in the 1960s in a rather curious way: in Britain, the two languages Coral 66 and RTL/2 were developed, but they had no real-time features. Finally, since 1969, also new real-time languages possessing real-time elements have been devised and later standardized, viz., PEARL in Germany, Procol in France (later replaced by LTR), and Ada in France and the U.S.A. A third group of universal, procedural programming languages that are often used for real-time programming purposes (although they contain, at most, rudimentary real-time features) includes C, Chill, Modula-2, and PL/M. The reasons for their wider application may be that they are better known and more easily available than genuine real-time languages.

In parallel to the three approaches outlined above, non-procedural languages and "fill-in-the-blanks" program generators have been devised for a number of particular application areas. Examples from the first group are Atlas for automatic test systems, Exapt for machine tool controls, and Step5 for programmable logic controllers. The program generators are generally vendor specific and contain preprogrammed function blocks for measurement, control, and automation that need to be configured and parameterized by the user. Program generators are available for a number of distributed process control systems and programmable logic controllers, and have proven to be relatively successful in their respective application areas.

Requirements for Hard Real-Time Systems

Per definitionem [14], real-time operation requires the availability of results within a given time span after the arrival of the input data. The timing constraints of hard real-time environments characteristically are tight and must be met under any circumstances: violation of these constraints at best will give rise to useless results and in general will endanger the controlled process and the safety of persons. Hence, the timeliness requirement is closely related to the reliability and security of operation. In addition, in order to design and to realize real-time systems that fulfill the above demands, the system behavior must be fully predictable. Predictability is a necessary precondition for the use of such a system in a safety-related application.

When compared with the requirements for conventional batch or time-sharing systems, the additional features that a user expects of real-time systems are all concerned with the dimension time. In contrast to the common practice prevailing in computer science, in real-time systems the dimension time deserves the highest attention and must be dealt with in an explicit way (cf. Chapter 2.1).

Thus, in these systems it is necessary to know a priori not only the execution times required by the different processes that are processed "simultaneously" but also their interdependence in time. The main reasons for this are:

- predictability of system behavior,
- supervision of timely task execution,
- application of feasible task-scheduling algorithms, and
- early detection and handling of transient overloads.

During the implementation phase of a real-time application, detailed knowledge of task execution times enables correct selection of that hardware configuration that will later yield the on-time performance specified in the design phase as a requirement. Real-time software must be guaranteed to meet its timing constraints [5, 7, 12, 24, 37, 51, 53, 56]. If full predictability of hardware and operating system behavior is provided, a schedulability analysis [53-55] based on the knowledge of task execution times can be carried out to ensure the timing correctness of real-time programs. Such an analysis and the predictability of the temporal system performance are prerequisites for the approval by pertinent authorities of real-time computing systems that are to be employed in safety-critical environments. In a fault-tolerant system, the execution of each task is supervised by the operating system in order to assure correct operation or to detect faults, respectively. Hence, each task is associated with a watchdog timer that requires the tasks overall run-time as a parameter.

As stressed above, a process control system employed in a hard real-time environment is expected to execute all tasks within predefined time frames, provided that doing so is actually possible. Algorithms that always generate appropriate schedules for all task sets executable under observation of given strict due dates are referred to as feasible, and several have been identified in the literature [26-31, 40]. In the state of the art of contemporary high-level real-time languages and operating systems, it is characteristic to note, that the presently employed scheduling methods, based on either fixed or on user-modifiable priorities, are not feasible.

The implementation of feasible scheduling algorithms, such as the deadline-driven algorithm that is optimal for single-processor systems, needs to be accompanied by language elements for specifying the due dates and by methods to determine total and residual run-times (or at least upper bounds thereof) required by tasks to reach their completion. Furthermore, if task run-times are a priori given, scheduling theory provides necessary and sufficient conditions which allow the early determination whether it will be possible to process a task set in time. Otherwise, parts of the workload have to be discharged. This can be achieved in an orderly and predictable manner if it is possible to state in the source program which tasks can be terminated or even replaced by ones with shorter run-times, if necessary in emergency or transient overload situations.

In order to improve the reliability of process control data processing, it is of utmost importance to increase the utilization of real-time high-level languages, because presently a large amount of the applications is still programmed in assembly languages. It is very hard or even impossible to check whether all above mentioned specifications are met in a system which is implemented in assembly language. A high order language must reflect the users' way of thinking, and the users generally are engineers and technicians, but not computer scientists. So the language features should be easily conceivable, safe to handle, and application oriented. Furthermore, wide use of dynamic language elements seems questionable and, also for security reasons, it must be possible to completely relinquish their use.

Survey of Contents

In the next section, we shall present an inventory and a comparison of the real-time elements provided by the main languages for this application area, viz. Ada, Industrial Real-Time FORTRAN, HAL/S, LTR, PEARL, PL/1, and Real-Time Euclid. In this context, we shall examine whether Ada, the most recent development, represents progress toward allowing the formulation of all problems in process control. Also, we shall take a closer look at PEARL, which was designed by process automation engineers and which differs from all other languages by having a concept for distributed systems. Comparing the facilities provided by these languages with the above-discussed requirements established by industrial environments for real-time data processing systems gives rise to a synopsis of important real-time features that are not yet or are only seldom realized. Therefore, in a later section we take remedial measures by defining and discussing appropriate new language constructs in a form independent of presently available languages. The chapter concludes by considering languages for programmable logic controllers (PLCs), which have not received any attention yet from computer science and engineering. The importance of PLCs for process automation is growing, and they are becoming versatile real-time systems in their own right. Therefore, such an interest appears to be well justified, especially since interesting developments are currently under way, including an international standardization effort: but there is also some duplication of work.

REAL-TIME FEATURES IN HIGH-LEVEL LANGUAGES: A COMPARISON

Selection of Reviewed Languages

Within the scope of this chapter, it is of course impossible to consider all languages aiming at applications in real-time processing. The selection of the languages to be reviewed here is determined by their dissemination and their suitability in industrial environments.

So most languages either are designed for special purposes or have only been implemented on a single computer model and have not found widespread usage, or have even been discontinued in an experimental stage. In [48], ALGOL68 was considered in a comparison of real-time languages. Although it allows parallel processing, we omit it here, since it does not provide real-time features. The older British developments Coral 66 and RTL/2 are designated as process control languages, but actually comprise only algorithmic elements; as far as real-time features, synchronization, and input/output (I/O) are concerned, these languages rely totally on calls of the underlying operating system. Therefore, we neither consider them here nor the following two languages: Concurrent Pascal and Modula. The latter are operating system implementation languages, and process control-oriented elements are missing. Modula in particular is not suited for industrial applications, since no I/O and timing facilities are incorporated and since it does not provide as language constructs the features necessary to model a set of tasks upon a technical process and to control this task set accordingly. Instead, Modula allows one to write machine-dependent peripheral driver routines and to utilize the users' own timing, synchronization, and resource-sharing mechanisms. Formulating real-time application programs in Modula would therefore require a considerable amount of non-problem-oriented work and would yield non-portable software. The above statements about real-time facilities hold even more for Modula-2 [57], especially with respect to tasking, since only the coroutine concept is implemented to express concurrency. There are many dialects of BASIC available that incorporate a minimum of real-time features aimed at scientific, not heavily used applications, where the control and evaluation of experiments is to be quickly programmed. Hence, these BASCI derivatives are also outside of our scope. Finally, there are the new languages Occam and Occam2, which are often considered to be real-time. They support the communication and processor configuration in Transputer-based distributed systems. Parallelism, however, can only be generated in a static way. Since the languages do not allow exception handling and only provide a non-deterministic guarded command feature for reacting to several possible events, they do not fulfill the predictability requirements of hard real-time systems.

Synopsis and Discussion of Real-Time Features Realized in Available Languages

The availability, and some further specifications of the real-time features of the seven languages under consideration here have been compiled in Table 1. The information used in this compilation was obtained from the references [1-4, 6, 15, 34, 36, 42, 46, 48, 50, 54, 55].

The first category of features listed in **Table 1** comprises elements required in process control applications but of a conventional character. This may cause contradiction as far as process I/O is concerned. Apart from their time scheduling, however, these operations are carried out by corresponding driver routines of the operating system in a manner similar to standard I/O.

In all the languages considered, parallel processing is organized on the basis of tasks or processes. These can be hierarchically ordered, as in Ada, HAL/S, and PEARL. Each language has as a foundation a different model of task states. However, it is not possible to interrogate as to which state a task is currently in. This missing feature is not really a restriction because, according to the dynamics of task execution, the task status delivered as a result of such an information service may have already changed before any reaction can be executed. Only Ada and PL/1 allow a determination of whether a task is active or already terminated. To control the execution of tasks and their inter-status transfer, tasking opera-

Table 1. Survey on Real-Time Elements in Ada, Industrial Real-Time Fortran, HAL/S, LTR, Pearl, Real-Time PL/1, and Real-Time Euclid

Category	Feature	Ada	Fortran	HAL/S	LTR	Pearl	PL/1	Euclid
Conventional elements	Bit processing	(n)(1)	y	y	y	y	y	y
	Reentrant procedures	y	n	y	y	y	y	y
	File handling	y	y	y	y	y	y	y
	Process I/O	y	y	n	y	y	y	y
Tasking	Hierarchy of tasks	y	n	y	n	y	n	n
	Task stati available	2	n	n	n	n	3	n
	Task scheduling	y	y	y	y	y	y	y
	Controllability of tasks	poor	complete	limited	limited	complete	poor	complete
	Exception handling in tasks	y	n	y	y	y	y	y
	Implied scheduling strategy	prio.,fifo	-	prio.	pre-emp. prio.	prio.	prio.,ca-dep.	pre-emp. deadline
	Usage of priorities	y	n	y	y	y	y	y
	Changeability of priorities	y(2)	n	y	n	y	y	y
Synchronisation	Semaphores	y(2)	y	n	y	y	n	n
	Further sync. means	rendezvous signal(2) shared variable	resourcemark	compool lock	blockstruct.	bolt blockstruct. implicitly	shared obj. lock	monitor event
	Resource reservation	y	y	y	y	y	y	y
	Resource stati available	y	y	n	y	n	y	y
	Resource alloc. strategy	fifo	ca-dep.	prio.	prio.	prio.,ca-dep.	prio.	deadline,prio.
	Deadlock prevention support	n	n	y	n	n	y	n
Events	Interrupt handling	y(1)	y	(y)	y	y	y	y
	Enable/disable interrupt	n(1)	y	(y)	(n)	y	y	y
Timing	Date/time available	y	y	y	(y)	y	y	y
	Cumulative run-time avail.	y(2)	n	(y)		n	n	n
	Forms of time scheduling	delay	various	various	delay cyclic	various	delay fixed date	various
	Timing control of synch. operations	y	n	n	n	n	y	y
Verification	Testing aids	raise exception call int. entries	set eventmark	simulation run-time deter. trace	mapping trace flag display	induce event trigger int.	n	raise except. assert

(1) Only indirectly possible in the final Ada version [3].
(2) Not any longer available in the final Ada version [3].

tions are provided as language features. In this respect PEARL, FOR-TRAN, and Real-Time Euclid offer a wide range of capabilities. In PEARL, these may even be scheduled. In Ada and PL/1, it is only possible to activate and abort tasks; in Ada, HAL/S and LTR, also a wait operation may be requested. Furthermore, HAL/S and LTR allow task activations to be scheduled for the occurrence of simple timing and other events. After the schedules for their initiation are met, the tasks compete for resources, especially for processors. All languages except FORTRAN imply that the appropriate dispatching decisions are then made on the basis of a strategy. The majority of languages uses task priorities that may be dynamically changed, but not in LTR. According to [1], an operating system underlying Ada programs may also utilize the first-in-first-out strategy. Only Real-Time Euclid employs the more appropriate concept of deadlines to schedule tasks.

All languages considered provide means for the synchronization of task executions and for the reservation of resources to be used exclusively. The most common synchronization feature is the semaphore. In addition, various other concepts are realized in the different languages. The availability of resources can be checked in all languages except HAL/S and PEARL. But even if a conditional lock of resources is avialable, the employed resource-allocation schemes are mostly either priority based or operating system dependent. Real-Time Euclid allocates resources by deadline or priority, depending on the nature of the resources. Only Ada systems work according to the first-in-first-out strategy. Only PL/1 has deadlock prevention as an objective of the language.

In all seven languages tasks may communicate with each other via single-bit messages called events. The handling of interrupts is also a common feature. However, these cannot be enabled or disabled in an Ada program.

As we shall see in the sequel, all languages are lacking in important timing features. The cumulative run-time of a task, which must be known to perform deadline driven scheduling, is available only in the preliminary version of Ada and in Euclid. For simulation purposes, this information is given by HAL/S measured in machine cycles. Whereas the capabilities of Ada, PL/1, and LTR for the time scheduling of task executions and operations are very limited, the former two languages and Euclid allow supervision of whether synchronization operations are completed within predefined time frames. The other languages and Euclid offer a wide range of time-scheduling features.

Real-Time Euclid is unique in our comparison of languages in that its task scheduling can be analyzed. This means that for a given program implemented in Real-Time Euclid, the user can always ascertain at compile time whether the program will meet its timing constraints. To allow such a static schedulability analysis, Real-Time Euclid has no constructs that take arbitrarily long to execute; the number of tasks in the system may not be arbitrarily high, and the timing and activation/deactivation constraints of each task are expressed explicitly.

Only HAL/S provides run-time determination and simulation facilities to aid in the verification of real-time software. For this purpose in four other languages, there is only the possibility of generating events under program control.

As far as actual implementations are concerned, the situation is quite different for the seven languages considered here. A vast number of Ada compilers is available, whose target systems include all major microcomputers; however, from its first implementations, Ada has had a bad reputation as a real-time language.

Various dialects of FORTRAN and PL/1 with a broad range of different and incompatible real-time capabilities are on the market. However, only the corresponding proposals for standardization [6, 36] that comprise experience with former realizations and provide the widest spectra of features are suitable for consideration in this language comparison. Of the subroutine package constituting Industrial Real-Time FORTRAN only subsets have so far been experimentally implemented. For extended PL/1, unfortunately, no implementations can be ascertained. HAL/S compilers exist for IBM/360 and /370 systems. LTR can be employed on SEMS's line of Mitra computers. Despite its good standing in this language comparison, LTR has been replaced by Ada for political reasons. PEARL programming systems are commercially distributed for some 50 different target computers, ranging from microcomputer to mainframe systems. PEARL systems are based on some 10 compilers and corresponding real-time operating systems providing the required features. Single-board microcomputer systems containing either complete PEARL programming environments or only object programs generated from PEARL sources are shipped in large quantities and are used in industrial, scientific, and military applications. Real-Time Euclid has been implemented at the University of Toronto (Canada). The system includes a compiler, a schedulability analyzer, and a run-time kernel. The target hardware is a distributed microprocessor system. The implementation and use of the system have taken place in a research environment only (cf. Chapter 2.2.) Thus, Real-Time Euclid is not available commercially.

A Closer Look at Ada and PEARL

According to the above availability considerations, Ada and PEARL are the only high-level real-time languages readily applicable in industrial control environments, since they have been implemented on a wide range of computers, including the major 16- and 32-bit microprocessor series (cf. Chapter 4.4) on which most contemporary process control computers are based. That is the reason to take a closer look at these two languages and discuss their merits and deficiencies.

Evaluation of Ada: The programming language Ada emanated from an effort by the U.S. Department of Defense (DoD) to make available a software-production tool suited to the implementation of embedded systems. Ada is supposed to save considerable software costs by replacing the vast variety of languages presently in use for DoD projects. The evaluation of these projects gave rise to requirements for a common DoD language that have been refined several times. As a result, Ada claimed to be the language of the 1980s and of being apt for the formulation of real-time applications. Owing to Ada's military background, one expects that it should provide remarkable properties with regard to safety, reliability, predictability of program behavior, and especially with regard to timing features, because military and technical processes are rather time sensitive and often require the observation of tight timing constraints with only small tolerances.

In the following discussion a series of shortcomings is mentioned that make it questionable whether Ada's claims can be maintained as far as real-time features are concerned. Although Ada was intended to support the development of large software systems in a well-structured way, experience shows that it "does not have the adequate capabilities to express and to enforce good system design" [18]. The reason for this may be that a language must reflect the users' way of thinking, and the users generally are engineers and technicians, but not computer scientists. So, the language features must be easily conceivable, safe to handle, and application oriented. Ada has disadvantages in this respect, because some of its constructs, although very elegant, are difficult to understand. Another important requirement for the design of a programming language is orthogonality. Unfortunately, orthogonality is often a contradiction to the user's way of thinking.

With regard to conventional elements, bit handling is only indirectly possible in Ada. The sole tasking elements provided are initiation and termination. Therefore, a provision for extensions of the language to allow for rapid mode shifting of tasks was demanded [35]. The only way to express time dependencies in Ada is to delay the execution of tasks by given time periods. On the other hand, absolute time specifications are impossible. Hence, the execution of tasks must be scheduled explicitly within their bodies in accordance with external demands whose arrivals are actively and if need be also selectively awaited. With an interrupt only a single task can be reactivated. The operations of enabling or disabling interrupts are not contained in this language. Furthermore, the prevention of deadlocks is not supported. Another design goal of Ada was to support distributable and dynamically reconfigurable applications. Since the language does not provide any related expressive means [35, 44], it is impossible to formulate distributed software within its framework.

Comparing Ada [3] with its preliminary version [1], which was published in 1980, one finds that its feasibility for real-time programming has actually been impaired by the elimination of several significant features:

- The initiate statement has been deleted.
- Tasks are immediately activated when their declarations have been elaborated.
- Tasks' once-assigned priorities may not be dynamically changed.
- Semaphores and signals, as predefined synchronization mechanisms, have been removed according to orthogonality.
- The cumulative processing time of tasks is no longer available.

Of course, the missing semaphores and signals can be implemented by means of the rendezvous concept in Ada. But such an implementation is a theoretical approach, because it requires additional tasks and results in low efficiency.

Remarkable in Ada are its elegant synchronization concept and the possibility of controlling whether rendezvous and synchronization operations take place within given time frames. Current experience [13] shows, however, that the rendez-vous takes too much time, to be carried through. On the other hand, the non-deterministic selection of an entry call by select statements, in case several rendezvous are simultaneously possible, introduces unpredictability into the program execution. The semantics of Ada imply that processors and other resources are dispatched according to the first-in-first-out strategy with regard to priorities. Therefore, when considering the above-mentioned shortcomings and the absence of most of the features (discussed below) that ought to be present in a contemporary real-time language, it must be concluded that Ada is less of a progress as far as its suitability for real-time applications is concerned.

The deficiencies of Ada summarized above have now been widely recognized. Some developers of large embedded real-time systems even expressed their concern in recent workshops that Ada cannot be called a real-time language. Nevertheless, Ada offers two approaches to overcome these problems. One is the very practical approach to access the underlying operating system as package specified in Ada but implemented in any other language. Unfortunately, this efficient approach is not portable. The other possibility is to build real-time packages, using Ada and its low-level constructs as implementation language. This approach, while portable, is not very efficient. To solve these real-time problems, a large number of change requests is being handled in the currently running language-revision process. According to the rules of ANSI and ISO, in 1988 the Ada9X project was started to revise the language, leading to a new standard. One goal is a substantial improvement of Ada to support the implementation of real-time and distributed systems.

Evaluation of PEARL. Although conceived a decade earlier, the **Process and Experiment Automation Realtime Language** (PEARL) provides much more expressive power for formulating real-time and process control applications than does Ada. The reason may be that PEARL was defined by electrical, chemical, and control systems engineers on the basis of their practical experience in industrial automation problems and after devoting special attention to the timing aspects of the applications. The development of PEARL already began in the late 1960s and was sponsored by the Ministry of Research and Technology of the Federal Republic of Germany. There are two levels of the language, a subset called

Basic PEARL and the superset Full PEARL, which were both adopted by the German Standards Institute (DIN). Evaluating Table 1 and the comparison above, PEARL appears to be the most powerful high-level real-time programming language for the automation of technical processes.

As far as its algorithmic part is concerned, PEARL has approximately the same facilities as Pascal, although the notation is different. It provides the additional data types CLOCK and DURATION and corresponding arithmetic operations. A PEARL program, which is the highest-level object defined in the language, is composed of modules. These modules may be separately compiled and linked together to build programs. This structure supports the modular composition of complex software systems. Each module may contain a system division and a number of problem divisions. In the system divisions, the hardware configuration is described with special regard to the process peripherals. User-defined identifiers are associated with hardware devices and addresses. Only these identifiers are referenced in the problem divisions in order to enhance the documentation value and the portability of PEARL programs. There must be at least one system division, but there may be several system divisions within one PEARL program. When moving to a different hardware, only the system divisions need to be modified. On the other hand, PEARL modules thus contain all information necessary to run them, thereby eliminating the need for the description of devices and files in the job control language of an operating system outside of PEARL. Besides high-level file-oriented I/O statements, PEARL also provides low-level constructs to exchange information with process peripherals and to handle interrupts, which are necessary for automation and control applications. Neither of these I/O statements, however, directly communicates with peripherals. Instead they communicate with virtual data stations or "dations" that are mapped onto real devices in the system divisions, where all implementation details are encapsulated for portability reasons. As already stated above, PEARL features the most comprehensive application-oriented range of elements for task scheduling and control, and for expressing the time behavior. In particular, with single-event schedules defined by

```
single-event-schedule::=
    AT clock-expression |
    AFTER duration-expression |
    WHEN interrupt-name
```

and general, cyclically repetitive schedules defined by

```
schedule::=
    AT clock-expression ALL duration-expression
    UNTIL clock-expression |
    AFTER duration-expression ALL duration-ex-
    pression DURING duration-expression |
    WHEN interrupt-name ALL duration-expression
        {UNTIL clock-expression | DURING dura-
        tion-expression}
```

the following five statements are provided for the control of task state transitions. A task is transferred from the dormant to the ready state by an activation, in the course of which the task's priority may also be changed:

```
[schedule] ACTIVATE task-name [PRIORITY posi-
tive-integer];
```

The inverse operation to this is the termination:

```
TERMINATE task-name;
```

Running or ready tasks can temporarily be blocked with the help of

```
SUSPEND task-name;
```

which is reversed by

```
[single-event-schedule] CONTINUE task-name;
```

Applied to the calling task itself, the last two operations are combined into

```
single-event-schedule RESUME;
```

All scheduled future activations of a task are annihilated by the execution of
```
PREVENT task-name;
```

In conclusion, it can therefore be said that PEARL combines the advantages of general-purpose and of special-purpose synchronous real-time languages [8], because it avoids many of the problems discussed above. Despite its merits, PEARL also has a number of deficiencies, as the interpretation of Table 1 reveals. The most significant of these is the lack of well-structured synchronization primitives with temporal supervision. The experience gained with the application of PEARL has also suggested some further enhancements of the present PEARL standards. The features to be included in this new version, called "PEARL 90" [52], comprise a few minor modifications of some non-real-time language elements, more powerful exception handling and a conditional REQUEST. With these features, a semaphore will be requested if and only if the semaphore is avialable. If the semaphore is occupied, the conditional REQUEST returns without blocking the calling task. An important property of this new construct is its uninterruptability by other PEARL tasks.

PEARL for Distributed Systems. Distinguishing it from all other available real-time languages, just recently a third part of PEARL was standardized in Germany [16], which allows for the programming of distributed applications and for dynamic system reconfiguration. This language extension "Multiprocessor-PEARL" contains elements for describing the hardware configuration of a multiprocessor system, i.e. of the nodes or "stations", of the physical communication network between them, and of the attachments to the peripherals and the technical process, and for describing the software properties, viz. of the distribution of software units among the nodes, of the logical communication channels and the corresponding transmission protocols as well as how to react on system malfunctions.

Multiprocessor-PEARL programs are structured by grouping modules to "collections", which are distributed to the single nodes in dependence on the system states. These collections are the reconfigurable elements for dynamic reconfiguration. Since collections may be moved between different processors, the system divisions describing the access to peripherals may not be contained in the collections. Instead, there is only one system division for each station in a given configuration. In addition, there is a global system division to enable the access to devices attached to other nodes. For the purposes of dynamic reconfiguration in response to changing system states Multiprocessor-PEARL provides configuration divisions and corresponding executable statements for loading and unloading of collections and for the establishment and the discontinuation of logical communication paths. The detailed discussion of these facilities is omitted here, because they are very similar to the reconfiguration language elements that will be introduced below.

The communication between collections is solely performed by message exchange employing the port concept. It avoids the direct referencing of communication objects in other collections and decouples the communication structure from the logic of message passing. Ports represent the interface between collections and the outside world. There are input and output ports and a collection may have several ones of them. In the problem divisions of collections messages are exclusively routed by mentioning ports. The logical communication paths between collections in a certain configuration are established by executable statements connecting ports of different collections with each other. One-to-many and many-to-one communication structures may be set up. The logical transmission links can be mapped to the physical ones in three ways, namely, by selecting certain lines, by specifying preferred lines, or by leaving the selection to the network operating system.

Three different protocol types are available to carry out communications, namely, the asynchronous "no-wait-send" protocol and the "blocking-send" and "send-reply" synchronous protocols. When the asynchronous protocol for sending a message is used, the message is passed to the network operating system for transmission and, if need be, buffering at the receiver. Applying the blocking-send protocol, after transmission the sender waits until it obtains an acknowledgement generated by the corresponding input operation in the receiving task. For the third type of communication protocol, a one-to-many connection is not possible. Here, the sending task waits until the receiving one has processed the message and returned a result. The send and receive operations can be endowed with timeout mechanisms if a synchronous protocol is selected.

NEW HIGH-LEVEL LANGUAGE AND OPERATING SYSTEM FEATURES ENABLING THE DEVELOPMENT OF FAULT-TOLERANT AND ROBUST REAL-TIME SOFTWARE

Synopsis and Discussion of Further Real-Time Features to be Implemented

When comparing the requirements for real-time languages and systems that were outlined above with the capabilities of existing languages, it becomes obvious that various elements especially important for the production of reliable software are still missing or are only rudimentarily present in available languages. We shall discuss these elements and give arguments for their necessity below. For reasons of easy reference, we also provide a survey listing of these desirable real-time features in Tab.2.

Application-oriented synchronization constructions
Surveillance of the occurrences of events within time windows
Surveillance of the sequences in which events occur
Timeout of synchronization operations
Timeout of resource claims
Availability of current task and resource stati
Inherent prevention of deadlocks
Feasible scheduling algorithms
Early detection and handling of transient overloads
Determination of entire and residual task run-times
Task-oriented look-ahead virtual storage management
Accurate real-time
Exact timing of operations
Dynamic reconfiguration of distributed systems when a failure occurs
Support of software diversity
Application-oriented simulation regarding the operating system overhead
Interrupt simulation and recording
Event recording
Tracing
Usage of only static features if necessary

Table 2: Desirable Real-Time Features

These requirements can be divided into three groups. The first of these includes constructs that either make the formulation of frequent applications easier or serve to supervise the stati of tasks and resources as well as the duration of synchronization and resource-claim operations. The second group consists of desirable operating system services that should be provided for the purpose of reliable and predictable software performance. To control these features, several language elements need to be introduced. Finally, the third group of features comprises software verification measures. Their utilization requires only a few control statements.

The leading idea behind all these proposals is to facilitate reliable, predictable, and fault-tolerant program execution by providing robust and inherently safe language constructs that generally aim to prevent software malfunctions. All this is a prerequisite for the safety licensing of software in hard real-time environments. This approach is oriented toward the objective of modern quality assurance, namely, *the systematic prevention of faults, not their detection.*

Despite their fundamental significance in real-time applications, almost none of the languages considered allows the specification of completion deadlines for task executions. However, as already stressed earlier, processors ought to be scheduled using algorithms capable of guaranteeing the observation of the tasks' deadlines. This goal cannot generally be achieved by dynamic priority schemes under control of the programmer, as supported by most languages and operating systems. Such scheduling algorithms also allow the early detection of whether it will be possible to process a task set in time. Otherwise, parts of the workload have to be removed. In order to do so in an orderly and predictable manner, the language should allow the specification of tasks that can be terminated, suspended, or even replaced by others with shorter run-times when an emergency or an overload situation requires it. The implementation of due-date-driven scheduling algorithms must be supported by language elements that allow the determination of the tasks' run-times, or upper bounds, and the updating of the residual run-times necessary before the tasks' final completion. In this connection, the stealing of memory cycles by direct memory access devices becomes a problem. However, in [25] it was shown that this problem can be solved in order to provide task scheduling schemes that meet the requirements of reliable and foreseeable software behavior in hard real-time environments.

To provide information about date and time, existing operating systems (and thus real-time languages as well) completely rely on interval
timers under control of a corresponding interrupt servicing routine.
Since the processor load of the servicing routine must be kept reasonably
small, the resolution of the time information remains rather poor. The
unpredictable operating system overhead introduces further inaccuracies
that become more intolerable the longer the systems run after restarts.
Therefore, highly accurate real-time clocks should be provided as hardware components (cf. Chapter 2.1). The attention of the supervising software will be initiated by an interrupt generated when actual time equals
the next critical moment that has been loaded into a comparison register.
Thus, the requirements of scientific and military applications with regard to very short time intervals can also be met. Provisions should be
made in the language to suspend tasks just before performing certain
operations, e.g., inputs. When the clock's interrupt signal can be directly
utilized to resume the execution of tasks waiting in such a manner, the
accurate timing of single operations is made possible. In this way, for example, measurements can be carried out at precisely equidistant or Gaus-
sian points.

Based on this timing hardware, several time-related surveillance features should be able to be expressed within a language that otherwise
would have had to be explicitly programmed. In control applications, a
guarantee that events occur within given time frames or in predefined
sequences is essential. As can be seen in Tab.1, Ada, PL/1, and Real-
Time Euclid already provide a timeout feature for synchronization operations. Moreover, the process of claiming resources and assigning them to
tasks must be supervised in a real-time environment. In this respect, in
[43] the following statement, which is certainly true for other languages
as well, is found: "Ada multitasking seems to assume that a calling task
requesting a resource can wait until that resource is available", i.e., in-
definitely.

When developing process control programs, not only the software
correctness, in the sense of mathematical mappings as in batch-process-
ing environments, has to be proved, but also its intended behavior in the
time dimension, and the interaction of concurrently active tasks need
verification. A step toward enabling the safety licensing of real-time programs (cf. Chapter 3.1) is to create the possibility for application-oriented
simulation. A test procedure, oriented toward hard real-time environ-

ments, must yield exact information as to whether time constraints can be observed and whether due dates can be granted, in distinct contrast to the prevailing "hope and pray" practice of assuming that "computers are so fast that they will easily master the given workload". Such a procedure would also allow determination of the adequate computer capacity needed for a certain application. In contrast to the language specification [36], which reads that state transitions (of tasks) are performed "instantly", i.e., in zero time, the processing time for the operating system overhead has to be taken into consideration in the course of a simulation. In the test phase, interrupts must also be simulated and protocoled to check whether a task set can cope with the requirements of a hard real-time environment. On the other hand, all events occurring during the execution of such a task set are to be recorded for subsequent error analyses. This feature is also useful in later routine application of a software package, because it can ease the post mortem error search. To guarantee a predictable program execution, every task should be fully expressible in terms of static language elements wherever necessary. It has already been mentioned that requested resources must be granted within predefined time frames. To give urgent tasks the possibility to elude busy resources or to initiate alternative actions, the present stati of resources must be available for questioning. Finally, it is to be stressed that real-time languages ought to make the occurrence of deadlocks impossible by employing an a priori prevention scheme.

Proposal of Additional Language Elements

Protection of Resources and Temporal Surveillance of Synchronization Operations: When several tasks access the same resource, most languages only provide very basic means such as semaphores and bolts, to secure data integrity [15]. Applying these to the locking of resources veils the nature of the operation to be performed, since there is no obvious and verifiable relation between resource and synchronizer. Furthermore, programming errors become possible by not releasing requested resources, that cannot be detected by the compiler, due to the missing syntactical connection between the mutually inverse synchronization operations. On the other hand, due to their complexity, sophisticated concepts like monitors or tasks to be accessed by rendezvous are also error-prone.

The access to shared objects can be protected with the help of implicit, "invisible" bolts to be generated by the compiler. To instruct it to do so in the case of shared basic objects, arrays, and structures, we introduce the optional attribute

SHARED

as part of the pertaining declaration syntax. As data types of shared variables, array elements, and structure components, respectively, only basic ones are admitted, because sharing others either leads to difficulties or is meaningless. For encapsulating the access to protected resources and to enforce the release of synchronizers, we introduce a LOCK statement that is similar to the structures considered in [6, 19, 20, 21] and that has a timeout clause, that was also proposed in [22] in a different context:

```
LOCK  synchronization-clause-list  [NONPREEMP-
TIVELY]
   [timeout-clause] [exectime-clause]
PERFORM statement-string
UNLOCK;
```

with

```
timeout-clause::=
   TIMEOUT {IN duration-expression | AT clock-
   expression}
   OUTTIME statement-string FIN ,
exectime-clause::=
   EXECTIMEBOUND duration-expression ,
synchronization-clause::=
   EXCLUSIVE(shared-object-expression-list) |
   SHARED(shared-object-expression-list)
```

The task executing a LOCK statement waits until the listed shared objects can be requested in the specified way. By providing a TIMEOUT attribute, the waiting time can be limited. If the lock cannot be carried through before the time limit is exceeded, the statements of the OUT-TIME clause will be executed. Otherwise, control passes to the statement sequence of the PERFORM clause as soon as the implied request operations become possible. The corresponding releases will be performed automatically upon reaching UNLOCK, or when terminating the construction with the help of the

```
QUIT;
```

statement. For reasons of program efficiency, seized resources ought to be freed as early as possible. To this end, the instruction

```
UNLOCK shared-object-expression-list;
```

can be applied already before dynamically reaching the end of the surrounding LOCK statement, where the remaining releases will be carried through. The optional exectime clause is introduced to enhance the predictability and safety of real-time systems. It limits the duration that a task is in a critical region. Thus, the possibility is prevented that programming errors resulting in infinite loops will be able to block the whole system. In order to handle a violation of a LOCK's execution time bound, a system exception is introduced. The optional attribute NONPREEMP-TIVELY serves to improve performance. By specifying it, the operating system is instructed not to pre-empt the execution of the LOCK statement due to the applied processor scheduling strategy. Thus, superfluous and time-consuming context-switching operations can be saved when a more urgent task requesting one of the locked resources commences execution before termination of the LOCK statement. Except for their appearance in synchronization clauses, shared objects may only be referenced within the framework of LOCK statements. They must be locked for exclusive access if the reference is used in any assignment context. Owing to its temporal supervision, the above-introduced LOCK synchronization statement, in combination with deadline-driven task scheduling, solves [26] the uncontrolled priority inversion problem [47] in a very simple and elegant way. In addition, deadlock prevention schemes can be very effectively based on it, as will be shown below.

Monadic Operators: In this section we shall define intrinsic functions providing status information on tasks and synchronizers. The functions yield results of type fixed. Given a certain model of task states and with an appropriate numbering of these states, the

```
TSTATE task-identifier
```

operator returns the number of the parameter's actual status.

In order to interrogate the stati of shared objects, we introduce the function

```
SYNC shared-object
```

which returns the values 0 or -1 in the unreserved or exclusive access states, respectively, or the number of tasks that have shared reading access.

Surveillance of the Occurrence of Events: To survey whether and in which sequence events occur, we shall propose in this section a new language feature. In this context, we use a wider notion of events to summarize

- interrupts,
- exceptions,
- time events,
- status transfers of synchronizers and tasks, and
- the assumption of certain relations of shared variables to given values.

These events may be stated according to the following syntax rules:

```
event::=
    WHEN interrupt-expression |
    ON exception-expression |
    AT clock-expression |
    AFTER duration-expression |
    status-function-call relational-operator
    expression |
```

```
shared-variable-reference relational-opera-
tor expression |
Boolean-type-shared-variable-reference
```

where status-functions are those introduced in the previous section. In the last three alternatives above, the events are raised when the corresponding Boolean expressions turn true. Now the new language element is defined by

```
EXPECT alternative-string FIN;
```

with

```
alternative::=AWAIT event-list DO statement-
string.
```

When the program flow of a task monitoring events reaches an EXPECT block, the expressions contained in the event specifications are evaluated and the results stored, and then the task is suspended until any one of the events mentioned in the AWAIT clauses occurs. Then the statement string following the associated DO keyword will be executed. In case several events listed in different AWAIT clauses occur together, the corresponding DO clauses will be performed in the sequence in which they are written down. When the operations corresponding to the just occurred event(s) have been executed, the task is again suspended awaiting further events. In order to leave the described construction after finishing a surveillance function and to transfer control behind the block end FIN, the

```
QUIT;
```

statement has to be applied. When this has been done, or when the construction has been otherwise left, there will be no further reaction to the events mentioned in the EXPECT feature. The applications, for which a scheduled synchronizer release statement was requested in [23, 49], can be programmed using the element described above. In particular, the EXPECT statement allows the formulation of surveillance execution and time behavior for I/O operations. Thus the EXPECT statement corresponds to the selective wait statement of Ada, but, being especially important for real-time systems, it has fully deterministic semantics.

Parallel Processing and Precedence Relations of Tasks Sets: Although parallelism can be expressed and realized with the task concept, it is sometimes necessary to request parallel execution of certain activities in a more explicit and specific way. Consequently, we introduce the following language feature:

```
PARALLEL activity-string FIN;
```

with

```
activity::=ACTIVITY statement-string.
```

Naturally, PARALLEL language elements may be nested and applied in sequence. Then, by including task-activation statements in the various ACTIVITY strings, they are particularly useful in expressing the bundling of task executions and in specifying precedence relations that hold between sets of tasks comprising arbitrary numbers of predecessor and successor tasks.

Additional Operating System Features to be Supported by Real-Time Languages

Inherent Prevention of Deadlocks: Employing the LOCK language element introduced earlier for claiming resources, the compiler can now verify the correct application of two well-known deadlock-prevention schemes, which may be requested by stating a corresponding pragma. According to the resource-releasing procedure, all required shared resources must be reserved en bloc before entering the critical region where they are used. This can easily be accomplished by mentioning them in the synchronization-clause-lists of the LOCK statement. To ensure a deadlock-free operation, the compiler only has to examine if no further resource requests appear within the PERFORM clause. In order to apply the partial ordering method, an ordering of the shared objects

needs to be declared. An appropriate facility, when this deadlock prevention scheme is to be applied, is introduced as

```
RESOURCE    HIERARCHY    sync-object    hierarchy-
clause-string;
```

with

```
hierarchy-clause::=›sync-object
```

Nesting of LOCK statements can then be allowed, as long as the sequence in which the resources are claimed complies with the predefined
hierarchical ordering.

Exact Timing of Operations: Although most languages allow the specification of time schedules for tasking operations, one cannot be sure
when they actually take place. Since this situation is unacceptable in
many industrial and scientific applications, a language element appears to
be necessary to request the punctual execution of tasking operations: i.e.,
the critical moments need to be actively awaited. For that purpose an optional

```
EXACTLY
```

attribute to be placed in time schedules will be used.

Application of Feasible Scheduling Algorithms: In the introduction it
has already been stressed , that processors ought to be scheduled employing procedures that can guarantee the observation of the strict deadlines
usually set for the execution of tasks in hard real-time environments.
This goal, however, generally cannot be achieved by priority schemes under control of the programmer, as supported by most operating systems
and languages. The implementation of feasible scheduling algorithms,
like the deadline-driven algorithm that is optimal for single-processor
systems, needs to be accompanied by language elements that can specify
the due dates and determine total and residual run-times (or at least upper

bounds thereof) required for task completion. Furthermore, such scheduling algorithms permit early detection of whether a task can be processed in time. If this is not possible, parts of the workload have to be discharged. In order to do so in an orderly and predictable manner, it should be possible to state in the source program which tasks could be terminated or at least replaced by others with shorter run-times, if an emergency or overload situation requires it. These requirements outline the objectives of this section.

We begin by replacing priority specifications with an optional deadline of the form

```
DUE AFTER duration-expression
```

in task declarations and in (re-) activation tasking statements. When the condition for a task's (re-) activation is fulfilled, this duration is added to the actual time yielding the task's due date. As an additional parameter, the deadline-driven algorithm requires the task's (residual) run-time, which is stated in its declaration in the form

```
RUNTIME {duration-expression | SYSTEM}
```

In general, the given duration can only be an upper bound for the required processing time. If this processing time is not known, by inserting the keyword SYSTEM here, the compiler is instructed to supply this time according to a method outlined below.

For each task three variables have to be allocated for use by the scheduling algorithm, whose task-control block may be denoted here by T. The due date is stored in

```
T.DUE
```

an object of type **clock**. The two other variables have the type **duration** and must be continuously updated while the task is being executed:

```
T.TIME
```

is initially set to zero and contains the accumulated execution time, whereas

T.RESIDUAL

initialized with the RUNTIME parameter, is decremented to provide the residual time interval needed to complete the task properly.

When the due dates and execution times of tasks are a priori available, using the following necessary and sufficient conditions [31], a feasible scheduling algorithm is able to detect whether a task set given at a certain moment can be executed to meet the specified deadlines.

For any time t, $0 \leq t < \infty$, and any task T with deadline $t_z > t$, let

$a(t) = t_z - t$ be its response time,
$l(t)$ the (residual) execution time required before completion, and
$s(t) = a(t) - l(t)$ its laxity (slack-time ,margin).

Then, necessary and sufficient conditions that a task set,   indexed according to the increasing response times of its n elements, can be carried through meeting all deadlines are,

a) for m=1, i.e. for single processor systems:

$$a_k \geq = \sum_{i=1}^{k} l_i, \; k=1,...,n, \tag{1}$$

b) and for m > 1, i.e., for homogeneous multiprocessor systems:

$$a_k \geq = \frac{1}{m} \left[\sum_{i=1}^{k} l_i + \sum_{i=k+1}^{n} max(0, a_k - s_i) \right], \; k=m,...,n-m+1, \tag{2}$$

$$a_k \geq = \frac{1}{n-k+1} \left[\sum_{i=1}^{k} l_i + \sum_{i=k+1}^{n} max(0, a_k - s_i) - \sum_{i=n-m+1}^{k-1} a_i \right], \; k=n-m+2,...,n. \tag{3}$$

For k=1,...,m-1 eq. (2) must be valid, except if there are j tasks with $a_k > s_i$ for $k < i \leq n$ and $j+k < m$ then

$$a_k \geq \frac{1}{k+j} \left[\sum_{i=1}^{k} l_i + \sum_{i=k+1}^{n} max(0, a_k - s_i) \right] \qquad (4)$$

must be fulfilled.

If at least one of the above inequalities is violated, a transient over-load situation must be handled. In order to do so in an orderly and pre-dictable manner, all interrupts are masked, all tasks are terminated, and the schedules for further activations of those tasks are deleted, whose declarations do not contain the

KEEP

attribute, introduced as an option. Then the remaining tasks, together with the emergency tasks scheduled for the event of an overload condi-tion, will be processed.

An overload, and the consequent discharge of load, may also arise as an effect of an error recovery. Since the recovery from an error occur-ring in a certain task requires additional time, the free task set may lose its feasible executability. The KEEP attribute provides the appropriate means to cope with such a situation and to implement a fault-tolerant software behavior.

The overload-handling scheme presented here is suitable for and ori-ented toward industrial process control applications. In contrast to this, the method outlined in [9] tries to cope with transient overload situations by sharing the load in a distributed system. Besides the method's disad-vantage that some tasks may still not be completed in time (or at all) and that this case is not handled, the method is unrealistic, since load-shar-ing is usually impossible due to the hard-wired connection of process pe-ripherals to distinct processors.

Run-Time Estimation of Tasks: In this section we shall discuss a way, to determine the execution time or, at least, an upper bound thereof, for each major executable statement of structured high-level real-time lan-guages. In combination, an estimation method is obtained that permits the algorithmic generation of overall execution time bounds for program units such as tasks and procedures. The initial ideas about such a method can be found in [29].

As long as the program flow is strictly sequential, no problem exists since only the processing times of the corresponding compiler-generated instructions need to be summed up. The language elements covered by this argument are assignment statements, in-line and access function evaluations, interrupt masking statements, i.e. ENABLE and DISABLE, branching to and from procedures, and operations invisible to the programmer that are caused by the block structure. Whereas run-time determination is exact for the above-mentioned features, in all other cases one cannot expect to obtain more than an upper bound by carrying out an appropriate estimation.

As far as IF and CASE statements are concerned, such an upper bound is yielded by adding the time required for the evaluation of the conditional expressions occurring in these statements to the maximum execution times of the two or more alternatives provided.

The other major program-control structure already causes difficulties, because the number of times the instructions in a repetition or iteration statement are performed generally depends on the values of variables and is, therefore, a priori not determined. In order to enable the run-time estimation for LOOP statements as well, we augment its syntax by the following clause:

```
MAXLOOP   fixed-literal   EXCEEDING   statement-
string FIN
```

If the number of iterations exceeds the limit set with this clause, the loop execution is terminated, and the statement string specified after the keyword EXCEEDING is carried out before control passes to the first statement behind the loop. At first, this restriction seems to be a serious drawback. But in actuality it enhances the reliability essential for real-time software, because with it faulty programming cannot lead to infinite loops and thereby to system hang-ups. The limitation of the number of loop repetitions for safety purposes was first demanded by Ehrenberger [17].

From the viewpoint of run-time determination the greatest concern is caused by the GOTO statement, that is added to its harmfulness resulting from the fact that its unrestricted use may produce difficult-to-survey and hence error-prone programs. This latter fact was perceived at an early stage and resulted in the development of structured programming, which revealed that in the majority of cases the application of jumps can be avoided, provided that appropriate structured language elements are available. Consequently, we can confine the usage of GOTO statements to the purpose of leaving loops, blocks, and other constructions, when some code is to be skipped. Thus, both the software reliability is enhanced and the run-time estimation is made possible, because jumps can be simply disregarded. The restricted use of the GOTO statement as outlined above can be enforced by the compiler.

The span between the time a synchronization operation requesting resources is reached and the time it is finally carried through is generally undetermined. However, an execution time bound for the LOCK statement introduced above can easily be obtained by adding the time required to claim the specified resources, the given waiting time, and the maximum execution times for the OUTTIME string on the one hand and for the PERFORM string including the release of the synchronizers on the other.

By combining the above-stated estimation techniques and applying them recursively, run-time bounds for program modules such as blocks, procedures, and task bodies can be determined. In particular, the method can also be applied to procedures that are part of a language's run-time package. Such procedures are invoked, e.g., by the I/O statements and the tasking statements; hence, run-time estimations can also be derived for these statements. From the view-point of run-time determination temporal suspensions of tasks do not need to be considered, since for the scheduler only the residual run-time required from a task's resumption to its final completion is significant. Statements simulating external interrupts or exceptions can be excluded in the context of run-time estimations, because they are only useful in the software-validation phase.

In order to take the run-time of exception handling into account, an exception handler is regarded as a procedure whose maximum run-time can be estimated. This run-time is added to the execution times of those parts of the task that are processed when the exception handler is invoked. The maximum task run-time without exception handling plus the maximum of all paths through the task that include calls of the exception handlers yields an overall upper bound for the task's execution time. According to [32], however, the necessity of exception handlers ought to be avoided wherever possible. This can be achieved by incorporating, into the application programs, appropriate checks before those statements whose execution might cause the invocation of an exception handler, and by providing suitable alternatives. This procedure is also advantageous from the standpoint of run-time determination.

A run-time estimation for an EXPECT block can be derived by considering the feature itself, as well as the preceding and succeeding parts of the surrounding task, to be separate tasks. First, the initial part of the task surrounding the EXPECT feature is performed; it finishes by activating all alternative actions of the EXPECT feature as separate tasks scheduled for the occurrence of the events specified in the respective WHEN clauses. Taking the maximum execution time of all alternatives present in the EXPECT block provides an upper bound for its processing time upon each activation or resumption, respectively. Finally, the task's continuation is scheduled for the end of the EXPECT element. As additional WHEN clauses also actions can be specified to be performed in case certain time limits are exceeded. Thus, utilizing the given time conditions, the overall run-time estimation of the task surrounding an EXPECT construct becomes possible.

The above discussion has shown that the exact calculation of a task's run-time will be an exceptional case. Hence, we have to content ourselves with estimations. But it will often be possible to improve the estimation of a task's residual run-time in the course of its execution, e.g., when control reaches a point where two alternative program sequences of different lengths join again. To this end, the statement

```
UPDATE   task-identifier.RESIDUAL:=duration-ex-
pression;
```

is introduced for setting the residual run-time to a new and lower value.

Owing to inaccuracies in the run-time estimation, overloads may be detected and handled, although they would not occur during the execution of the corresponding free task sets. This is the price that must to be paid for the predictable and safe operation gained. The rationale for the provision of the UPDATE statement was to minimize the occurrence frequency of the mentioned effect. By the thinking criteria of classic computer science, the low processor utilization achieved with this run-time estimation technique may be considered a serious drawback. For embedded real-time systems, however, suboptimal processor utilization is totally irrelevant because costs have to be seen in the framework of the controlled external process and its safety requirements. Taking into account the costs of a technical process and the possible damage that a processor overload may cause, the cost of a processor is usually negligible. Hence, processor utilization is not a feasible design criterion for embedded real-time systems.

Having derived run-time estimation techniques above, now we shall outline their practical implementation in close co-operation with a compiler for a high-level real-time programming language. The implementation is realized in the form of a pre-processor working on the source code and a post-processor interpreting the compiler-generated assembly language code. The pre-processor is essentially a control-flow analyzer determining all paths leading through software modules such as tasks or procedures. In software verification methodology such tools are employed for the static analysis of software structures [11, 45]. The analyzer decomposes the body of a module into single entry single exit sections whose execution times can be calculated or estimated according to the rules given above, once the object code is available. Furthermore, all control paths leading through a module,> as well as the linear code sequences constituting each path, are identified by the analyzer. The source code of the module is finally annotated by specially labeled comment lines marking the boundaries of the sections and code sequences thus identified and stating the estimation rules to be applied. The extended source code is subsequently subjected to the compiler. Like other comments, the annotations are disregarded by the compiler, but are copied, as comments again, at their respective positions into the generated assembly language code modules. Here they are interpreted by the post-processor, which first calculates the execution times of all marked linear sequences by adding the execution times of the single machine instructions contained in them. Based on these and the library-provided run-time bounds for invoked standard procedures, the application of the specific estimation rules mentioned in the various annotations finally yields an upper bound for a module's execution time.

Support of Task Oriented Virtual Storage Management: In [26], nearly optimal look-ahead algorithms for virtual storage administration were given. These employ the codes of entire tasks as paging elements and are closely related to deadline-driven scheduling. The idea behind these algorithms is that deadline-driven scheduling orders the elements of free task sets according to increasing response times. Since the tasks are processed in this sequence, an ordering of the corresponding storage accesses to program code and data by increasing forward distance is also implied. This observation suggests basing a virtual storage administration scheme on task objects as paging elements, thus representing an application-oriented realization of working sets. It is even possible to take none-ready tasks into account, since, for utilization in a more sophisticated look-ahead algorithm, information is available in process control systems that specifies when temporarily non-ready tasks will be (re-) activated. When this occurs, these tasks may supersede others, that became ready at earlier points in time.

As was shown in [26], the above-mentioned information can be extracted by considering buffered task activations and the schedules for task activations and continuations. The calculation of the required parameters is easily possible for time dependent schedules only. To enable a similar determination of future task (re-) activations in the case of interrupt-driven schedules, the user must supply an average occurrence frequency for each of them. This can be achieved by augmenting the interrupt definition in the system division with a corresponding optional attribute:

```
INTERVAL duration-literal.
```

In PEARL, for instance, some features can be utilized to provide further directives for storage management. System data and shared objects with GLOBAL scope should be placed together with the supervisor in a non-paged storage area. The same semantics may be assigned to the attributes RESIDENT and REENT of tasks and procedures, respectively. The MODULE concept in connection with permanent residency can be employed to gather shared variables and procedures as well as heavily used small tasks in one page.

Dynamic Reconfiguration of Distributed Systems: One objective for the utilization of a distributed computer system is to provide the ability to react to erroneous states of some of its components. To this end, the available hardware redundancy can be employed. It necessitates the capability of dynamic reconfiguration, i.e., an automatic migration of software modules from a malfunctioning unit to an operational one. However, it should be kept in mind that the applicability of dynamic reconfiguration to maintain the operation of a process control system is relatively limited, because control applications are heavily I/O oriented and relocation of software to other processors is meaningless when the peripherals hard-wired to a faulty processor can no longer be used.

In order to apply the concept of dynamic reconfiguration, a language construct is needed that specifies how software modules are assigned to the various processors initially, and how this assignment is to be rearranged when different faults occur. Such a configuration statement is introduced as follows:

```
CONFIGURATION
    initial-part
    [reconfiguration-part-string]
ENDCONFIG;
```

with

```
initial-part::=load-clause-string
```

```
reconfiguration-part::=
    STATE Boolean-expression
        remove-clause-string
        load-clause-string
    ENDRECONF;
```

```
load-clause::=LOAD task-identifier TO processor-identifier;
```

```
remove-clause::=REMOVE task-identifier FROM processor-identifier;
```

The configuration statement contains an initial part and may contain reconfiguration parts. The assignment of tasks to those processors that are present in the system for undisturbed normal operation is specified in the initial part by a sequence of LOAD clauses. For an error condition represented by a Boolean expression in the STATE clause of a reconfiguration part, the latter determines how the software is to be redistributed when the error occurs. This is carried out by specifying which tasks are to be removed from certain processors and which tasks are to be loaded to other ones with the help of REMOVE and LOAD clauses.

Providing Graceful System Degradation by Utilizing the Concept of "Imprecise Results": Recently, the "imprecise computation approach" was proposed [10, 38, 39, 41] to provide flexibility in scheduling. In a system that supports imprecise computations, intermediate results produced by prematurely terminated server tasks are made available to their client tasks. By making results of poorer quality available when the results of desirable quality cannot be obtained in time, real-time services, possibly of degraded quality, are provided on a timely basis. This approach makes it possible to guarantee schedulability of tasks while the system load fluctuates.

In order to utilize the imprecise computation approach, real-time tasks need to have the "monotone property": the accuracy of their intermediate results is non-decreasing as more time is spent to obtain the results. The result produced by a monotone task when it terminates normally is the desired result and is called the *precise* one. External events such as timeouts and failures may cause a task to terminate prematurely. If the intermediate result produced by a server task when it terminates prematurely is saved and made available, a client task may still find the result usable and, hence, "acceptable". Such a result is then called *imprecise*. The imprecise computation approach makes scheduling hard real-time tasks significantly easier for the following reason: to ensure that all deadlines are met requires only that the assigned processor time of every task, i.e., the amount of time during which the processor is assigned to run a task, is equal to or larger than the amount of time required for the task to produce an acceptable result, which is called its "minimum execution time". The scheduler may choose to terminate a task any time after it has produced an acceptable result. It is not necessary to wait until the execution of the task is finished. Consequently, the amounts of processor time assigned to tasks in a valid schedule can be less than the amounts of time required to finish executing the tasks, i.e., their "execution times".

As already mentioned above, the prerequisite for employing the imprecise computation approach is that real-time tasks possess the monotone property. Iterative algorithms, multiphase protocols, and dynamic programming are examples of methods that can be used to implement monotone tasks. Unfortunately, however, tasks with the monotone property are seldom found in process control applications and, therefore, the very promising concept of imprecise results cannot be utilized in its current form. Therefore, we present here a modification that makes the concept feasible for control and automation applications. In order to detect and handle errors on the basis of diversity and to cope with transient overloads, the altered concept is employed to provide graceful degradation of a system in response to the occurrence of faults.

Transient Overloads: Despite the best planning of a system, there is always the possibility of a transient overload of a node due to an emergency situation. To handle such a case, many researchers have considered load-sharing schemes that migrate tasks between the nodes of distributed systems. In industrial process-control environments, however, such schemes are generally not applicable, because they only hold for computing tasks. In contrast, control tasks are highly I/O bound, and the permanent wiring of the peripherals to certain nodes makes load-sharing impossible. Therefore, we present here a fault-tolerant scheme that handles overloads by degrading the system performance gracefully and predictably.

In addition to the above described overload-handling scheme that uses the KEEP attribute for tasks, we now introduce a more sophisticated method based on the concept of imprecise results. The method's basic idea is that the programmer provides for each task alternatives for their executable parts, i.e., the task bodies. To this end, the run-time attribute of a task declaration may be given in a third form:

```
RUNTIME SELECTABLE
```

Such a task provides alternatives with different run-times for its executable part. These are made available for selection by the task scheduler in the framework of the following language construct:

```
TASK_BODY alternative-string FIN;
```

with

```
alternative::=
    ALTERNATIVE_WITH_RUNTIME {duration-expres-
    sion | SYSTEM};
    statement-string
```

The run-time parameters of each alternative are specified in the same form described earlier for tasks. The compiler sorts the alternatives in decreasing order of their run-times. Thus, when a task with this feature is activated, the task scheduler can select for execution the first task-body alternative in the ordered list that allows feasible executability. Since the order generally conforms with decreasing quality and accuracy of the results produced by a task, this scheme represents a realization of the requirement of graceful degradation of system performance when a transient overload occurs.

The usefulness of the above-outlined approach can be shown by considering the following example: In process control applications, a frequently occurring task is the periodic sampling of analog or digital external values. In the presence of a transient overload, for one or a few measuring cycles, a good estimation of an external value can be provided to client tasks by extrapolating the present value from the most recent measurements. Since no driver programs need to be executed, the task's execution time is reduced to a very small fraction. Nevertheless, acceptable results are delivered and, thus, a system break-down is prevented.

Diversity-Based Error Detection and Handling: Software diversity can be used to detect and cope with programming errors. In general, however, this approach is rather costly, because several different versions of a program module need to be executed either in parallel or sequentially. Here, savings can be achieved by employing the concept of imprecise results again. Instead of having different modules, which are to yield the same results, only one module is provided to determine the precise results. These are compared with estimations, i.e., imprecise results, generated by less complex and faster-to-execute software modules. This is performed in the framework of the following language structure:

```
DIVERSE
    alternative-string
ASSURE statement-string
FIN;
```

with

```
alternative::=ALTERNATIVE statement-string.
```

Each alternative constitutes a diverse software module. When all occurring alternatives are executed, the comparison of their results is carried through in the ASSURE clause. Its function is to verify the precise results. If, however, an error is detected, imprecise results can be assigned to the result variables, which will be used in the continuation part of the program. The rationale behind this approach is that the processes of obtaining the approximations are more reliable, since the corresponding software modules are less complex.

Naturally, the above described semantics could also be implemented with if-then-else statements. The proposed language feature, however, provides the possibility for the compiler and operating system to distribute the single alternatives to different processes and, thus, to allow their parallel execution.

Considering again the example of the last section, an extrapolation routine could be used for verifying a sampled external value. If the measured and the extrapolated values only differ within a predetermined margin, the ASSURE clause will release the sampled value for further usage in the program. Otherwise, an error in the external data source, the process peripheral, or an invoked software module is detected. For further calculations, it is preferable to use the approximation, that was obtained with higher reliability.

In the form described above, the concept of imprecise results can be employed to provide software diversity at relatively low cost combined with graceful degradation in the presense of errors.

Software Verification Features

Tracing and Event Recording: For the purpose of controlling whether a real-time software package works in the intended way, it is necessary to record intermediate results and the occurrence of events influencing the single activities. The first feature to be provided in this context is tracing, which is not typical for process control systems. Hence, we can refer here to other languages like FORTRAN [36] and LTR [42], where statements were defined instructing the compiler to generate additional code for writing specified traces into certain files. In order to avoid frequent changes in the source code, a compiler option appears helpful that selects whether the tracing-control statements are to be considered or to be treated as comments. When the behavior of a real-time system is to be understood, it must be known when the events determining the state transfers of tasks have occurred. The kind of events to be considered here are

- interrupts, exceptions, and changes of masking states,
- state transfers of tasks and synchronized variables,
- attainment and actual execution of tasking and synchronization operations.

These events, or specified subsets thereof, should be recorded on a mass storage device during routine operation as well, in order to enable the post mortem analysis of software malfunctions. When simulations are carried out in the test phase, such files represent the corresponding output and require no further specification.

Restriction to Static Language Features: In hard real-time environments the application of dynamic language elements appears questionable, since it introduces unpredictability with respect to capacity and time requirements, making the work of a scheduling algorithm impossible. Therefore, for the sake of reliability, the use of variable array dimensions and of recursive procedure calls should, at least optionally, be suppressed.

Concluding Remarks

In the past, and especially in the case of Ada, concessions have been made when designing real-time languages in order to enable their implementation under existing operating systems. This contradicts a feasible proceeding, because operating systems ought to support the language features, in an inconspicuous manner, and to bridge the gap between language requirements and hardware capabilities. Hence, the development of a process control system should commence with the definition of a suitable language incorporating a host of features enhancing software reliability. Then the hardware architecture should be designed, enabling the implementation of the language as easily and efficiently as possible, and thus keeping the operating system and its overhead relatively small. In the case where implementation and operating efficiencies of a real-time system are in conflict with the efficiency and the safety of the process to be controlled, the latter naturally must be valued more highly.

In the section above, some shortcomings of presently available real-time languages with regard to process control applications have been indicated, and various real-time features have been proposed to overcome these drawbacks, making it possible to express all process control programs fully in terms of a high-level language. The proposals emphasize providing language constructs that are inherently safe and that facilitate the development of fault-tolerant and robust software.

As shown above, there is no language available yet, that fully meets the requirements of hard real-time systems. Also, the constructs described here have not yet been incorporated into an environment that can be used to implement industrial automation systems. Among the available languages, PEARL seems to have the smallest number of deficiencies. Therefore, it would be most cost effective to add the features proposed here to PEARL, both because the necessary changes in the compilers and in the associated run-time systems would be relatively limited and because only a small number of highly portable PEARL compilers is available. Unfortunately, PEARL is not widely accepted outside of Germany. On the other hand, the definition and implementation of a new language containing the above-specified real-time constructs will be very expensive. In addition, it must be noted that the time needed to bring a new language to a broad audience takes at least five years. Pascal, C, and Ada needed eight to ten years from their beginnings to their public acceptance. Although PEARL would be the better basis for a clean prototype implementation of the enhancements, a first prototype could most rapidly be realized in Ada by means of packages, i.e., by the existing concept for language extensions. Of course, such an approach has deficiencies with respect to performance and the checking of the dynamic behavior at compile time. But the advantage of such a solution would be very early availability. Thus, the implementors of real-time systems could already begin to benefit from the favorable properties of these constructs, e.g., predictable behaviour, no deadlocks, less synchronization errors, etc. With the practical experiences gained, an increasing demand for these features would result, forcing the language designers to incorporate them into the next generation of real-time languages, which will hopefully be "real" ones.

LANGUAGES FOR PROGRAMMABLE LOGIC CONTROLLERS

Programming languages and generally computer-aided software development tools for **programmable logic controllers** (PLCs) are presently only available in vendor-specific form. The commonly used programming methods can be subdivided into two main groups: a textual one based on instruction lists, i.e., low-level machine-specific programming languages similar to assembly languages, and a semi-graphical one employing ladder diagrams. The latter representation is a formalization of electric circuit diagrams to describe relay-based binary controls. If other operations than binary ones are to be programmed, however, the ladder-diagram language already must be extended by special-purpose function elements.

The International Electrotechnical Commission (IEC) has worked out a very detailed draft on the definition and standardization of languages for the formulation of automation projects [33]. The languages are apt for all performance classes of PLCs. Since they provide a range of capabilities that is larger than would be necessary for just covering the classical application area of PLCs, viz. binary processing, they are also suitable for the front-end part of distributed process control systems. This, however, does not hold for their communication features and the user interface level.

The draft defines a family of four system independent languages, two textual and two graphical ones:

IL	Instruction List
LD	Ladder Diagram
FBD / SFC	Function Block Diagram / Sequential Function Chart
ST	Structured Text

The IEC claims that the languages are equivalent and can be transformed into one another. It is the goal of this standardization effort to replace the programming in machine, assembly, and procedural languages by employing object-oriented languages with graphical user interfaces. Although it still contains the low-level Instruction List language, the draft emphasizes high-level graphical languages for function charts incorporating the basic analog functions. The Function Block Diagram language has been derived from diagrams of digital circuits, in which each chip represents a certain module of the overall functionality. The direct generalization of this concept leads to functions that are depicted by rectangular boxes and that may have inputs and outputs of any data type. Functions and function blocks may perform binary, numerical, analog, or character string processing. The schematic representation of logical and functional relationships by symbols and connecting lines provides easy conceivability, in contrast to alphanumerical representations. However, also schematic representation turns out to be difficult to understand when the complexity increases. A function diagram is a process-oriented representation of a control problem, independent of its realization. It serves as a means of communication between different interest groups concerned with the engineering and the utilization of PLCs, which usually represent different technical disciplines. The Function Block Diagram language is supplemented by another graphical component: the Sequential Function Chart language. The representation method of sequential function charts can be considered as an industrial application of the general Petri-net concept, utilized to formulate the coordination and cooperation of asynchronous sequential processes.

The draft distinguishes between functions and function blocks. A function is an object that generates for given inputs one function value at its output, using a certain algorithm. The language provides elementary Boolean and arithmetic functions that can be combined to yield new derived library functions. These are stored under identifying symbols. Function blocks may have more than one output. Furthermore, and also in contrast to functions, upon multiple invocation with equal input values, function blocks may yield different outputs. This is necessary to be able to express internal feedback and storage behavior. **Fig. 1 and 2 give** more examples of how the internal structure of a function block can be represented at various levels of detail. The figures reveal that the Function Block Diagram language requires only four different structural elements:

1. function (block) frames, i.e. symbols,
2. names,
3. connectors, and
4. connections.

The system independent software engineering for PLCs is carried out in two steps:

1. set-up of a function (block) library, and
2. interconnection of function (block) instances.

In the second of the above-mentioned steps, the interaction of function block instances is determined in the form of a function block diagram in order to yield the solution to an application problem. To carry out this second, constructive step of PLC programming, i.e., the linking of function block instances, the user invokes, places, and interconnects predefined functions and function block instances from his library. Such a library consists of standard functions, that are universally applicable for automation purposes and are usually provided by the system vendor, and of project-specific elements compiled by the user himself. New library entries can be generated either by defining a new function (block) directly from scratch or by editing and modifying an already existing one. Once a library has reached a certain size, the second method is employed more frequently. The interconnection of function (block instance)s is carried out interactively by pointing to the respective inputs and outputs on the screen and by selecting a polygonal line connecting feature of the applied CAD system.

This section outlines the present status of the IEC standardization effort and discusses possible further developments emphasizing the aspect of computer aided tools for planning purposes.

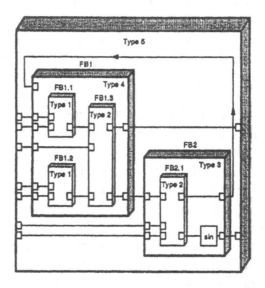

Figure 1: Function blocks (FB) with 3 levels according to IEC 65A

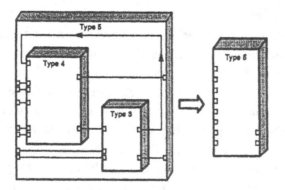

Figure 2: Function block with two levels of detail according to IEC 65A

Survey and Evaluation of the Proposed IEC Languages

As already mentioned above, the IEC draft [33] defines two low-level and two high-level PLC programming languages and provides a unified environment for their utilization. The reason for this approach is to enhance the acceptance of the forthcoming standard: the majority of potential users has a background as electrical engineers or electricians and, therefore, will initially prefer to use the two low-level languages, which are very similar to the vendor-specific languages presently in use. Ladder diagrams are in widespread use among people concerned with the power part of process installation and instrumentation. These diagrams represent an abstraction and formalization of electrical current flow diagrams and, although not standardized yet, have reached a stable state of their development with only slight variations between the different vendors' versions. In contrast to this, instruction list languages are very vendor specific, because they are nothing other than assembly languages of the processors utilized in PLCs. To ease the migration from such a language to the IEC standard, the draft provides the language Instruction List, which is an assembly language for a hypothetical processor possessing an accumulator and using a symbolic single-address scheme.

Main Common Features of the Four IEC Languages: In the four languages proposed by the draft [33], there is a unified form for identifiers, single and multi-element variables, keywords, numerical and character strings, and time literals, as well as elementary and derived data types comprising various Boolean, integer, real, string, and time formats, and enumerated types, subranges, structures, and arrays, respectively. Furthermore, the declaration, specification, and initialization of objects in the various program organization units can only be carried out with standardized language constructs. The program organization units defined in the standard are function, function block, and program. When executed, a function yields as its result exactly one data element, which may be multi-valued. Functions do not contain internal state information: i.e., the invocation of a function with the same arguments always yields the same result. Functions and their invocations can be represented either graphically or textually. The draft states a wide range of intrinsic functions, which are also common to all four languages. Some of them are extensible, i.e., they are allowed to have a variable number of inputs, and are considered to apply the indicated operation to each input in turn. Examples for extensible functions are the Boolean And and Or. The available intrinsic functions can be subdivided into the following groups: type conversion, numerical functions including logarithmic and trigonometric operators, bit and character string processing, selection and comparison functions, and, finally, operations on time data. The second type of program organization unit, viz. the function block, yields one or more values upon execution. Multiple, named instances, i.e., copies, of a function block can be created. Each instance has an associated identifier and a data structure containing its outputs, internal variables, and possibly input variables. All the values of the output variables and the necessary internal variables of this data structure persist from one execution of the function block to the next. Therefore, invocation of a function block with the same arguments does not necessarily yield the same output values. Only the input and output variables are accessible outside of a function block instance, i.e., the function block's internal variables are hidden from the outside. Function block instances can be created either textually or graphically. As for functions, the draft predefines a number of function blocks providing bistable elements, edge detectors, counters, timers, and message transfer and synchronization operators. The execution of program units can be controlled by employing the task concept or sequential function chart elements. As in the real-time languages considered above, a task represents a thread of code that is to be executed in parallel with others. It is capable of invoking the execution of a set of program organization units, either on a periodic basis or upon occurrence of the rising edge of a certain Boolean variable. The general form of a task is similar to that of a function block. It provides a priority input to determine the order of task processing in case there is a conflict.

For the purpose of performing sequential control functions, the draft defines sequential function chart elements. These represent a feature unique in programming languages. The elements serve to partition the PLC program organization unit types program and function block into a set of steps and transitions interconnected by directed links. Associated with each step is a set of actions, and with each transition is associated a transition condition. A step is a situation in which the behavior of a program organization unit with respect to its inputs and outputs follows a set of rules defined by the actions associated with the step. A step is either active or inactive. At any given moment, the state of a program organization unit is defined by the set of active steps and by the values of its internal and output variables. As outputs, each step has a Boolean variable indicating its activity status, and a variable of the type time that specifies the time elapsed during its execution. The initial state of a program organization unit structured with **sequential function chart** (SFC) elements is represented by the initial values of its internal and output variables, and by its set of initial steps, i.e., the steps that are initially active. A transition represents the condition whereby control passes from one or more steps preceding the transition to one or more successor steps along the corresponding directed links. Each transition has an associated transition condition, which is the result of the evaluation of a single Boolean expression. One or more actions may be associated with each step . A step with zero associated actions is considered as having the function of waiting for a successor transition condition to become true. An action can be a Boolean variable, a piece of program code expressed in one of the four languages, or again a sequential function chart. Optionally, an action can specify a Boolean feedback variable, which can be set by the action to indicate its completion. Associated with each action is a Boolean action control variable whose value depends on the actual time and the qualifier selected for the action. A list of the different qualifiers is given below. An action is executed continually while this Boolean variable is true. Upon its transition to false, the action is executed one final time. Upon initiation of a program or function block organised as a sequential function chart, its initial steps are activated. Then, evolution of the active states of steps takes place along the directed links when caused by the clearing of one or more transitions. A transition is enabled when all the preceding steps connected to the corresponding transition by directed links are active. The clearing of a transition occurs when the transition is enabled and when the associated transition condition is true. The clearing of a transition simultaneously leads to the activation of all the immediately following steps connected to the corresponding transition by directed links, and to the deactivation of all the immediately preceding steps.

Specific Language Constructs: Now we want to mention briefly those features that are present in only one of the four languages:
Most operators of the *IL* language have the form

```
result:=result op operand,
```

i.e., the value of a result variable, or hypothetical accumulator, is replaced by its current value operated upon by the operator with respect to the operand. The classes of standard operators available in IL are load/ store, Boolean set/reset, elementary Boolean and arithmetic functions and comparisons, as well as instructions for jumping and subroutine control.

The second textual language, *ST,* has a Pascal-like syntax and functionality. In particular, it features general mathematical expressions including function invocations and the following statement types: assignments, function block invocations, premature exits from function (block)s and iterations, if and case selections, and for, while, and repeat iterations. The draft's graphic languages are used to represent the flow of a conceptual quantity through one or more networks representing a control plan, i.e.,

- "Power Flow" analogous to the flow of electric power in an electromagnetical relay system, typically used in relay ladder diagrams;
- "Signal Flow" analogous to the flow of signals between elements of a signal processing system, typically used in function block diagrams; and
- "Activity Flow" analogous to the flow of control between elements of an organization, or between the steps of an electromechanical sequencer, typically used in sequential function charts.

The lines connecting the graphic language constructs determine the execution control flow.

A program in the *LD* language enables a PLC to test and modify data by means of standardized graphic symbols. These symbols are laid out in networks in a manner similar to a "rung" of a relay ladder logic diagram. LD networks are bounded on the left and right by power rails.

The main language elements of LD are links, static and transition-sensing contacts, and momentary, latched, retentive, and transition-sensing coils.

Finally, the language *FBD* only knows functions, function blocks, and tasks as its elements, which are interconnected by signal flow lines.

Critique of the IEC Proposal: A serious shortcoming of the IEC draft as it stands now is that none of its four languages provides symbolic I/O capabilities such as using the concept of logical units. Instead, I/O ports can only be addressed directly by specifying their respective hardware addresses. Taking into consideration the fact that control programs are highly I/O bound, the standard's final version ought to contain appropriate facilities for external data interchange. Otherwise, system-independent programming of PLCs will still be impossible in the future. Another drawback is that the cyclic execution of control actions is part of the sequential function chart semantics. As a consequence, the duration of the single process states is implementation dependent and not under program control. This limitation must be removed in order to provide a truly system-independent software-development tool and to not restrict future developments and new implementations. Considering that the four languages of the IEC draft are aimed toward the control of industrial automation applications, it is very surprising how poor the available timing-control features are. The only means of time control are to delay actions and to schedule tasks cyclically. There are, however, no capabilities for absolute timing and for the temporal supervision of activities. Finally, the competition of tasks for requested resources is resolved on the basis of priorities. While easily implementable, this concept it totally inappropriate for industrial control purposes. As was already stressed above, processors ought to be scheduled employing procedures capable of guaranteeing the observation of the strict deadlines usually set for the execution of tasks in real-time environments.

Sequential Function Charts

Theoretical Background: The mathematical model of sequential function charts has been derived from the well-known theory of Petri-nets (cf. Chapter 3.1). A sequential function chart is a directed graph defined as a quadrupel

$$(S,T,L,I)$$

with

S = $\{s_1,...,s_m\}$ a finite, non-empty set of steps,

T = $\{t_1,...,t_n\}$ a finite, non-empty set of transitions,

L = $\{l_1,...,l_k\}$ a finite, non-empty set of links between a step and a transition or a transition and a step, and finally

I $\subset$ S the set of initial steps.

The sets S and T represent the nodes of the graph. The initial steps are set at the beginning of the process and determine the initial state. Actions are associated with each step and are being executed while a step is set. Each transition is controlled by a Boolean condition: if the preceding step is set and (in the Boolean sense) the condition turns true, the subsequent step is set and the preceding one is reset.

Hence, a sequential function chart directly and precisely provides an answer to the question: "How does the system react if a certain step is set and the subsequent transition condition will be fulfilled?" The end of a step is characterized by the occurrence of the process information, which fulfills the condition for the transition to the following step. Consequently, steps cannot overlap. Actions can be initiated, continued, or terminated during a step.

Coordination of Asynchronous Sequential Processes: The representation method of sequential function charts can be considered as an industrial application of the general Petri-net concept, utilized to formulate the coordination of asynchronous processes. The main elements of sequential function charts are

- steps,
- transitions,
- actions, and
- connections that link steps and transitions with one another.

They are employed under observation of the following boundary conditions:

- any step may be associated with one or more actions, and
- there is a transition condition for any transition.

A control application is statically represented by a sequential function chart. Through its interpretation, observing certain semantic rules, the dynamic aspect of the described control procedure can be revealed.

A sequential function chart generally consists of steps, which are linked to other steps by connectors and transitions. One or more actions are associated with each step. The transition conditions are Boolean expressions formulated as function block diagrams or in ST.

At any given point in time during the execution of a system,

- a step can be either active or inactive, and
- the status of the PLC is determined by the set of active steps.

A Boolean variable X is associated with each step, expressing with the values "1" or "0" its active or inactive state, respectively. A step remains active and causes the associated actions as long as the conditions of the subsequent transitions are not fulfilled. The initial state of a process is characterized by initial steps, which are activated upon commencement of the process. There must be at least one initial step.

The alternating sequence step/transition and transition/step must be observed for any process, i.e.,

- two steps may never be linked directly, i.e. they must be separated by a transition, and
- two transitions may never be linked directly, i.e. they must be separated by a step.

A transition is either released or not. It is considered to be released when all steps connected to the transition and directly preceding it are being set. A transition between steps cannot be performed, unless

- it is released, and
- its transition condition has turned true.

The elaboration of a transition causes the simultaneous setting of the directly following step(s) and resetting of the immediately preceding ones. Transitions that are to be executed concurrently must be synchronized.

Actions are graphically represented by rectangles, which are connected to the steps initiating them. Each rectangle is internally subdivided into three areas, the first of which is called action qualifier. It contains a short character string characterizing the action:

N non-stored, unconditional
R reset
S set / stored
L time limited
D time delayed
P having pulse form
C conditional

Combinations of these codes are possible; they are interpreted from left to right.

CAE Tools for PLC Software Development

Presently, tools are being developed to support the system-independent high-level graphical programming of PLCs. These tools are based on the textual and the graphical high-level languages as defined in the IEC standardization proposal [33] Since the degree of programming detail achievable in the graphic FBD/SFC language is limited, the structured text language ST is used for the formulation of a library of project specific software modules in the form of function blocks that contain, and at the same time hide, all implementation details. These modules are

194

then utilized and interconnected in the graphical languages FBD/SFC for expressing solutions to automation and control problems. Thus, the advantages of graphical programming, viz. orientation toward the engineer's way of thinking, inherent documentation value, clarity, and easy conceivability, are combined with those of textual programming, viz. unrestricted expressibility of syntactic details, of control structures, of algorithms, and of time behavior.

Below and in Fig. 3, we give a survey of the functionality of these tools. They are aimed at supporting methodical and structured top-down design, the design in the conceptual phase, and the set-up of a library of well-tested standard modules for a project or a class of projects. At the same time they represent very powerful tools for rapid prototyping.

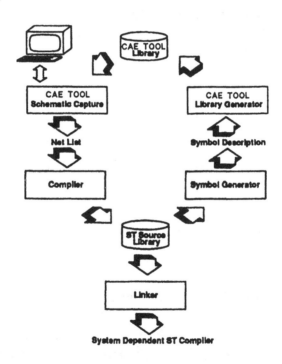

Figure 3: Flow chart of the CAE tool set

When a new software is to be developed, the programmer uses a CAE tool to set up his drawings of function blocks and sequential function charts. In particular, he fetches appropriate graphical objects from his library, places them in a worksheet, and links connectors to these objects, according to his logic. After the drawing process is complete, the worksheet is stored and then submitted to a utility program of the CAE tool, which generates lists of all objects and all interconnection nodes occurring in the drawing. These lists constitute a textual representation that is fully equivalent to the contents of the considered worksheet. They are then submitted to a second postprocessor, viz. a compiler generating code in the IEC language ST. In particular, the compiler produces complete program units with the declarations of input, output, and local variables, of tasks, and of instantiated function blocks, with calling sequences of functions, function block instances, and tasks, and with the encoded description of steps, transitions, and actions. The text of these program units is stored in a library, which, of course, can also be filled by a text editor with hand-coded modules. The library serves as input for a final postprocessor, i.e. a linker, that generates "runable" programs by including, from the library, the source codes of all functions and function blocks,⟩ that are invoked in a module, and any subprograms thereof, that is specified by the user as a main program. Such programs are runable in the sense that they completely describe a control problem. Naturally, they still need to be translated into the machine language of a specific PLC. When functions and function blocks are directly coded in the language ST and stored in the text library, it is desirable to have a tool available that automatically generates from their source codes the corresponding graphical symbols to be used in the drawing process. Thus, a second editing step to set up entries for the graphical library of the CAE tool is avoided. To this end, a further tool interprets the name and the input and output declarations of a function or function block in order to generate the textual description of an appropriately sized graphic symbol. The output of this tool is finally subjected to a utility of the CAE system, which translates the descriptions into an internal form and places this form into its component library.

Acknowledgment: For their help and contributions in preparing this chapter we should like to express our appreciation to Prof. A. Alvarez, Dr. G. Berry, Dr. R. Henn, Prof. H. Rzehak, Dr. G.-U. Spohr, and Dr. A. Stoyenko.

REFERENCES

[1] "Preliminary Ada Reference Manual", and: J.D. Ichbiah et al.: "Rationale for the Design of the Ada Programming Language". *ACM SIG-PLAN Notices 14,6* Parts A and B, June 1979

[2] "Ada Reference Manual" (July 1980). In: *H. Ledgard: Ada - An Introduction.* New York-Heidelberg-Berlin: Springer-Verlag 1981

[3] "The Programming Language Ada Reference Manual", American National Standards Institute, Inc., *ANSI/MIL-STD-1815A-1983.* Lecture Notes in Computer Science 155, Berlin-Heidelberg-New York-Tokyo: Springer-Verlag 1983

[4] U. Ammann: "Vergleich einiger Konzepte moderner Echtzeitsprachen". *6. Fachtagung der GI über Programmiersprachen und Programmentwicklung.* Darmstadt 1980. pp. 1-18. Informatik-Fachberichte 25. Berlin-Heidelberg-New York: Springer-Verlag 1980

[5] T. Baker: "Implementing Timing Guarantees in Ada". *Proc. 4th IEEE Workshop on Real-Time Operating Systems,* Cambridge, MA, July 1987

[6] R. Barnes: "A Working Definition Of The Proposed Extensions For PL/1 Real-Time Applications". *ACM SIGPLAN* Notices 14, 10, 77-99, October 1979

[7] G. Berry, S. Moisan, J.-P. Rigault: "Esterel: Towards a Synchronous and Semantically Sound High Level Language for Real-Time Applications". *Proc. IEEE Real-Time Systems Symposium,* December 1983

[8] G. Berry: "Real Time Programming: Special Purpose or General Purpose Languages". *Proc. 11th IFIP World Computer Congress,* San Francisco, 28 August - 1 September 1989

[9] S. Biyabani, J.A. Stankovic, K. Ramamritham: "The Integration of Deadline and Criticalness Requirements in Hard Real-Time Systems". *Proc. 5th IEEE/USENIX Workshop on Real-Time Software and Operating Systems.* pp. 12-17. Washington, D.C., 12-13 May 1988

[10] J.Y. Chung, J.W.S. Liu, K.J. Lin: "Scheduling periodic jobs using imprecise results". *Technical Report No. UIUCDCS-R-87-1307,* Department of Computer Science, University of Illinois, Urbana, IL, November 1987

[11] D.L. Clutterbuck and B.A. Carré: "The verification of low-level code". *IEE Software Engineering Journal,* 97-111, May 1988

[12] B. Dasarathy: "Timing Constraints of Real-Time Systems: Constructs for Expressing Them, Methods of Validating Them". *IEEE Transaction on Software Engineering,* Vol. 1, No.1, pp.80-86, January 1985

[13] N.W. Davis et al.: "Practical Experiences of Ada and Object Oriented Design in Real Time Distributed Systems". pp. 59-79. In: *A. Alvarez (Ed.): Ada: the design choice. Proc. Ada-Europe International Conference.* Madrid, 13 - 15 June 1989. Cambridge: Cambridge University Press 1989

[14] DIN 44300: Informationsverarbeitung. October 1985

[15] DIN 66253: Programmiersprache PEARL, Teil 1 Basic PEARL, Vornorm, Juli 1981; Teil 2 Full PEARL, Norm, Oktober 1982

[16] DIN 66253 Teil 3: Programmiersprache PEARL - Mehrrechner-PEARL (PEARL for distributed systems). Berlin: Beuth-Verlag 1989

[17] W. Ehrenberger: "Softwarezuverlässigkeit und Programmiersprache". *Regelungstechnische Praxis 25*, 1, 24-29, 1983

[18] T. Elrad: Comprehensive Race Control: "A Versatile Scheduling Mechanism for Real-Time Applications". pp. 129-136. In: *A. Alvarez (Ed.): Ada: the design choice. Proc. Ada-Europe International Conference.* Madrid, 13 - 15 June 1989. Cambridge: Cambridge University Press 1989

[19] P. Elzer: "Ein Mechanismus zur Erstellung strukturierter Prozessautomatisierungsprogramme". *Proc. Fachtagung Prozessrechner 1977,* Augsburg, March 1977. Informatik-Fachberichte 7, pp. 137-148. Berlin-Heidelberg-New York: Springer-Verlag 1977

[20] P. Elzer: "Strukturierte Beschreibung von Prozess-Systemen". *PhD Thesis. Reports of the Institute for Mathematical Machines and Data Processing,* University of Erlangen-Nuremberg, Vol. 12, No. 1, Erlangen, February 1979

[21] P. Elzer: "Resource allocation by means of access rights, an alternative view on realtime programming". *Proc. 1980 IFAC-IFIP-Workshop on Realtime Programming,* pp. 73ff. Oxford-New York: Pergamon Press 1980

[22] A. Fleischmann, P. Holleczek, G. Klebes and R. Kummer: "Synchronisation und Kommunikation verteilter Automatisierungsprogramme". *Angewandte Informatik 7,* 1983, 290-297

[23] A. Ghassemi: "Untersuchung der Eignung der Prozessprogrammiersprache PEARL zur Automatisierung von Folgeprozessen". PhD Thesis. Universität Stuttgart 1978

[24] V. Gligor, G. Luckenbaugh: "An Assessment of the Real-Time Requirements for Programming Environments and Languages". *Proc. IEEE Real-Time Systems Symposium,* December 1983

198

[25] W.A. Halang: "On Methods for Direct Memory Access Without Cycle Stealing". *Microprocessing and Microprogramming 17,* 5, May 1986

[26] W.A. Halang: "Implications on Suitable Multiprocessor Structures and Virtual Storage Management when Applying a Feasible Scheduling Algorithm in Hard Real-Time Environments". *Software - Practice and Experience 16(8),* 761-769, 1986

[27] R. Henn: "Deterministische Modelle fuer die Prozessorzuteilung in einer harten Realzeit-Umgebung". *PhD Thesis.* Technical University Munich 1975

[28] R. Henn: "Zeitgerechte Prozessorzuteilung in einer harten Realzeit-Umgebung". *GI - 6. Jahrestagung.* pp. 343-359. Informatik-Fachberichte 5. Berlin-Heidelberg: Springer-Verlag 1976

[29] R. Henn: "Ein antwortzeitgesteuertes Unterbrechungswerk - Auswirkungen auf Betriebssystem und Programmstruktur". *Proc. GMR-GI-GfK Fachtagung Prozessrechner 1977, Informatik-Fachberichte 7,* pp. 345-356, Berlin-Heidelberg-New York: Springer-Verlag 1977

[30] R. Henn: "Antwortzeitgesteuerte Prozessorzuteilung unter strengen Zeitbedingungen". *Computing 19,* 209-220, 1978

[31] R. Henn: "Feasible Processor Allocation in a Hard-Real-Time Environment". *The Journal of Real-Time Systems 1,* 77-93 (1989)

[32] IEC Standard 880: "Software for computers in the safety systems of nuclear power stations". Geneva 1986

[33] International Electrotechnical Commission, Technical Committee 65: Industrial Process Measurement and Control, Subcommittee 65A: System Considerations, Working Group 6: Discontinuous Process Control, Working Draft "Standards for Programmable Controllers", Part 3: Programming Languages, IEC 65A(Secretariat)90-I, December 1988

[34] A. Kappatsch: "Full PEARL Language Description". *PDV-Bericht KFK-PDV 130.* GfK Karlsruhe 1977

[35] J. van Katwijk, H. Toetnel: "Language Extensions to Allow Rapid Mode Shifting in the Ada Programming Language". pp. 26-36. In: *A. Alvarez (Ed.): Ada: the design choice. Proc. Ada-Europe International Conference.* Madrid, 13 - 15 June 1989. Cambridge: Cambridge University Press 1989

[36] W. Kneis (Ed.): "Draft Standard on Industrial Real-Time FORTRAN". International Purdue Workshop on Industrial Computer Systems. *ACM SIGPLAN Notices 16,* 7, 45-60, 1981

[37] R. Lauber: "Prozessautomatisierung I". Berlin-Heidelberg-New York: Springer-Verlag 1989

[38] K.J. Lin, S. Natarajan, J.W.S. Liu, T. Krauskopf: "Concord: a system of imprecise computations". *Proc. 1987 COMPSAC Conference*, Tokyo, October 1987

[39] K.J. Lin, S. Natarajan, J.W.S. Liu: "Imprecise results: utilizing partial computations in real-time systems". *Proc. IEEE Real-Time Systems Symposium*, San Jose, CA, December 1987

[40] C.L. Liu, J.W. Layland: "Scheduling Algorithms for Multiprogramming in a Hard-Real-Time Environment". *JACM 20*, 46-61, 1973

[41] J.W.S. Liu, K.J. Lin, S. Natarajan: "Scheduling real-time, periodic jobs using imprecise results". *Proc. IEEE Real-Time Symposium*, San Jose, CA, December 1987

[42] LTR Reference Manual. "Compagnie d'informatique militaire, spatiale et aeronautique", Velizy, October 1979

[43] L. MacLaren: "Evolving Toward Ada in Real Time Systems". *ACM SIGPLAN Notices 15*, 11, 146-155, November 1980

[44] A. Di Maio et al.: "DRAGOON: An Ada-Based Object-Oriented Language for Concurrent, Real-Time, Distributed Systems". pp. 39-48. In: A. Alvarez (Ed.): *Ada: the design choice. Proc. Ada-Europe International Conference.* Madrid, 13 - 15 June 1989. Cambridge: Cambridge University Press 1989

[45] "Malpas" (Malvern Program Analysis Suite). Rex, Thompson & Partners Ltd., Farnham, England

[46] P.M. Newbold et al.: "HAL/S Language Specification". *Intermetrics Inc.*, Report No. IR-61-5, November 1974

[47] R. Rajkumar, J.P. Lehoczky: "Task Synchronization in Real-Time Operating Systems". *Proc. 5th IEEE/USENIX Workshop on Real-Time Software and Operating Systems.* pp. 18-22. Washington, D.C., 12-13 May 1988

[48] R. Roessler, K. Schenk (Eds.): "A Comparison of the Properties of the Programming Languages ALGOL68, Camac-IML, Coral 66, PAS 1, PEARL, PL/1, Procol, RTL/2 in Relation to Real-Time Programming". *International Purdue Workshop on Industrial Computer Systems.* Physics Institute of the University of Erlangen. Nuremberg 1975

[49] R. Roessler: "Betriebssystemstrategien zur Bewältigung von Zeitproblemen in der Prozessautomatisierung". *PhD Thesis.* Universität Stuttgart 1979

[50] A. Schwald, R. Baumann: "PEARL im Vergleich mit anderen Echtzeit-sprachen". Proc. "Aussprachetag PEARL". *PDV-Bericht KFK-PDV 110.* GfK Karlsruhe, März 1977

[51] P. Sorenson: "A Methodology for Real-Time System Development". *Ph.D. Thesis,* Department of Computer Science, University of Toronto, 1974

[52] K. Stieger: "PEARL 90. Die Weiterentwicklung von PEARL". In: *R. Henn, K. Stieger (Eds.): PEARL 89 - Workshop über Realzeitsysteme.* pp. 99-136, Berlin-Heidelberg: Springer-Verlag 1989

[53] A. Stoyenko: "Real-Time Systems: Scheduling and Structure". *M.Sc. Thesis,* Department of Computer Science, University of Toronto, 1984

[54] A. Stoyenko, E. Kligerman: "Real-Time Euclid: A Language for Reliable Real-Time Systems". *IEEE Transactions on Software Engineering,* Vol. 12, No. 9, pp. 941-949, September 1986

[55] A. Stoyenko: "A Real-Time-Language With A Schedulability Analyzer". *Ph.D. Thesis,* Department of Computer Science, University of Toronto, 1987, Tech. Report CSRI-206

[56] S. Ward: "An Approach to Real Time Computation". *Proc. Texas Conf. on Computing Systems,* November 1978

[57] N. Wirth: "Programming in Modula-2". Berlin-Heidelberg-New York: Springer-Verlag 1982

7

COMPARISON OF SYNCHRONIZATION CONCEPTS
A DISSENTING VIEW

K. F. Gebhardt

Berufsakademie Stuttgart
Rotebühlplatz 41
D-7000 Stuttgart 1

INTRODUCTION

The following article is a contribution that presents a somewhat dissent-ing view to the ongoing, almost ideological, dispute between different real-time concepts and standards as realized in a language or an operating system.

Computer sciences, particularly when addressing questions of *performance, practicality, taste and even esthetics,* are outside the realms of scientific proofs. As has been demonstrated in the past, plain force - institutional and economic is often the strongest if not the only argument in such matters. This is true for real-time processing as well, where systems can be safety-relevant, where many applications are of industrial and military character, and where discussions tend to get emotional money and morals being at stake. Despite this clearly non-scientific situation, there is, literally speaking, *progress* in the computer sciences not unlike in philosophy and in the arts, in the sense that one cannot regress behind major insights or breakthroughs or ignore certain thoughts but not in the sense that unrefutable scientific findings have been detected that would provide a solid foundation for a theory of real-time processing as such.

In this chapter, after a discussion of the meaning of the term synchronization, the following section will present for real-time systems the pivotal synchronization concepts - *semaphores and rendezvous* . An extended example then will show how to formulate programs for cooper-ating and competing processes with these concepts. The last section will give a few suggestions on how to use these concepts in the world of com-mercially available real-time languages or operating systems.

WHAT IS MEANT BY SYNCHRONIZATION?

A real-time system has to deal with states of external processes, trying to execute code in order to reach certain internal process states before or after certain external process states are reached. To control a process generally poses no problem if the progress of the external process can be arbitrarily delayed or accelerated by the real-time system proper. However, most external processes move uncontrollably from state to state, such as chemical reactions. Neither man nor a real-time system can influence their progress without having to take a significant and normally undetermined dead time into account. In such a case, we need some independent process to measure the progress of external process states in order to be able to design an orchestrated sequence of external and internal process states. As a measure, we can make use of any independent periodic process such as the movements of the sun, a pendulum, or a quartz crystal. The practical realization of such an independent process is a clock which really is nothing more than a means to measure time (as Einstein said: "Zeit ist, was meine Uhr anzeigt").

Synchronization via time would in theory be a useful technique if it wouldn't imply a *deterministic* behavior of the external process. But, by definition, deterministic processes do not have to be controlled. Real-time systems, on the other hand, are needed and made to control external processes whose behavior generally cannot be predicted precisely and are thus essentially *non-deterministic*. Only the controlling real-time system has to be deterministic, i.e., it has to react to situations presented by the external process in a foreseeable way.

In practice, therefore, synchronization via schedules or time specifications over longer than absolutely necessary periods should be avoided and are in fact dangerous because of the non-deterministic nature of almost all technical processes. In that sense we agree with Turski, who in essence suggests that "time is irrelevant for real-time systems" (cf. Chapter 2.1).

Apart from these theoretical caveats, there are some almost unsurmountable practical problems with the time process itself: it takes about one microsecond for a signal to travel 300 meters. Since a modern microprocessor, running at ›20 MHz, can execute several instructions in $1\mu s$, several such microprocessors, when not in immediate physical neighbourhood, *cannot* be synchronized, using time as a synchronization means (if not an until now hypothetical general clock with a precision that is «$1/20\mu s$ is available (as suggested, for example, in Chapter 2.1)).

For example, we want to synchronize a distributed system consisting of two computers physically located 300 m apart, to exchange sensible data via shared memory on a pure time-synchronization basis with the following scheme: computer 1 uses the shared memory every 100 µs for 50 µs starting at time zero and computer 2 uses the shared memory every 100 µs for 50 µs starting 50µs later. Obviously, very soon both computers would work with inconsistent data.

Time-synchronization elements are nevertheless offered by real-time languages such as PEARL. They are appealing to novice users because of PEARL's easy to apprehend scheduling syntax, e.g.:

```
AT 12:00:00 ALL 7 SEC DURING 2 HRS ACTIVATE MYTASK;
```

It turns out that this is one of the most errorprone features of the language! If MYTASK, being typically non-deterministic, takes longer than seven seconds, the schedule cannot be met and the whole program will crash because in PEARL, for good reasons, tasks must not be activated twice before completion.

Although PEARL and other real-time software try to take into account the timing of external events, up to now most languages and operating systems do not even try to take into account time information about internal processes (with the remarkable exception of Euclid (cf. [1] and Chapter 2.2)), which makes synchronization via time even more unreliable.

To summarize it with an example: Trains are *primarily* run not by the timetable, but by signals! We can use timetables or schedules only as a rather coarse planning tool. The ultimate sequence of events, i.e., the exact "before, after, or together", has to be determined by other synchronization mechanisms (whereby the word "synchronization", whose Greek origins imply a reference to time, must be used for lack of a more suitable alternative).

SYNCHRONIZATION CONCEPTS

In the sequel, we will concentrate our discussion on *semaphore* and *rendezvous* as the fundamental synchronization concepts which do not use time. Comments on *bolt*-variables and *monitors* will be given after the extended example has been presented.

Definitions: In order to effectively deal with the complexity of real-time systems - viz. many different things going on simultaneously and usually getting in each others way - one models these systems as a set of cooperating and/or competing processes or tasks running (quasi-) in parallel. To control the interaction between these processes, two concepts - *semaphores* and *rendezvous* - are used in the majority of real-time applications.

Semaphores: Introduced by Dijkstra [2] to guarantee mutual exclusion of critical program regions, semaphores are global, non-negativ integer variables (s) which, after initialization, can be manipulated by only two atomic (non-interruptable) functions - p(s) and v(s) (cf. Chapter 2.1, 4.1, and 4.6). These functions are defined as follows:

```
p(s):  if s>0
               then s:=s-1
               else  suspend calling process
                     and put it into waiting queue
                     for s

v(s):  if the waiting queue for s is not empty
               then  activate  exactly  one  of  the
                     waiting processes
               else  s:=s+1
```

As defined above, this is an example for a *counting* semaphore. If the semaphore can take on only the values 0 and 1, then it is a *binary* semaphore. In this case, "s:=s+1", in the definition of v(s), has to be replaced by "s:=1". In addition to the original Dutch names "passeren" and "verlaten", numerous other names are used for the operations p(s) and v(s) :

```
p - wait   - REQUEST - take - Sperren   - Passieren
v - signal - RELEASE - give - Freigeben - Verlassen
```

In the following examples, processes are sometimes shown in parallel columns just to indicate (quasi-)simultaneous execution and to save space, but not to indicate any time-relations between statements. In particular, statements in a row are not necessarily executed synchronously.

The standard application of the semaphore is the mutual exclusion of critical regions by sandwiching those regions between p(s) and v(s):

Initialize semaphore crit to crit=1, allowing only one process being in the critical region (crit=k would allow up to k processes to be in the critical region.)

```
process A            process B            process C
...                  ...                  ...
p(crit);             ...                  ...
critical region;     ...                  ...
v(crit);             ...                  p(crit);
...                  p(crit);             critical region;
...                  critical region;     v(crit);
...                  v(crit);             ...
...                  ...                  ...
```

The first process reaching p(crit) will set crit to zero. Any process reaching p(crit) later will be suspended until the first process - on eventually leaving the critical region by executing v(crit) - will either activate one of the suspended processes or will set crit back to 1 if no process waits to enter its critical region. As the extended example below will show, semaphores can be used to control more complex situations of process-interaction as well.

The semaphore - being a global variable - is based on a shared memory concept which has to be controlled by one centralized institution. This is difficult to implement in a distributed environment though. Further, the functioning of the above mutual exclusion scheme depends on every process being loyal in using the semaphore operations. One has to make sure - eventually over years of software maintenance - that the semaphores are used in the same consistent way in every module and at any time throughout the life cycle of the program.

Rendezvous: The rendezvous concept introduced by Hoare [3] is based on message passing concept. It is the synchronization mechanism of Ada. The original idea was, to synchronize two and only two processes in order to perform one activity together, possibly excluding any third process from this activity. A simple example - the standard client-server situation - in Ada-like syntax will illustrate the usage of the rendezvous:

```
process server          process client1          process client2

...                     ...                       ...
...                     server.crit (pa);         ...
...                     ...                       ...
accept crit (pa) do ...                           ...
   critical region;    ...                        server.crit(pa);
end crit;               ...                        ...

...                     ...                       ...
```

The process server offers an "entry" procedure crit. In the declaration part of the task server - not shown in the example - the entry procedure crit is marked by the keyword "entry"; in the task body it is marked by the keyword "accept". It can be used by other processes in the following way: The statement accept crit(pa) and the procedure call server.crit(pa), respectively, are synchronization points where server and client wait for each other. The server waits at accept crit(pa) until a client has reached server.crit(pa). The client waits at server.crit(pa) until the server has reached accept crit(pa). If more clients than one reach their server.crit(pa)-calls they are placed in a FIFO-waiting queue.

After synchronization, the server executes the body of the procedure crit with the parameters (pa) and the priority of the client while the client is suspended. After the execution of crit and after eventually returning parameters (pa), the client is allowed to proceed with its code.

The next client in the queue will be served when the server - typically looping - has reached the statement accept crit(pa) again.

The parameters (pa) of the procedure crit are used to exchange messages between processes.

The accept statement can only wait for one procedure call. The select statement introduces the possibility of waiting for more than one call. One example is the mutual exclusion of different activities (e.g., crit1 and crit2). With the select-feature, alternative entry routines are offered. Furthermore, these alternatives can be opened or closed depending on the value (true or false) of a condition (guard):

```
process server          process client1  process client2

...                         ...                 ...
select when condition1 =>   ...                 ...
  accept crit1(pa) do       ...                 ...
    critical region1;       ...           server.crit2(pa);
  end crit1;                ...                 ...
or when condition2 =>       ...                 ...
  accept crit2(pa) do       ...                 ...
    critical region 2;      ...                 ...
  end crit2;        server.crit1(pa);           ...
else                        ...                 ...
  statements;               ...                 ...
end select;                 ...                 ...
...                         ...                 ...
```

The select statement functions as the synchronization point at the server side where it is determined which accepts are open by evaluating all conditions. After this, any change in the conditions of the select statement cease to effect the status (open or closed) of an accept. The server is suspended if no call for an open accept is pending.

The first client requesting any open accept alternative is served. If more clients are requesting services, a FIFO-waiting queue is opened for each accept in the select block. How the system makes a choice between the queues is undefined in the sense that the Ada-programmer should not make any assumptions about a queuing strategy (e.g., FIFO, LIFO, or random). The optional "else" branch is used to execute statements - typically, for some kind of "maintenance or bookkeeping" - if no rendezvous is possible, i.e., if no client is in any of the waiting queues.

An Extended Example

The two robots Nina&Boris are supposed to screw together an un-specified number of bolts and nuts. We want to program these robots so that one robot takes a nut, the other one a bolt from the conveyor belt using the supplied routines get_nut(robotname) and get_bolt(robotname); further, that they position nut and bolt - position_nut(robotname), position_bolt(robotname) - and wait for each other to start turning the bolt or the nut, respectively.

The turning of the bolt, respectively nut is continued until each robot has the "feeling" of the required tightness -turn_right_until_tight (robotname)). Finally, one robot lets the product go - let_go(robotname) and the other robot puts the product away - put_away(robotname), maybe into a basket.

For geometric reasons, only one robot can get to the conveyor belt. All the other operations can be done in parallel without risking a collision between the robots. The robots' "boss" naturally wants to see a maximal production-rate. Therefore, it has to be taken into account that the positioning of the bolt takes relatively longer than the positioning of the nut. Finally, to "please" the goals of the robots' boss, we try to let the two robots work at the same time as much as possible.

First solution: Using semaphores and PEARL-like syntax, we synchronize the robots (comments in /* ... */):

```
MODULE (NINABORISMAIN);

SYSTEM;

PROBLEM;
  DECLARE sbelt SEMA RESET(1);        /* semaphore for      */
                                      /* mutual exclusion   */
                                      /* at the conveyor    */
                                      /* initialized to 1   */
  DECLARE sboris SEMA PRESET(0);      /* semaphores for     */
                                      /* cooperation        */
  DECLARE snina SEMA PRESET(0);       /* of Nina and Boris  */
                                      /* initialized to 0   */

nina : TASK PRIORITY 2; /* gives orders to robot Nina     */
  REPEAT
    REQUEST(sbelt);          /* p-operation  mutual exclusion */
    get_nut('Nina');         /*              at the conveyor  */
    RELEASE(sbelt);          /* v-operation  belt             */
    position_nut('Nina');

    RELEASE(snina); REQUEST(sboris);/* possibly waiting      */
                                    /* for Boris             */
    turn_right_until_tight('Nina');
```

```
  RELEASE(snina); REQUEST(sboris); /* possibly waiting */
                                   /* for Boris to let go */


    put_away('Nina');
  END;
END;


boris : TASK PRIORITY 2;/* gives orders to robot Boris */
  REPEAT
    REQUEST(sbelt);        /* mutual exclusion          */
    get_bolt('Boris');     /* at the conveyor           */
    RELEASE(sbelt);        /* belt                      */


    position_bolt('Boris');


    RELEASE(sboris); REQUEST(snina); /* possibly waiting */
                                     /* for Nina         */
    turn_right_until_tight('Boris');
    REQUEST(snina);        /* possibly waiting for Nina */
                           /* to be done turning */
    let_go('Boris');
    RELEASE(sboris);/* allows Nina to put the product away */
  END;
END;


START_TASK : TASK PRIORITY 1;
  ACTIVATE nina;         /* tasks nina and boris are   */
  ACTIVATE boris;        /* getting started            */
END;


MODEND;
```

The mutual exclusion at the belt is fairly easy to be understood because semaphores are made for situations like that. But it requires some concentration - one has to go back and forth between tasks nina and boris, eventually put down a scenario - in order to understand the RELEASE()-REQUEST()- constructs. Adding a third or more robots would complicate the situation further. Because of their unstructured or non-local behavior, semaphores are often called the "gotos" of real-time programming.

By the way, the problem above can be solved using only two semaphores where one has to be a counting one. But then the correctness of the program becomes even less obvious to understand.

The synchronization of the two robots using the rendezvous is shown in Ada-like syntax (comments are marked by a double dash --):

```
procedure ninaboris is

-- declaration of tasks
   task nina is        -- gives orders to robot Nina
    entry ready_to_turn;-- Nina offers two rendezvous
    entry try_to_let_go(robot:String);  -- to the outside
   end nina;

   task boris is        -- gives orders to robot Boris
   end boris;           -- Boris offers nothing
                        -- to the outside

   task belt is         -- for mutual exclusion at conveyor
                        -- belt an extra
                        -- task - a belt server -
                        -- is needed
    entry try_to_get_nut(robot: String);
    entry try_to_get_bolt(robot: String);
   end belt;

--implementation of tasks
   task body nina is
   begin
    loop
     belt.try_to_get_nut("Nina");-- task belt responsible
                                 -- for mutual exclusion
      accept ready_to_turn; -- rendezvous with Boris, this
                            -- accept serves just as  syn-
                            -- chronisation point,
                            -- no parameters are passed,
                            -- no procedure body
      turn_right_until_tight("Nina");
      accept try_to_let_go(robot) do-- serializing let_go
        let_go(robot);              -- by rendezvous with
      end try_to_let_go;            -- Boris
      put_away("Nina");    -- Nina puts product away,
                           -- so Boris can already start
                           -- grabbing the bolt
    end loop;
   end nina;
```

```
task body boris is
begin
  loop
    belt.try_to_get_bolt("Boris"); -- task belt respon-
      -- sible for mutual exclusion
    nina.ready_to_turn;      -- rendezvous with Nina
    turn_right_until_tight("Boris");
    nina.try_to_let_go("Boris");-- serializing let_go
      -- by rendezvous  with Nina
  end loop;
end boris;


task body belt is     -- the two critcal belt procedures
begin                 -- are managed at the same place
  loop
    select
      accept try_to_get_nut(robot:String) do
        get_nut(robot);
      end try_to_get_nut;
    or
      accept try_to_get_bolt(robot:String) do
        get_bolt(robot);
      end try_to_get_bolt;
    end select;
  end loop;
end belt;
  -- main program ninaboris
begin    -- together with the main program ninaboris
         -- all declared tasks - nina, boris, and belt -
         -- are started (without explicit statement)
  null;  -- nothing done by the main program
end ninaboris;
```

For the mutual exclusion at the conveyer belt, an additional task - a belt server - is required. This task has to guarantee that nothing collides at the conveyer belt and has the advantage that the critical regions are concentrated at one location. No matter how many robots are eventually added to the system getting bolts, nuts, and other things at the conveyer belt, such extensions can be handled in a consistent manner.

As a matter of fairness, it should be left entirely to the reader to decide what concept will lead to readable, well maintainable, and easily extensible software .

But we can't help making at least a few remarks: First, one preliminary question to think about: In order to manage a library, do we give every user a key to the library, or do we employ a librarian and leave the key with her/him...?

Do we need semaphores? Semaphores can be emulated using the rendezvous-concept [4]. For some people, the fact that Ada abandoned semaphores is a deficiency of the language. For us it is a sign of the quality of Ada and thus a serious argument against languages using semaphores. Offering only the rendezvous-concept, the language becomes more orthogonal and safer to use, not allowing an errorprone feature.

Is the rendezvous too complex? As the example demonstrates, the rendezvous is easy to understand and safe to use. For the programmer it is certainly not complex. Therefore, the complexity prejudice with respect to the rendezvous-concept must come from the compiler constructors. Indeed, it can take about 100 times more code to implement the rendezvous-concept than the semaphore-concept. But the rendezvous with all its facettes is a much more powerful tool than the semaphore.

Is the rendezvous too slow? The complexity of a code bears less on the speed of execution than the quality of the algorithm (e.g., more complicated sorting algorithms are often faster). But still it is true that the rendezvous is usually much slower than semaphore-operations, mainly because a procedure call is involved. But in order to serialize one critical activity with semaphores, one would like to use a procedure call anyway in order to avoid code repetition. This would slow down the semaphore procedure as well. There is probably still room to optimize rendezvous implementations, however. In standard applications such as mutual exclusions it may be feasable to emulate the rendezvous concept with semaphores. As the example in Appendix A demonstrates, then three times as many semaphore operations are necessary as with a straightforward semaphore implementation of a mutual exclusion. However, considering the added safety of the rendezvous, this would be a factor one should be able to live with.

Bolt-variables: To handle the reader-writer-problem, the bolt-variable and some auxiliary operations (RESERVE, FREE for the writers, ENTER, LEAVE for the readers in PEARL-Syntax) had to be invented [5], because the solution of this problem using semaphores "is very messy" [4]. Appendix B shows the definition of the reader-writer-problem and how it can be solved with the rendezvous-concept in a more elegant and flexible way by installing a transaction monitor.

Monitors: The monitor is a structured synchronization tool, localizing critical regions but it is still based on a shared memory concept [6]. It can easily be implemented using the rendezvous-concept, so there is no incentive to go into the semantics of the involved condition variable and corresponding operations [4]. Furthermore, the use of monitors can be quite complex as a comparison of Appendix B of this chapter and the monitor solution in [4] of the reader-writer-problem would show.

What language to use?

The software market offers two dominant approaches to real-time programming: Either, one can use a language with more or less implemented real-time features (Ada, PEARL, Modula, Forth, Euclid, Occam), or one can take a real time operating system and use a non real time language (C, FORTRAN, Pascal, BASIC), realizing real-time features by system calls.

With respect to the first alternative, if one is still convinced of semaphores, there is PEARL which offers semaphores and bolt-variables as synchronization tools. But PEARL has some annoying properties, like the repetition of record definitions in every module where the record is used, creating serious modularity problems, or the unreadable case-statement, and a few other things which make programming unpleasant in our view. Although there are language extensions for distributed systems (cf. Chapter 4.1), they are practically not easily available so that connectivity - not to speak of portability - may constitute a problem. Outside of W.-Germany PEARL is largely unknown.

On the other hand, if one is convinced that the rendezvous is a much superior and safer concept than semaphores, the only choice right now would be Ada, preferably a bare machine Ada, since only Ada offers at the moment the rendezvous-concept. With Ada should come a Ada Programming Support Environment (APSE) which is difficult to get at the moment. Further, a good Ada compiler is expensive and not available for every machine or target system, maybe due to the complexity of the compiler. Connectivity and realizing I/O is often a problem as well.

Other real-time languages, measured by distribution and maturity, are presently only of academic interest.

The majority of real-time developers, if not otherwise forced by the US DoD and NATO, obviously uses some real-time operating system running on the target processor, typically a single board computer, and write programs in C, FORTRAN and Assembler on a workstation, preferably in a cross-development-environment using her/his customized environment. This way, there usually arise no problems connecting the real-time system to a higher computer system, running a time-sharing operating system. The disadvantage of this approach is that there are many different real-time operating systems on the market which usually offer more or less sophisticated semaphores as the only means of synchronization.

With UNIX seemingly emerging as the operating system and development environment of the near future and with C++ being an object oriented extension to C emerging as the language of the near future, the logical real-time extension is Concurrent C++ [7]. In hard real-time applications, Concurrent C++ would run on top of a real-time operating system or in a bare machine mode on a target-CPU. Concurrent C is indeed an appealing extension. Besides the rendezvous-concept Concurrent C offers the ability to parameterize process types, offers clauses for selectively accepting transaction (Concurrent C's term for Ada's rendezvous) requests, asynchronous transactions, and transaction pointers. Therefore, offering a superset of the concurrency features of Ada and much better object oriented features, we expect Concurrent C++ to emerge as the language for virtually all applications (real-time, non-real-time, industrial, business, and even artificial intelligence). Appendix C gives an impression of how Concurrent C++ works.

As long as Concurrent C++ is not yet available for one's own real-time operating system or target machine, we recommend to use C - preferably already C++ just because the compilers (typechecking, data abstraction facilities) enforce better code, even if no object oriented features are used - and to try to implement synchronization using rendezvous emulations as shown in Appendix A.

Appendix A - Rendezvous Emulation

The example below shows how the rendezvous-concept can be emulated using semaphore operations.

Initialize semaphore a = 1; semaphore s = 0; semaphore c = 0; Define some shared memory to exchange messages between processes;

```
Task:Server      Task:Client1       Task:Client2   Task:Clients...

loop             ...                ...            ...
p(s);            ...;               ...            ...
read message     p(a);              ...            ...
  from           write message to   ...            ...
shared memory;   shared memory;     ...            similar;
do various       v(s);              ...            ...
critical         p(c);              ...            ...
  activities;                       similar;       ...
write message to read message from  ...            ...
shared memory;   shared memory;     ...            ...
v(c);            v(a);              ...            ...
end loop;        ...                ...            ...
```

Despite still being unstructured, such an emulation has at least the advantage of collecting the critical activities at one place. All possible clients have to adhere rigorously to the prescribed semaphore sandwiching-scheme. The message from the client to the server might contain information about the selection of critical code so that select-features can partly be emulated.

This example shows again how hard to read semaphore synchronizations are.

Appendix B - Transaction Monitor

Reader-writer-problem: Writing and reading processes are mutually exclusive. But more than one process can read at the same time. Only one process can write at the same time. While at least one process is still reading, all other processes that want to read are not allowed to read before a writer can write again.

This problem can be solved with the rendezvous by installing a process as transaction monitor emulating the bolt-variable or monitor concept in Ada-like syntax. (For each bolt variable one transaction monitor has to be installed. Eventually, this could be facilitated in Ada or Concurrent C by defining the transaction monitor as a task type.)

```
procedure reader_writer_problem is

-- declaration of tasks
   task transaction_monitor is
       entry ENTER;   -- Bolt variable vocabulary is used.
       entry LEAVE;
       entry RESERVE_FREE;-- 2 functions of bolt-variable
   end transaction_monitor;-- concept merged into one.

   task type readers is   -- Tasks declared as types!
   end readers;

   reader: array(1..10) of readers; --  array of tasks
                            -- of type readers

   task type writers is
   end writers;

   writer: array(1..10) of writers;-- declaring array
                            -- of tasks of type writers
-- implementation of tasks

   task body transaction-monitor is
   begin
      readers: integer;
      readers:= 0;
      loop
        select
          accept ENTER do readers:=readers+1; end ENTER;
        or
          accept LEAVE do readers:=readers-1; end LEAVE;
        or when readers=0 =>
          accept RESERVE_FREE do
            critical_code_of_the_writers;
          end RESERVE_FREE;
        end select;
      end loop;
   end transaction_monitor;
```

```
task body readers is
begin
    ...
    transaction_monitor.ENTER;
    critical_code_of_the_readers;
    transaction_monitor.LEAVE;
    ...
end readers;

task body writers is
begin
    ...
    transaction_monitor.RESERVE_FREE;
    ...
end writers;

begin-- main program of reader-writer-problem
      -- the transaction monitor task and
      -- the arrays of reader- and writer-tasks are all
      -- activated
    null; -- nothing done by the main program
end;
```

The rendezvous approach is much more flexible than the bolt-variable concept. If we want to make sure that the shared memory is initialized by first allowing a writer to write, the accept RESERVE_FREE ... - possibly with a special initialization code - can be inserted before the loop. Or if we want to limit the number of consecutive reads to prevent "starvation" of the writers, this can be easily accomplished by introducing another counting variable.

Appendix C - Extended Example in Concurrent C++

We present the extended example in Concurrent C++ to give a short impression of the syntax with respect to some of the synchronization concepts (comments are introduced by a double slash //):

218

```
// ninaboris
// declaration of processes as types
process spec ninatype() {// gives orders to robot of type nina
    trans void ready_to_turn();// Nina offers two transactions
    trans void try_to_let_go(char*robot);  // to the outside
};

process spec boristype(ninatype partner) {
// gives orders to robot of type boris. Takes a process ID "partner" of type
// ninatype as parameter.
};  // boristype offers nothing to the outside

process spec belttype() {
// for mutual exclusion at conveyor belt an extra task - a belt server - is
// needed
    trans void try_to_get_nut(char *robot);
    trans void try_to_get_bolt(char *robot);
};

belttype belt;
// belt is globally defined as process of type belttype
// implementation of process types
process body ninatype() {
    for(;;) {
        belt.try_to_get_nut("Nina");// task belt is responsible for
                                        // mutual exclusion
        accept ready_to_turn();// rendezvous with Boris, this
                                    // accept serves just as  synchroni-
                                    // sation point, no parameters are
                                    // passed, no procedure body
        turn_right_until_tight("Nina");
        accept try_to_let_go(char *robot) {// serializing
            let_go(robot); //let_go by rendezvous  with Boris
        }
        put_away("Nina");
// Nina puts product away, so Boris can already start grabbing the bolt
    }
}  // end body ninatype
```

```
process body boristype(ninatype partner) {
    for(;;) {
        belt.try_to_get_bolt("Boris");  // task belt is responsible
                                        // for mutual exclusion
        partner.ready_to_turn();        // rendezvous with partner
        turn_right_until_tight("Boris");
        partner.try_to_let_go("Boris");
                    // serializing let_go by rendezvous with partner
    }
} // end body boristype

process body belttype() {
// the two critcal belt procedures are managed at the same place
        for(;;)
        select {
            accept try_to_get_nut(char *robot) {
                get_nut(robot);
            }
        or
            accept try_to_get_bolt(char *robot) {
                get_bolt(robot);
            }
        }
} // end body belttype
main() {// program ninaboris
belt = create belttype();
// a process of type belttype is created with resulting ID belt, process belt is
// also activated at this point
ninatype nina; // declaration of process nina as ninatype
nina = create ninatype(); // creation and activation of process nina
boristype boris; // declaration of process boris as boristype
boris = create boristype(nina); // creation and activation of
// process boris whereby the ID of process nina is passed as a parameter
 } // end ninaboris
```

Processes in Concurrent C++ have to be declared as types and they can take among other parameters also processes as parameters. Therefore, if we had not chosen to keep the structure of the example in parallel with the Ada implementation, a nicer and more general design would have been possible. Especially silly is the hardcoding of the names "Nina" and "Boris".

REFERENCES

[1] A. D. Stoyenko, "A Schedulability Analyzer for Real-Time Euclid", *Proceedings of the IEEE Real-Time Systems Symposium* 1987

[2] E.W. Dijkstra, "Cooperating Sequential Processes", in *"Programming Languages"*, F Genuys, Ed., Academic Press 1968

[3] C.A.R. Hoare, "Communicating Sequential Processes", *CACM* 21, 666, 1978

[4] M. Ben-Ari, *"Principles of Concurrent Programming"*, Prentice-Hall, 1982

[5] W. Werum and H. Windauer, *"PEARL Process and Experiment Automation Realtime Language"*, Vieweg & Sohn, 1978

[6] P. Brinch Hansen, "Concurrent Programming Concepts", *ACM Computing Surveys* 6, 223, 1973

[7] N. Gehani and W.D. Roome, *"The Concurrent C Programming Language"*, Prentice-Hall, 1989

Acknowledgments: I thank Wolfgang Helbig for his critical suggestions and for introducing me to Ada, Henry Vogt for showing me Concurrent C and many other forerunning developments and Bernd Schwinn for instructive coffee breaks.

8

REAL-TIME DATABASE SYSTEMS

Hans Windauer
Werum GmbH
D-2120 Lüneburg

INTRODUCTION

When is a database system considered a "real-time database system"? Up to now there has been no scientific definition. The minimum requirements for a data base system to carry the label "real-time" could be summarized tentatively as follows:

- The throughput of data must stay in synch with the input, which for database system basically means that such operations as `insert`, `select`, `update` and `delete` records must be executed within predefined time frames.
- A further requirement for a database system to qualify as a "real-time" system would be that such operations be performed "sufficiently" fast.

Of course, this definition is incomplete because the input data rate can differ even within one field of application (e.g. data acquisition from an experimental source) from 1 KB/sec to several 100 KB/sec. In server/client configurations in local area networks (LAN), for example, the number of database operations to be performed by the server can be 10 per second (e.g. for 20 clients) or 20 per second (e.g. for 40 clients) etc. Where is the line drawn between normal requirements and real-time requirements?

Instead of solving this definition problem, we want to present three application projects, which are representative for certain classes of typical real-time applications using databases. Then requirements for real-time database systems which could be used in such real-time applications will be described.

THREE EXAMPLES FOR REAL-TIME APPLICATIONS USING DATABASE SYSTEMS

Data Acquisition from Experiments

During the next Spacelab mission D-2, the data of the medical experiments will be transmitted to earth and stored in a database system in real-time. Thus the medical experts can evaluate the data using comfortable database functions while the mission is still in progress.

Approximately 200 measurement channels produce up to 100 KB data per second (e.g. electrocardiogram, ECG). About every four hours a 1.2 GB disk is filled by the database server on a VAX computer. While these data are copied into an archive on optical disk, the server fills a second 1.2 GB disk with the data of the next 4-hour period, and so on. These devices work alternately during the entire mission (8 - 10 days).

The medical experts use VAX workstations, which are connected to the server via Ethernet. They use SQL to select channel-oriented records with lengths of up to 640 KB.

The database server is located on a VAXstation 3500; the workstations are also VAXstations 3500 or VAXstation II/GPX, and the operating system is VAX/VMS.

Process Monitoring and Documentation

In an oil refinery, a PC is installed to acquire data from approximately 1000 measurement points every 5 minutes, 24 hours a day, 7 days a week. The data are displayed on monitors as color graphics or curve diagrams. They are stored in a database. These data are the sole source which the administration uses to bill its customers. Since the production of the data cannot be repeated, the data base is installed twice, on two mirror disks. Thus the database must write twice for `insert`, `update` and `delete` operations. The records are 300 bytes long and contain one key each.

Every night between midnight and 1.30 a.m. the PC has to generate 60 production reports with a total of 300 pages. An average of 400 `select` operations are necessary in the database to generate one page.

Hence, the database system has to perform within 90 minutes:

120,000	`read with key`	(for the production reports)
18,000	`read with key`	(for update, 90 : 5 = 18)
36,000	`write with key`	(mirror disk!)

174,000 database operations

In other words, between midnight and 1:30 a.m., the database system has less than 31 msec for one database operation (the application also needs some time). Naturally, the update of the measurement data has higher priority than the display or generation of the production reports.

The PC consists of an Intel 386 processor with 16 MHz, 8 MB main memory and two 260 MB disks. The operating system used is OS/2.

Acquisition and Storage of Production Quality Data

A company produces air bags for cars and delivers them to automobile factories. For liability reasons, the factories only accept air bags which have a "quality passport." This document must contain all data which describe the quality and the production process of a specific air bag. These data must also be stored on optical disks for 10 years.

For acquisition, visualization, documentation, and storage of data, a local area network with five PCs is installed in the plant where the air bags are produced. Three PCs (clients) acquire the data from 48 interactive terminals and several production machines. They visualize the data and send it to the database server, which is installed twice as a "double server" on the two remaining PCs. The two PCs work as master and slave. If the master fails, the slave takes over the server function. Therefore, during normal operation every operation on the master is also performed on the slave.

The time constraints are essentially characterized as follows: a cycle consists of three select operations and one update operation with records of 200 bytes length, with 2 keys each. The size of the corresponding database tables is 4 MB. A cycle must be executed in less than 220 msec. The database server is installed on 2 IBM PS/2 model 80's (386 processor) with the operating system OS/2. The clients are IBM model 60 (286 processor) with OS/2. The LAN is an IBM Token Ring.

REQUIREMENTS FOR REAL-TIME DATABASE SYSTEMS

Performance

As mentioned above, all input data must be entered in real-time, i.e. no data must be lost. The examples show that, even in PC applications, access times of 30 to 50 msec are required. These times closely correspond to the average disk access times. Because PCs with OS/2 have a transfer unit of 1 KB (i.e. one write operation to disk transfers 1 KB), higher performance for random access is only possible with small records and large buffer systems in the main memory (cache for the database).

More powerful - and more expensive - computers permit a higher performance of the database system. This is shown in the first example; there the transfer unit is 64 KB.

In general, a real-time database system should be adaptable to the specific limitations of the computer and operating system used. Tuning parameters are, for example, the transfer unit and the number and size of buffers.

In real-time computing systems it is necessary to know a priori the time required by the different parallel processes, i.e. the tasks, to complete their execution for the following main reasons [1]:

- predictability of system behavior
- supervision of timely task execution
- application of feasible task scheduling algorithms, and
- early detection and handling of transient overloads.

Therefore, a real-time database system should guarantee certain access times. The database operations can be seen as functions which are performed by the calling task (procedure call) or by another task on the same or on a different computer (remote procedure call). In any case the execution of these functions requires time. Of course, these access times depend on the hardware, the operating system, the values of the tuning parameters and - last but not least - the database layout of the specific application. Records that are 100 bytes long can be handled faster than records of 2 KB.

The best way to describe this requirement is to say that a real-time database system must be able to guarantee the access times that are measured and considered sufficient in a certain hardware and software environment which is relevant for the application. Normally, a simulation of the real environment before project start is too expensive and time consuming. Therefore, prototyping with measurements for critical system parts is required. Here the use of database systems has been proven to be very helpful.

The property of guaranteed access time, however, assumes that the data base system does not need to be reorganized. This property is also necessary in "nonstop" operations such as in the above example.

Multi-tasking Access with Priorities

A real-time database system must support the "preemptive scheduling with priorities" of the multi-tasking operating system. The second example - the oil refinery - shows this clearly: during the generation of the production reports, the update of the measurement data must not be delayed. Therefore, the task executing the update operations must be able to interrupt and to surpass the task executing the select operations for the production reports.

In general, a real-time database system must permit tasks with higher priorities to surpass tasks with lower priorities within the database system

Locking Mechanisms and Transactions

Locking of records must be organized and performed automatically by the data base system. Only those records which are accessed for update or deletion should be locked.

Many real-time applications (including those mentioned above) operate with single records with no relation to other records. Therefore, a real-time data base system should offer "single actions" which perform a database operation (insert, select, update or delete) for one record independent of any transaction mechanism. This can increase the performance.

Of course, there are other real-time applications, such as in computer-aided manufacturing, where transactions are necessary. Here a transaction is a sequence of database operations in one or several database sets which is executed by the database system either completely or not at all. In keeping with the arguments for multi-tasking accesses with priorities, transactions should also be handled according to their priorities (which could be the same as the executing task). If two transactions want to lock the same record, the transaction with the lower priority should be reset to its starting point to avoid deadlock situations. Records already locked by this transaction must be released.

Recovery

Normally, a database system uses checkpoints to protect the consistency of its user and administration data. At checkpoint t_i the contents of changed main memory buffers are written to the disk so that user and administration data are consistent on the disk. If the computer fails before reaching the next checkpoint t_{i+1}, the database system automatically resets at checkpoint t_i. Insertions, updates and deletions after t_i are lost.

In order to avoid such a loss of data, a real-time database system should offer a logfile on disk, where important insertions, updates and deletions are recorded. Now - after a restart at the latest checkpoint t_i - all insertions, updates and deletions recorded in the logfile should be repeated automatically by the database system. Also, transactions which have been completely executed between t_i and the failure should be automatically repeated by the database system after restart.

Thus the database system can reach the same (consistent) state as at the time of failure without loss of data.

The measures described thus far prevent a loss of data if the CPU fails, but not if the disk fails. To prevent failure in applications with high data safety requirements (see second example), the database system should be able to store the data twice on mirror disks - a "master" disk and a "slave" disk. The data is read from the master disk and written to both master and slave disk. If the master disk fails, the slave disk becomes the master disk. After the repaired slave disk is restarted it must be updated. Whether the update is performed when the database system is operating or stopped depends on the application.

The best solution is a hardware solution where master and slave disks are synchronized automatically during normal operation or after a failure and restart. Analogously this is also true for dual computer configurations. However, in many cases they are still unavailable or too expensive. Then it is helpful if the database system supports a dual computer configuration such as in the third example, where the database system is installed on both computers. One system is the master, and the other is the slave. During normal operation, the master system sends change orders (resulting from `insert`, `update` and `delete` operations) to the slave system, where they are also executed. If the master system fails, the slave system becomes the master system. After the failed system is restarted, it must be updated from the master system in the same way as for mirror disks.

Storage

As shown in the first and third example, real-time database systems should offer the possibility to write important data into an archive. The writing is done with low priority during normal operation. An inventory of the archive should automatically be generated.

Server/Client Configurations in Local Area Networks

Real-time applications may be realized with computer networks (like the third example), or they may be parts of computer networks - e.g., a real-time data acquisition system has to deliver selected data via LAN to a higher level control system. Therefore, real-time database systems should be able to work in LANs according to the server/client model.

Evidently, the performance of such a database system also depends on the communication system (hardware and software) which is used for the communication between the database server and its clients. Lower levels (such as in the ISO model, cf. Chapter 3.6), e.g.. level 2, offer faster communication than higher, more comfortable levels.

It is important that the priorities of the accessing tasks are considered by the server - even if these tasks are located on client computers.

It is necessary to exchange information (orders, state information) when certain events happen in CAM applications consisting of a LAN with an area control system and several cell controllers. Creating a new order or reaching a certain work state, is such an event. It may be helpful if a central real-time database system is used for the information, and if this database system has a trigger mechanism which allows messages for specific clients to be scheduled, depending on certain operations in the database. These messages, which can include records (e.g. orders), are automatically sent to the specified clients when the specified operation takes place. Thus the clients can wait for a message instead of periodically looking for new orders. This "message routing mechanism" can reduce the operations on the LAN and the database server drastically.

Diskless Database Systems

On large 32 bit computers with correspondingly wide data busses it is already possible to use real-time database systems which store the whole database in the main memory. Of course, they can achieve very short access times (less than 3 msec), which are sometimes necessary, such as in the generation and distribution of electrical energy.

In connection with the introduction of 32 bit processors and operating systems in the PC world, it will soon be necessary for real-time database systems to be able to function on a PC without disk. These PCs will stand alone, e.g., for NC applications or will be used as diskless nodes in LANs of CAM applications.

Especially in the latter case, the real-time database systems should have automatic interfaces to other standard relational database systems located on larger computers with disk in the LAN.

EXPERIENCES WITH A REAL-TIME DATABASE SYSTEM

The requirements listed above, are the result of many discussions with software engineers realizing real-time applications. They are especially deduced from experiences with the development and use of the real-time database system BAPAS-DB (Basis for Process Automation Systems) which was also used for the three examples mentioned [BAPAS-DB is trademark of Werum - see References]

Flexible Buffer Systems

As mentioned above, the performance of a database system depends on the hardware, operating system and database layout. BAPAS-DB offers a very flexible concept for the configuration of the database system according to special requirements. All database tables are accessed via a standard buffer system. The numbers and lengths of the buffers are set during installation. However, it is also possible to create additional special buffer systems with another number or length of buffers. Now a database table with an arbitrary record length (e.g. 64 KB) can be assigned to a special buffer system with 64 KB long buffers. Or a database table can be assigned to a special buffer system that is half as big as the table. Thus approximately 50% of the operations for this table occur only in the main memory.

Main Memory Tables

BAPAS-DB also permits the creation of database tables which are located only in the main memory ("main memory tables"). From the user's or programmer's point of view, these tables are accessed in the same way as the other tables. Internally, however, these main memory tables are handled by special access strategies which act directly on the tables without considering buffer mechanisms. This concept of main memory tables has already been used in several applications to realize diskless real-time database systems. There the systems are initialized by streamers which are taken away when the system begins operation.

Special Access Strategies

The performance of real-time applications and the productivity of the software development can be increased by using special access strategies within the data base system. BAPAS-DB offers not only standard strategies such as B-Tree and Hashing strategies but also a strategy IND-SQL-AU, which permits a record to be selected by key, resulting in the delivery of its values of the last 30 minutes (as an example). This is very helpful and efficient if the record represents a measurement point in a plant and the accessing task wants to display the curve of the measured values of the last 30 minutes.

During Spacelab mission D-2, a special access strategy is used for data acquisition. When collecting the data in large main memory areas, the strategy adds information about application-dependent access paths. Areas that are full are written to the disk. The user selects records by means of an extended SQL interface. This interface is the same as the interface used for gain access to other tables using other strategies (e.g. Hashing or B-Tree).

It is important that such a special access strategy substantially increases the performance. On a VAXstation 3500 with 16 MB main memory, two 1.2 GB disks and VMS operating system, it permits the insertion of up to 433 KB/sec.

REFERENCES

[1]   W.A. Halang, (1989) "Real-Time Programming Languages", In: M. Schiebe, S. Pferrer (Eds.), *Echtzeitverarbeitung*, Proc. 10. Workshop, Ulm, 10.-12. October 1989.

[2]   P.C. Lockemann, M. Adams, M. Bever, K.R. Dittrich, B. Ferkinghoff, W. Gotthard, A.M. Kotz, R.P. Liedtke, B. Lüke, J.A. Müller, (1985) "Anforderungen technischer Anwendungen an Datenbanksysteme", In: Blaser A, Pistor P (Ed), *Datenbanksysteme für Büro, Technik und Wissenschaft*, Proc. GI-Fachtagung, Karlsruhe, March 1985, Informatik Fachberichte 94, page 1 - 26, Springer-Verlag, Berlin Heidelberg New York Tokyo, 1985

[3]   G. Stumm, W. Wildegger, E. Mikusch, "Einsatz eines Echtzeit-Datenbanksystems für die Abspeicherung von Experimentdaten im Rahmen der Spacelab Mission D-2", In: Rzehak H (Ed.), *Echtzeit '91,* Proc. Congress, Stuttgart, June 1991.

[4]   V.H. Tristram, "Local Area Network and Computer Aided Manufacturing" - Gedanken zum praktischen Einsatz, IBM Deutschland GmbH, Munich, 1987.

[5]   Werum, BAPAS-DB, "The Open Realtime Database System for Industrial Purposes", Werum, Lüneburg, 1989.

Remark: The portable BAPAS-DB database system has been developed by Werum GmbH, Erbstorfer Landstr. 14, D-2120 Lüneburg. It is currently installed in over 750 industrial installations, mainly in process control systems and CAM applications.

9

MICROPROCESSOR ARCHITECTURES: A BASIS FOR REAL-TIME SYSTEMS

Thomas Bemmerl

Lehrstuhl für Rechnertechnik und Rechnerorganisation
Institut für Informatik der Technische Universität München
Arcisstr. 21
D-8000 München 2

INTRODUCTION AND MOTIVATION

Four years after the "birth" of the microprocessor in 1971, Fortune Magazine wrote the following about the effects of this technology on human life:

"The microprocessor is one of those rare innovations that simultaneously cuts manufacturing costs and adds to the value and capabilities of the product. As a result, the microprocessor has invaded a host of existing products and created new products never possible before."

The ever growing market for microprocessor applications demonstrates that this statement is still true today. Microprocessor-based systems enter new application areas and win new market shares. These applications very often can be characterized by a system structure in which the microprocessor is integrated and embedded (*embedded system*) into a technical process. Because of this integration of the embedded system into the real-world process, the processor used has to fullfill the time constraints of the technical process. Therefore, one of the most innovative application areas for today's microprocessors is real-time systems. This leads us to the main topic of this chapter:

What architectural concepts are offered by today's and future microprocessors, and how useful are they in real-time environments?

When answering these questions, we restrict the class of microprocessors analyzed to the *high-end general-purpose processors*. The author believes that this constraint does not limit the general acceptability of the conclusions. First, the major developments in new microprocessor architectures are dominated by this class of processors. Secondly, the technology of low-end microprocessors (microcontrollers), very often used for smaller real-time systems (embedded systems), is influenced significantly by the high-end general-purpose processors.

Most computer systems used in (hard) real-time environments are based on microprocessors. The reasons for this are the low space and power consumption compared to processors implemented in non-VLSI *(Very Large Scale Integrated Cicuits)* technology. Therefore, the restriction of this discussion about microprocessors does not limit the generality of the topics covered. In particular, all future processors will be implemented in VLSI technology and therefore will represent microprocessors.

DEVELOPMENTS IN SILICON TECHNOLOGY - THE DRIVING FORCE

There is no doubt that one of the driving forces of microprocessor development in recent years has been silicon technology. This development has been accompanied by the integration of architectural minicomputer and mainframe concepts into VLSI technology. Today's microprocessors are comparable to mainframes in the 1970s, integrated on one VLSI chip. The most significant parameters for the evolution of silicon technology are

- the increase of integration density
- the development of price per bit and
- the growth of the die size

Using the microprocessor families from Intel Corporation as an example, Fig. 1 demonstrates the increase of the number of transistors per chip. Comparing the first microprocessor 4004 from the year 1971 with the 32-bit microprocessor 80386 announced in 1985 yields a very interesting result: in 14 years, the number of transistors per chip was increased by a factor of 120 on an eight-times-larger die.

Processor	Year	Devices	Technology	Physical Address Range	Add Time (us)
4004	1971	2300	PMOS	4K	10.8 (4 Bits)
8008	1972	3500	PMOS	16K	20.0 (8 Bits)
8080	1974	5000	NMOS	64K	2.0 (8 Bits)
8085	1976	6000	NMOS	64K	1.28 (8 Bits)
8086	1978	29000	NMOS	1M	0.375 (16 Bits)
80286	1982	130000	NMOS	16M	0.25 (16 Bits)
80386	1985	275000	CMOS	4096M	0.125 (32 Bits)
80486	1989	>1000000	CMOS	4096M	(?)

Figure 1: Increase of integration density

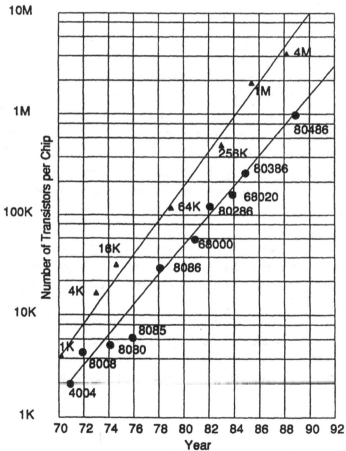

Figure 2: Growth of integration density for memory and micro-processors

Fig. 2 explains the developments in integration density from a more general point of view. The integration density of memory chips is plotted separately from microprocessor components. For dynamic RAM (Random Access Memory) chips, the integration density was raised by a factor of 1.5 per year, whereas the microprocessors have increased their integration density only by a factor of 1.35 each year. The reason for this difference is the more rectangular geometry of memory chips and therefore their simplified design and production. Today's densest dynamic RAM chips consist of a capacity of 4 MBit, whereas the largest microprocessors consist of approximately 1.2 million transistors. In the near future we will see RAMs with 16 MBit per chip and microprocessors with up to 5 million transistors.

Another significant parameter for the development of silicon technology is the price of integrated circuits. The development of new application markets has increased the number of pieces sold, which consequently has dropped the price. On average, the price for memory chips has been reduced by 40% per bit and year; the price for microprocessors only has dropped 25% per year. This continuing price reduction makes VLSI technology even more interesting for new application areas, thus forming the price/application cycle. Apart from price reductions and increases in integration density, researchers have developed new manufacturing techniques and tools to increase the die size of VLSI chips. The die size was increased on average by a factor of 1.15 per year. Based on these developments of die size, integration density, and price, the following three trends may be predicted for the future:

1. Performance increases and geometry reductions based on silicon technology improvements cannot be continued forever. However, for the near future these concepts will lead to acceptable improvements.
2. The continuing price reduction will drive the applicability of microprocessor technology in new application areas.
3. These trends can only be continued by increasing the number of pieces sold or by reducing the development/production costs.

EVOLUTION OF THE PROCESSOR ARCHITECTURE

Before starting with discussions on architectural concepts, it is necessary to define the term *processor architecture*. There is no unique definition of this term known from the literature. In this chapter we define the architecture of a processor to be the semantic interface between software and hardware. Therefore the definition of a processor architecture covers the following issues:

- instruction set
- register model
- data types and data formats
- memory model
- exception handling
- interrupt handling

Ten years ago the microprocessors lagged behind the minicomputers and mainframes with respect to functionality and performance. Current microprocessors (Motorola 68040, Intel 80486, Sun SPARC, MIPS R3000) are functionally equivalent if not superior to mainframes and minicomputers. New application fields for microprocessors, the increase of die size, and further improvements of design and production techniques will extend the success of the microprocessors into the future. Fig. 3 demonstrates faster improvements in functionality of microprocessors compared to mainframes. More and more concepts invented first for traditional mainframes (cache, virtual memory management, etc.) will migrate into the world of the microcomputer in the future. Further, microprocessors of the future will offer new architectural concepts and therefore will represent the driving force in the development of improvements in computer architecture in general.

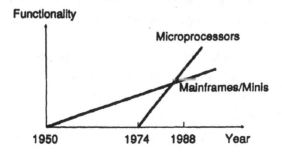

Figure 3: Functionality improvements of microprocessors and mainframes

Instruction Sets

In the 1970s, the development of new instruction sets for computer architectures stagnated. Microprocessors were designed and implemented with copies of the instruction sets known from the minicomputer world for years (Digital Equipment VAX, Motorola 680xx, Intel 80x86). With the advent of the 1980s, these very conservative developments were complemented by several new research projects. The three most well-known computer architecture projects of that type have been

- the IBM project 801 [12]
- the RISC I project at the University of California in Berkeley headed by D.A.. Patterson [11]
- the MIPS project at Stanford University initiated by J. Hennessy

The idea that was common to all three projects became famous as the so-called RISC idea (*Reduced Instruction Set Computer*) [5]. The instruction and address modes offered by a microprocessor have been restricted to those that are from a statistical point of view used most in application programs. Consequently, several architectural concepts typical for the RISC approach have been developed. Based on these concepts, various companies implemented early products using RISC designs (Hewlett Packard Spectrum, Fairchild CLIPPER, MIPS R2000, AMD 29000, Sun SPARC, Motorola 88000) [5]. The basic RISC concepts that emerged were:

1. Nearly all instructions are to be executed in one clock cycle.
2. The maximum clock rate of microprocessors is limited by the length of the critical path. RISC designs try to reduce the clock cycle time by shortening this critical path. This is done by adding to the instruction cycle only those functions that increase the overall performance of the processor.
3. The silicon available is used to optimize the price/performance ratio. Very often this means spending transistors for on-chip cache memories or performance-increasing concepts instead of special-purpose features.
4. RISC designs try to optimize pipeline execution within the processor.

5. The interface between hardware and software puts more burden on the software. One slogan often used for RISC designs is, "Never do at run-time what you would be able to do at compile time."

The basic RISC concepts can be used to derive parameters and attributes typical for RISC-type microprocessors. These attributes and parameters are well suited to characterize RISC microprocessors:

1. Arithmetical and logical operations are confined to registers. *A load/store architecture* is used, which means that only load and store instructions may access memory and I/O.
2. RISC processors offer many general-purpose registers on the processor chip (lower limit: approximately 32).
3. As the name indicates, the number of instructions in RISC designs is reduced. In available RISC processors, the number of instructions offered to the programmer extends from 39 for RISC I to 120 for IBM801.
4. To minimize instruction decode time, only instruction formats with fixed lengths are used.
5. Delayed branch techniques are used to hide control and data-flow interruptions due to pipeline flushes.
6. RISC processors try to use as little microcode as possible. This is mostly a consequence of the reduced instruction set and therefore simple decoding.

In conclusion, the RISC philosophy has already influenced and will strongly continue to influence microprocessor architectures in the future. But it is worthwhile to mention that the processor core discussed in the basic RISC concepts is only part of a complete microprocessor. Future developments of microprocessors also have to take care of support for floating-point arithmetic units, multiprocessor systems, operating systems, memory management, and development tools. These support functions have to be combined efficiently with the basic RISC idea.

Instruction sets for microprocessors widely used in real-time systems are conceptually identical to those described above. In particular, in real-time programs bit-manipulating instructions and boolean operations are necessary to control peripheral devices (AM29000, Intel 80960).

Register Models

Typical run-time characteristics of application programs play a major role for the design of register models. Programs written in high-level programming languages (C, Pascal, Modula, Fortran, etc.), according to the block structure of these languages, exhibit a block-oriented dynamic behavior. Normaly, after every 20 to 50 instructions of these programs, a *procedure* call or *return* is executed. Therefore, the performance of procedure calls and returns has a major impact on the overall performance of a microprocessor. Today's microprocessors address this issue by optimizing the procedure entries and exits with specific register models . The most well-known register models are:

- Register Windowing (RISC I, Sun SPARC)
- Stack Cache (AMD 29000) [6]
- Frame Cache (Intel 80960)

Fig. 4 gives a graphical explanation of these three register models. All three approaches are based on two preconditions: the *locality of reference* to stack frames during procedure entries/exits; and the assumption that on average most procedures create only *small* stack frames.

The optimization of procedure calls and returns is particularly important for real-time systems to ensure short response times. Disadvantages of the described procedure entry/exit schemes with respect to real-time systems are the exceptions to the scheme described above.

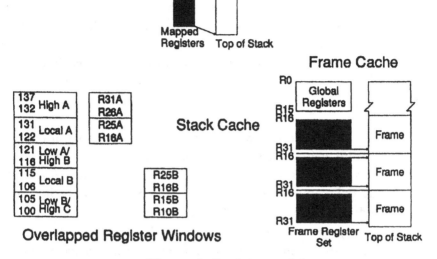

Figure 4: Register models

The Interrupt System

Real-time systems put a high burden on the interrupt system of a processor because of their time-related interaction with the external world (technical process). Conventional interrupt systems of microprocessors have been implemented with external, parameterized, and software-programmed interrupt controllers. The priority management of these external chips was not correlated to the internal execution priorities of the processors. Today's and future microprocessors associate the interrupt priorities with internal processor priorities. This is done by integrating the interrupt controller into the microprocessor. Interrupt systems of future processor architectures will handle interrupts from different sources within a multiprocessor and will be able to remember pending interrupts.

Microprocessors best suited for real-time systems can deal concurrently with a large number (250 and more) of interrupts from different sources. In addition, these processors offer a very short interrupt response time and can handle high interrupt rates. In particular, so-called microcontrollers contain integrated interrupt controllers.

Operating System Support

Microprocessor-based systems are used in increasing numbers for multitasking and/or multiuser environments. These applications require the development of highly sophisticated multitasking, multiprogramming, and multiuser operating systems for these architectures. In order to simplify these developments and to increase performance, the microprocessors have to offer operating system support within the architecture.

Virtual Memory Management: In virtual memory management, the virtual program address space is separated from the processor's physical address space. The motivation for this technique is multiuser/multiprogram mode and memory utilization. The mapping of virtual addresses to physical addresses is done in today's microprocessors with external or internal *memory management units* (MMUs). These MMUs implement three memory management techniques already known from the mainframe world:

- segmentation (Motorola 68010, Intel 80286)
- paging (Intel 80386, Motorola 68030)
- segmentation and paging (Intel 80386, AMD 29000, Motorola 88000)

The paging scheme is explained in **Fig. 5**. Via various address translation tables, the virtual address is mapped step by step onto a physical address. All future microprocessors will offer memory management units on the chip (Intel 80486, MIPS R3000, AMD 29000).

For processor architectures mostly used in real-time environments, the following three features of the MMU seem to be typical:

1. a flat address space
2. a simple address translation scheme
3. a fixed translation of virtual to physical addresses during the total execution of a program (no swapping)

Reasons for these features are once more the response times necessary and their predictability in real-time systems.

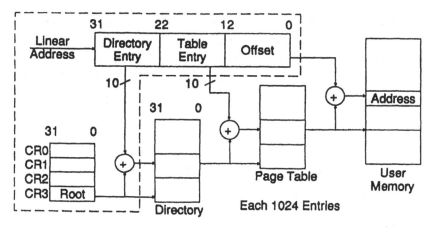

Figure 5: Memory management schemes (paging)

Memory Protection: The availability of memory management in addition to the address translation is also a prerequisite for the implementation of memory protection. Memory protection is necessary in microprocessor systems for the following reasons:

- support for simplified debugging and testing by run-time checks and exception handling
- protection of the operating system from erroneous application programs
- protection of the operating system and application against illegal access or even destruction

The schemes used for memory protection are adopted from the minicomputer and mainframe world once more:

- The access to main memory pages is checked via the address-translation tables of the MMU available already.
- The introduction of privilege levels with different access rights is used to protect different system components against each other.

The trend in microprocessor architectures goes toward an integration of memory protection into the processor chips. Processors used mostly for real-time environments tend to offer only two privilege levels (user/supervisor mode).

Multitasking Support: As already mentioned, new operating systems for microprocessors are significantly determined by their multitasking capabilities. To support this development, new microprocessors offer more and more multitasking support integrated in VLSI technology. In today's microprocessors, two levels of on-chip multitasking support can be characterized:

1. Some processors offer the notion of tasks and explicit context switches implemented in hardware (Intel 80386, Weitek 32100).
2. Other manufacturers have extended this concept and have implemented complete scheduling algorithms in hardware (Inmos T800).

Typical real-time systems have a large number of tasks running concurrently and a high rate of context switches. Therefore, efficient implementations of context switches in hardware are a basic prerequisite for short response times in real-time systems.

Floating-Point Support

The definition of the IEEE 754 floating-point standard has had a great influence on the development of floating-point coprocessors (Motorola 68881, Intel 80387, Weitek 1167). But the implementation of floating-point performance within an extra chip, separated from the CPU, leads to some disadvantages:
- The extra floating-point chip demands additional space and power.
- The protocol necessary between the floating-point chip and the CPU causes a problem due to performance losses.
- New developments of either the CPU or the floating-point chip may raise the problem of generation inconsistencies and incompatibilities.

Because of these problems, future processor architects will try to integrate the floating-point chips into the CPUs (Intel 80860, 80486, Motorola 88000). Apart from the integration of floating-point performance into the processors, it would also be worthwhile to transfer other special-purpose functionalities into the CPU (e.g., signal processors).

Debugging and Monitoring

The increasing complexity of microprocessor applications has had a major impact on the development of software for these applications (cf. Chapter 3.1). In particular, within projects based on embedded systems (cf. Chapter 2.2), software development has become the greatest effort. To deal with this software complexity, the software engineer needs adequate tools, which have to be supported by the processor architecture. For some years now, microprocessors have offered features for single stepping and breakpointing. These rudimentary features are no longer sufficient, and future microprocessors will integrate more sophisticated development supports:

- event trigger logic on instructions and data (Intel 80386, Motorola 68851, National Semiconductor 32082, Intel 80860)
- breakpointing and tracing based on different types of events
- new levels of granularity for the definition of events (threads, high-level languages)
- logical and time related sequential event combination

For real-time systems and multiprocessors, the integration of these features into the processor chips is very important. Without these hardware-based monitoring techniques, it is impossible to do nonintrusive monitoring of dynamic activities. Future development tools using hardware monitors and in-circuit emulators may only be implemented based on these technologies.

Special-Purpose Processors

A controversial question is the relationship between special-purpose and general-purpose microprocessor architectures. Classes of special purpose microprocessor architectures are

- language-specific architectures (LISP, Prolog, Smalltalk)
- application-specific machines (signal processing, image processing)
- architecture specific processors (systolic arrays, data flow machines)

As in the past, it can be expected that general-purpose architectures also will dominate in the future because of their greater flexibility. Additional reasons for this statement are compatibility, portability, and the increasing research and development costs necessary for developing new microprocessor architectures. Special-purpose processors are well suited to add additional functionality to the general-purpose CPU.

PERFORMANCE-INCREASING CONCEPTS AND PROCESSOR IMPLEMENTATION

This section adresses mainly performance issues of microprocessor designs. Architectural concepts are discussed that have been integrated into recent microprocessor designs for performance increases. More implementation-oriented questions of these architectural concepts will be addressed in addition, i.e. the ability to implement these features in VLSI technology.

The problem of measuring the performance of computers is well known in the field. Nevertheless, everyone is sure that microprocessors have become much faster over the years. Based on measurements in MIPS (a poorly standardized measurement metric), an analysis of available microprocessors demonstrates that the performance of microprocessors has been increasing by a factor of 2.25 each year in the past. Apart from improvements in silicon technology and the architectural concepts described in the last section, these performance increases are mainly due to cache-memory integration, on-chip pipelining, and improved processor implementation techniques.

Pipelining

Increasing performance by using parallelism in the shape of pipelining is a technique often used in computer design. In recent microprocessor architectures, this concept is used to execute multiple phases of different instructions in parallel. This concept only works perfect, when the sequential-execution flow is not disturbed by jumps, branches, procedure calls/returns and interrupts, which cause a pipeline "flush". In order to keep the number of necessary pipeline flushes as low as possible, various solutions have been developed [9, 14]:

- delayed execution of jumps by using code reorganization (delayed branch)
- doubling of the first pipeline stage
- prediction of the probability of branches based on heuristics (branch prediction)
- using code reorganization and adequate register allocation to avoid pipeline flushes (condition code scoreboarding, register scoreboarding)

A trend that can be recognized in processor design is that (apart from pipelining) future processors will offer much more sophisticated concepts for executing more than one instruction in parallel. Superpipelined, superscalar, LIW (Long Instruction Word), and VLIW (Very Long Instruction Word) computers as well as "out-of-order execution" are initial steps in this direction.

With respect to real-time systems, it is necessary to remember that at high interrupt rates and context switch rates , the performance of pipelined systems will decrease substantially.

Cache Memories

Another popular concept for performance increases in computer architecture is cache memories. In recent microprocessor designs, this architectural concept was adopted from the mainframe world and integrated into the processor chips (on-chip caches). Cache memories are used to avoid memory-access latency. Apart from their normal function, on-chip cache memories are a precondition for *branch prediction*. The efficiency of cache memories measured in hit rates depends heavily on intelligent prefetching strategies [2]. On-chip instruction cache memories (typical data from measurements gathered 1990) offer the following efficiency:

- 512-Byte cache: approximately 80% hit rate
- 2-KByte cache: approximately 95% hit rate

With an increasing number of transistors available for microprocessors, data-cache memories will be available on-chip also. The minimum size for data caches is about 4 KByte to get hit rates comparable to the ones described above for instruction caches. On-chip data and instruction caches are often separated. The problem with on-chip caches is the high number of transistors necessary for adequate hit rates. For the implementation of a 16-KByte on-chip cache with 16-Byte linesize, approximately 1 million transistors are required.

Real-time systems are influenced in two ways by on-chip caches:

1. The high interrupt and context switch rate often causes useless entries in the cache and consequently a low hit rate.
2. Very important routines of real-time programs may fit completely into on-chip caches and therefore can be executed without misses. By freezing the cache, the hit rate may be increased.

Modular Designs

In the introduction to this chapter it was already mentioned that the complexity of new microprocessor generations can only be managed with improved design and production concepts. One of these new concepts is a higher modularity of new designs. The idea here is to structure the processor design into well-defined reusable modules. Obviously this strategy has the following advantages:

- Submodules are independent and can be developed independently of each other.
- New processors may be combined easily based on already existing modules from different previous designs.

- Depending on the available integration density, a processor can be implemented either on one or on more dies.
- The modules can be tuned independently of each other with respect to different requirements.
- Application specific or special-purpose modules can be added easily.

Although adequate interface definitions between modules are difficult, future processors will be designed based on this design method. The consequence for real-time systems can be that future "real-time" processors are modifications of "standard" microprocessor families. An integration of these microprocessor cores with application-specific integrated circuits (ASICs) is a feasible goal.

Testability

The ever higher integration and complexity of microprocessors dramatically increases the efforts necessary for chip testing. Problems in this area arise from three sources:

1. The numbers of test patterns necessary for complete testing cannot be computed even by the fastest test machines.
2. It becomes increasingly complicated to access internal resources (cache, translation lookaside buffer, etc.) from outside the chip because of pin limitations.
3. Testing at the board level becomes more difficult, because of higher board integration, surface-mounted-device techniques, and multilayer boards.

One solution for these problems is the integration of specific test circuits and test algorithms into the processor chip. These tests are initiated from the outside via additional pins, and the results may also be checked via these additional interfaces (boundary scan).

INFLUENCES ON SYSTEM ARCHITECTURE

The adequate system design and architecture of a microcomputer system is as important as the microprocessor itself. Major system parameters and architectural system concepts are the processor bus (cf. Chapter 3.5), multiprocessor structures, the coprocessor, and reliability issues (cf. Chapter 4.1). In this section, we will discuss the relationship between processor architecture and these system concepts.

Processor Bus

The performance bottleneck of today's microcomputer systems is very often not the CPU, but the interface between processor and memory the processor/memory bus. Instructions cannot be delivered by the memory system as quickly as the processor would be able to execute them. Typical features of today's processor buses are

- synchronous and demultiplexed operation
- burst mode and pipelining
- dynamic and on-line bus sizing for I/O devices
- separated instruction and data buses

One solution to overcome the bandwidth problem in future generations of processors will be the use of processor buses with more than 32 Bit (e.g., 64 or 128 Bit).

Multiprocessor Architectures

As people become more aware of the fact that performance increases based only on silicon-technology improvements cannot be extended forever, more effort is being spent on research in parallel computing.

Available multiprocessor architectures can be categorized into three classes:

1. Homogeneous parallel computers based on identical microprocessors in each processor node and nonbus-based interconnection networks (e.g., hypercube, tree, array) [4]

2. Loosely coupled multiprocessors based on standard bus systems (VMEbus, Multibus I, II, Futurebus) [3]
3. Tightly coupled multiprocessors with specific proprietary buses and global shared memory [8]

Because of the physical limitations of performance increases in silicon technology, all three classes of parallel architectures will become more important in the future.

Typical real-time systems will be implemented mainly as heterogeneous multiprocessor systems of types 1 and 2.

Coprocessors

Special-purpose tasks can be handled in a microcomputer system very elegantly with coprocessor technology. The coprocessor offers a major contribution to the performance of the system by off-loading the CPU. Well known types of coprocessors are

- network coprocessors
- direct memory access controllers
- graphics coprocessors
- floating point coprocessors

The interface between the CPU and the coprocessor can be implemented either tightly coupled (CPU specific) or loosely coupled (CPU independent). Based on the modular design methodology described above, the integration of coprocessors as reusable, independent modules becomes feasible.

Reliability and Fault Tolerance

The last architectural system feature, which becomes more and more important, is the integration of fault-tolerant concepts into processors. The design of fault-tolerant system structures is motivated by the decreasing reliability and availability of microcomputer systems with increasing number of components. The integration of fault-tolerant con-

252

cepts into VLSI technology is used to overcome this problem. Recent concepts are based on the integration of master/checker circuits into CPUs (see **Fig. 6**). In particular for real-time systems, the availability of fault-tolerant concepts is very important, because of the very often security-sensitive applications of embedded systems.

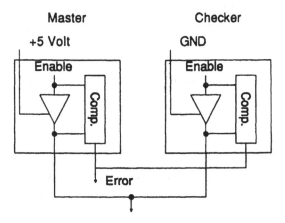

Figure 6: Master/checker circuit

EVOLUTION OF APPLICATION AND MARKET SEGMENTS

To recall the importance of the microprocessor for the development of economy and human life, an overview of the most important applications and market segments of this technology is now given. On one side, the microprocessor is increasing in importance in traditional computer market sectors, such as

- personal computing
- workstations
- office computing
- supercomputing

In these application areas, the microprocessor has primarily influenced the price/performance ratio. Apart from these traditional market areas, the microprocessor has invaded more and more application areas. In particular, market segments in which traditional computers are not attractive because of their space/power consumption and their price/performance ratio. Examples of these innovative application areas are

- automobiles
- technical medicine
- graphics
- image processing
- mechanical engineering

- expert systems
- signal processing
- animation
- simulation
- industrial automation

In many of these applications, the microprocessor has been integrated into an external technical process and therefore is being used within a real-time system.

REFERENCES

[1]   T. Bemmerl, "Realtime High Level Debugging in Host/Target Environments", *EUROMICRO '86*, Venice, Sept. 1986

[2]   T. Bemmerl, W. Karl, P. Luksch, "Evaluierung von Architekturparametern verschiedener Rechnerstrukturen mit Hilfe von CAE-Workstations", *ITG/GI-Fachtagung "Architektur von Rechnersystemen"*, München, FRG, März 1990

[3]   T. Bemmerl, A. Löw, "Vergleich und Bewertung von Bus-Systemen", *Elektronik*, Franzis-Verlag, Febr. 1989

[4]   T. Bemmerl, P. Schuller, "Monumental - Eine Einführung in Parallele Supercomputer", *Elektronik Informationen*, Würzburg, Sept. 1989

[5]   A. Bode, "RISC-Architekturen", *Reihe Informatik*, BI-Verlag, Band 60, 1988

[6]   D. Ditzel, R. McLellan, "Register allocation for free: The C machine stack cache", *Proc. Symp. on Architectural Support for Programming Languages and Operating Systems (ACM)*, pp. 48 - 56, 1982

[7]   J. Hennessy, et al. "Hardware/software tradeoffs for increased performance", *Proc. Symp. on Architectural Support for Programming Languages and Operating Systems (ACM)*, pp. 2 - 11, 1982

[8]   R.H. Katz, et al. "Implementing a cache consistency protocol", *Proc. 12th Annu. Symp. on Computer Architecture (ACM)*, pp. 276 - 283, 1985

[9] J.K.F. Lee, A.J. Smith, "Branch prediction strategies and branch target buffer design", *Computer, vol. 17,* no. 1, pp. 6 - 22, 1984

[10] M.S. Papamarcos, J.H. Patel, "A low-overhead coherence solution for multiprocessors with private caches", *Proc. 11th Annu. Symp. on Computer Architecture (ACM)*, pp. 348 - 354, 1984

[11] D.A. Patterson, C.H. Sequin, "RISC I: A reduced instruction set VLSI computer", *Proc. 8th Annu. Symp. on Computer Architecture (ACM)*, pp. 443 - 457, 1981

[12] G. Radin, "The 801 minicomputer" *Proc. Symp. on Architectural Support for Programming Languages and Operating Systems (ACM)*, pp. 39 - 47, 1982

[13] J.E. Smith, J.R. Goodman, "Instruction cache replacement policies and organizations", *IEEE Trans. Comput.*, vol. C-34, no. 3, pp. 234 - 241, 1985

[14] J.E. Smith, A.J. Pleszkun, "Implementation of precise interrupts in pipelined processors", *Proc. 12th Annual Symp. on Computer Architecture (ACM)*, pp. 36 -44, 1985

10

BUSES IN REAL-TIME ENVIRONMENTS

F. Demmelmeier
SEP Elektronik
Ziegelstr. 1
D-8153 Weyarn

INTRODUCTION

The term "bus" has two meanings: 1) It refers to an common electric pathway connecting different parts of a computer (contrary to individual connections); 2) it refers to a network topology, with several network nodes sharing a single physical channel (contrary, for example, to star-like or ring topologies). In the following section the first definition is used.

Buses have a key position in modern computer architectures. They allow the transport of data on all levels of communication. At the bottom line, internal buses are used to transfer data within the computer or central processing unit (CPU) and thus are an integral part of the computer architecture; on a higher level, buses link the CPU with peripheral components and connect computers with each other. Buses are present as part of data processing hardware, such as office equipment, scientific instruments and they transfer data in airplanes (cf. Chapter 5.2) and cars.

Internal buses and system buses which connect peripheral components like logical devices and physical devices are the domain of parallel bus systems which allow high transfer rates over short distances; serial buses are used to interconnect peripheral components at lower speeds but across longer distances. Thus, the collection of process-data, the sharing of resources such as the communication between computers and peripherals in different offices, and the data transport within networks, particularly over long distances, are almost exclusively based on serial bus systems.

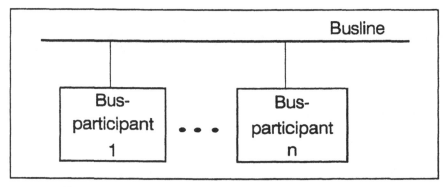

Figure 1: Bus as a common communication path

Since participants in bus systems are characteristically switched together via a common means of connection (cf.Fig. 1), transfer rules must be established in order to ensure proper communication between all participants. These rules encompass all layers of communication, from mechanical coupling to actual data transmission.

Because bus systems have a key position in all kinds of data communication in digital systems, they influence their timing characteristics profoundly. The internal structure of a computer system is demonstrated in Fig. 2. Here, the close interactions between the CPU, the application software and the peripherals is emphasized. Naturally, a close dependence exists between the timing of the CPU, the software, the data transmission, and the processing in the peripheral component:

The rate of processing of a CPU which is measured in *million instructions* per second (MIPS) puts a upper limit to the basic performance of a real-time system. Through special architectures that are partly parallel and partly pipelined, basic performances of up to 100 MIPS are available in modern CPUs.

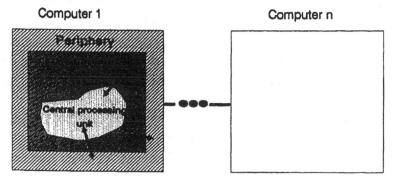

Figure 2: Central processing unit/software/periphery

The supervisory software - the *operating system* (OS) - transforms the performance of the CPU into the computer performance available to the user's application program. Its timing characteristics depends predominantly on the architecture of the underlying OS (eg. Real-Time OS vs. standard Time-Sharing Systems, cf. Chapter 2.3).

Application programs which are built on top of the operating-system software, are indirectly affected by the timing of the data transport between the CPU and peripheral components.

Interfaces between the computer and the user or technical process influence the timing characteristics of the overall system, depending further on the absence or presence of peripheral intelligence. A direct dependence on the peripheral equipment's timing behavior on basic performance is examplified by the so-called *DMA-Mode* (Direct Memory Access Mode), where the interface is granted access to the computer's memory thus bypassing the cpu which may even be temporarily cut off from access to memory during DMA-transfer.

When more than one computer is connected via a network, the situation becomes even more complex.

In the following, the relative merits of parallel and serial bus systems are briefly discussed, and then the timing characteristics of serial bus systems are shown.

PARALLEL BUSES

The characteristic feature of parallel buses is a parallel transfer pathway, together with the mechanical and electrical connectors and also the relevant transfer protocol. Signals are transferred via a set of conducting elements, which can be arranged in different ways, e.g., wires in a cable, etched conductors, and flat cables. Such buses absorb and emit high amounts of electrical interference also between bus lines which has to be taken into account together with proper choice and balancing of the connecting parts.

The number of bus participants is principally not limited, but a new one can only be added if the electrical, mechanical, and logical environment allows this. Obviously, only one participant can be active on one bus, but an unlimited number of passive participants can be there simultaneously. The signal lines of parallel buses may be divided into different categories, such as the address bus, the data bus, and the control bus. Between the address bus and the data bus can be further differentiated so that time-dependent lines can be attached to different categories. The result is called the *multiplex* bus.

Another important bus characteristic is whether its operation mode is *synchronous* or *asynchronous*. A synchronous mode means that procedures involving the bus are delimited by a time frame, so that there is no possibility that one participant can jeopardize the bus timing. In the asynchronous mode, one bus participant can interrupt the bus traffic deliberately until the operations belonging to this participant have been carried out. However, the price for this flexibility is that in asynchronous mode, one failing bus participant can bring the whole bus operation to a halt.

Parallel bus systems are obviously faster because they don't need to waste time to serialize parallel data. They are thus used when a high to very high data transfer rate is necessary. This advantage is diminished, however, by interference problems and by a certain inflexibility because its communication format is set up once and for all by its number of lines.

In Fig. 3 the parameters of some important parallel buses are shown. Parameters included here are the transfer protocol, the address bus width, the data bus width and the possibility for a multi-master operation (more than one bus-master on the bus), the number of bus interrupts and the maximum data transfer rate (cf.[1]).

	Q-Bus	Multibus	VME-Bus	SCSI-Bus	IEC-Bus
Communication	asynchronous multiplex	asynchronous no multiplex	asynchronous no multiplex	synchronous or asynchronous	asynchronous multiplex
Addressbusbits	22	20 standard 24 extended	24 standard 32 extended		8
Databusbits	16	8/16	16 standard 32 extended	8 +Parity	8
Multimaster	yes	yes	yes	yes	yes
Bus arbitration levels	1	1	4	1	1
Interrupt levels	4	8	7	1	1
Fault signals	DC-Fail AC-Fail Paritycheck		Systemfault Busfault		
Maximum Datarate in MB/s	2 (3)	5	24	asynchron 1,5 synchron 4	1

Figure 3: Characteristics of parallel bus systems

SERIAL BUSES

A serial bus is characterized by a single transmission line to connect the bus participants. The data transfer on the bus line is carried out in a serial manner. Existing information, which is normally parallel in a digital system, must be serialized before it can be brought onto a serial bus. The serialization must be delimited by a begin and end identification and the corresponding hardware must have facilities to control access to the bus. As a consequence serial transfer is normally slower. Further, in serial transmission, the transfer time of a message grows with the number of bits transferred in a linear way, whereas in a parallel bus it grows in steps, i.e., only with the integer number of words transferred. This disadvantage can often be compensated for by the important feature that single-line connection between participants can be interference-free and be carried over much longer distances. Thus an increase in the reliability of data transfer can be achieved. Of course, the connectorn technology plays a major role, since connectors are always a bottleneck in the transfer technique. The transfer of data and synchronization information via one line is possible through a special code. The clock information contained in the original signal will be mixed with the data information and then, as a combined data stream, transferred onto the bus. The choice of the coding procedure for a serial bus is determined by considering the type of connection used.

Due to the need to control longer distances, electrical-potential differences between the individual bus participants may have to be taken into account. To overcome this problem technically, all bus participants must be connected via glass fibers or by "free of direct current" technology in which case it is necessary to ensure a direct current free data stream by the type of code used.

In the course of the development of bus networks, an immense number of topological variations have been created. Typical examples are the bus topology or the ring topology (see below).

Shown in Fig. 4 is a classification of serial buses by distance and speed. The basic performance of a bus system will be measured by the bit transfer rate. The spectrum reaches from 1 kbit/sec to a few 100 Mbit/sec.

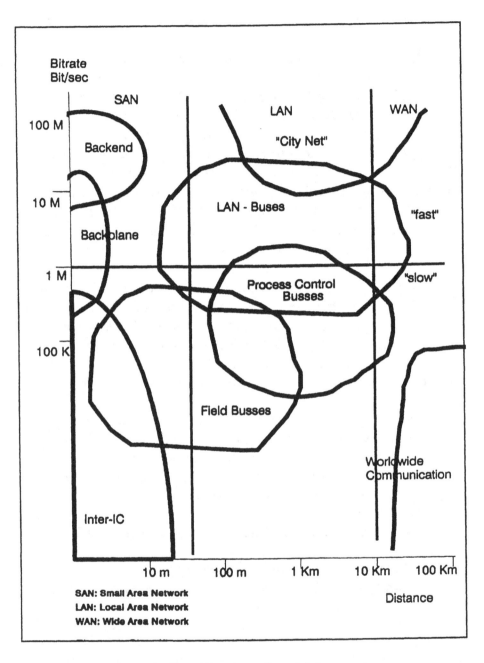

Figure 4: Classification of serial bus systems

Distances, which are bridged through bus systems on bounded media, range from a few centimeters up to the kilometer scale. Serial buses for short distances include the inter-IC-buses, which are used for communication between modules within computer systems. They reach a capacity up to a few 100 kbit/sec. The so-called *backplane buses*, which exist quite often as serial buses within parallel bus systems, connect distances of up to 50 cm and have a speed of up to 10 Mbit/sec. The *back-end buses* are for communication between the CPU and very fast peripheral components, e.g., drive systems. They are designed for high speed and have transfer rates up to a few 100 Mbit/sec. The most important application area of serial buses concerns the *field buses*, the *process-control buses*, and the *LAN buses* (local-area network buses). The classification of these buses is done with different kinds of emphasis which varies between high speed, high reliability, connection of long distances, and low costs.

On the level of the LAN-Buses, the *Carrier Sense Multiple Access with Collision Detect (CSMA/CD) buses* and the *oken passing buses* represent two popular, standarized channel allocation procedures. The demand for process-control buses which is widely identical with the demand for LAN buses has lead to the dominant position of the Ethernet bus (as a representative of the CSMA/CD world) and the token passing bus.

Field-bus systems, which link the small and smallest connecting devices of technical processes and their associated information to a major processing unit, are at the moment in different standardization stages. No single system on a wide basis will make a breakthrough, since requirements with reference to reliability, speed, and distances diverge strongly.

As shown in **Fig. 4**, the area of long-haul communication consists mainly of point-to-point-connections.

To fulfill the different demands regarding information transport and exchange, bus hierarchies have been created. **Fig. 5** shows typical hierarchical levels present in the area of materials-processing. On the level of factory planning, the main computer (in connection with other computers) must exchange a large volume of data. With reference to the transmission system and system administration, real-time behavior plays no major role. However, real-time behavior on the level of process-control buses and field buses is important. The overall timing is governed by the signals of the technical process. The computers that are used, together with their bus systems, have to consider the time demands in all subcomponents. These subcomponents are the computers themselves, the operating system, and the communication facilities.

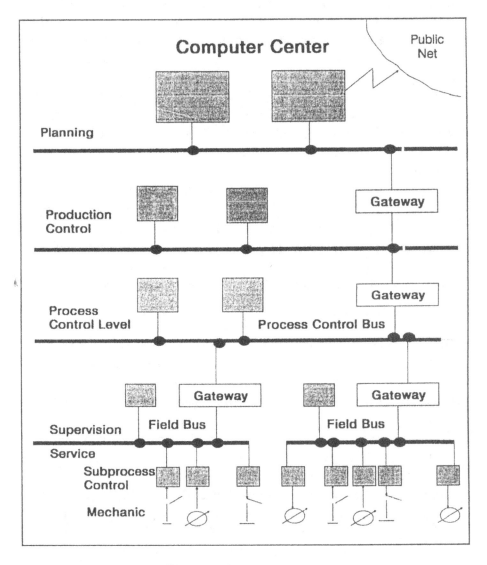

Figure 5: Bus hierarchies

As carriers of information between different subcomponents, bus systems are the backbone of any data exchange. Apart from the rules necessary to achieve a bus-like operation between participants, further buse-wide rules may also exist. Some of these may even involve the user's "point of view" when the bus systems are exchanging information between two participants.

Due to the need to maintain physical connection and the correct transfer of information, a number of protocol layers are defined in the ISO/OSI-reference model for the communication of open systems. Fig. 6 shows the levels 1 to 7. Whereas levels 1 and 2 belong to the bus system and, in the case of technical application, exist as hardware modules, levels 4 to 7 are realized in software exclusively. Level 3 is not necessary for bus systems, since each user can "hear" all the information on one bus system. The effort to implement all seven layers of the ISO-reference model is enormous and requires a very high processing rate. The timing behavior of communication between two participants is then mainly determined by communication software. Therefore, in real-time systems, it is desirable to use only levels 1, 2, and 7 (cf. also discussion on this topic in Chapter 3.6). This means that the intention of the ISO-reference model, to let communication systems of different manufacturers communicate with each other, will be lost because of the demand for real-time capability, at the expense that compatibility between systems will no longer be guaranteed. This is one reason for the still existing functional and topological variety of bus systems which seems to encompass new ones even, which have appeared on the market recently.

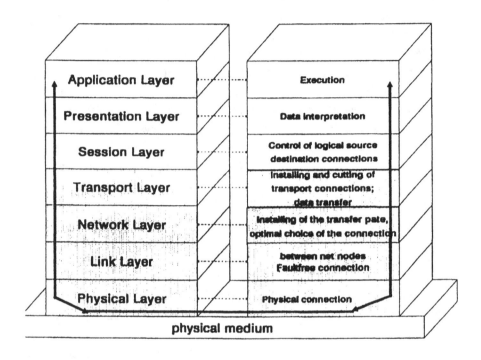

Application Layer	Execution
Presentation Layer	Data interpretation
Session Layer	Control of logical source destination connections
Transport Layer	Installing and cutting of transport connections; data transfer
Network Layer	Installing of the transfer pate, optimal choice of the connection
Link Layer	between net nodes Faultfree connection
Physical Layer	Physical connection

physical medium

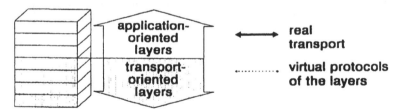

application-oriented layers

transport-oriented layers

← → real transport

.... virtual protocols of the layers

Figure 6: ISO/OSI reference model for open systems

ACCESS PROCEDURE AND TIME BEHAVIOR OF SERIAL BUSES

On a bus, as already mentioned, each participant can passively be con-nected. The main criterion for communication is, which participant owns the bus, i.e., which one has the "right to talk" on the bus and for how long.

To assign the right to talk, the following methods are possible:

- central distribution of the right to talk
- cyclic distribution of the right to talk
- rotating right to talk
- spontaneously usurped right to talk

In the following, the access methods will be described briefly by a circle of discussants, representing the bus participants. In principle, access to the common medium (the bus transmission) is symbolized by the situation in which one participant talks and the others listen. If more participants are equivalently connected, an information loss will occur as soon as two talk at the same time. The granting of the right to talk in the discussion circle also symbolizes the access rules and therefore the protocol. The access rules define the time conditions of the bus systems. Time conditions mean the waiting time that accumulates until one par-ticipant has the right to talk and therefore receives access to the bus. Naturally, in a real-time bus, each participant should be guaranteed a maximal waiting time (Fig. 7).

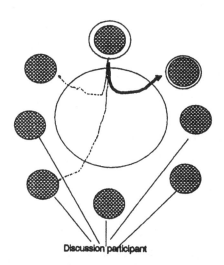

Figure 7a: Central allocation of the right to speak

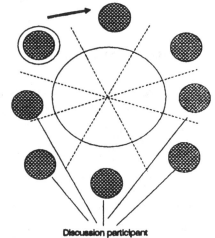

Figure 7b: Cyclic allocation of the right to speak

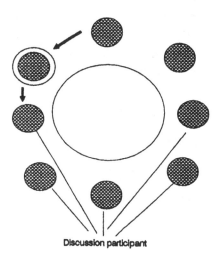

Figure 7d: Rotating right to speak

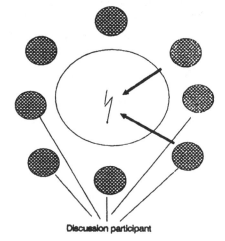

Figure 7c: Spontaneus usurpation of the right to speak

Every piece of information transferred on a bus system is divided into

- header information
- the data to be transferred
- the protection code

The header information and the security information symbolize the protocol information.

With reference to access methods, two groups exist: The non-contention oriented groups and the contention oriented groups. In the first there is a dominant participant, the so-called *bus master* or *discussion leader.* In the second group, there is no bus master and every participant has to compete for access to the bus.

Central control of the Right to Speak: Master Process (Fig. 7a)

This is a centralized polling method, where the discussion leader is the bus master. Only the bus master grants the right to talk. A central timing control also exists. Each participant will be asked by the bus master in turn or by use of a list whether a transfer wish exists. In this way, each participant is certain to receive the right to talk within a certain time slot. The message lengths define the waiting time for the other participants (see Fig.8). The existing basic transfer capacity of the bus systems cannot reach 100%, because administrative information has to be exchanged, i.e., the right to talk and the relevant feedback.

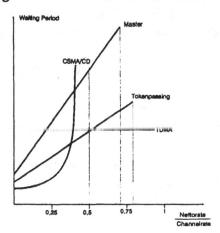

Figure 8: Waiting periods for participants of bus-systems

Cyclic distribution of Right to Speak: Time Division Multiplex Access (TDMA) (Fig. 7b)

This is a distributed polling method in which the total time frame is distributed evenly (Time Division) among the existing participants. Each participant is alotted a certain time slot (Multiplex Access) during which he should send some information to the bus system; if one participant cannot send any information, then time will be wasted. If there is more information than can be sent to the bus during the time available, the participant is responsible for dividing information into different time slots. The waiting time for each participant is constant. In case of failure, a central element comes into play that stops the whole communication. The administrative effort of this method is relatively low, so that the maximal possible transfer capacity can nearly be reached.

Rotating Right to Speak: Token-Passing (Fig. 7c)

Here participants agree upon a certain order that the passing of the right to talk (a bit pattern, called *token*) will follow. After the initialization of the whole discussion circle (or the bus system), one participant has the token, which he can use to gain access to the bus or which he can pass on to the following participant. Therefore, the topology of a token passing bus will be a logical ring. When a message is sent, participants will check its destination as it passes by. Thus participants are responsible for the identification and acceptance of messages as well as for the passing on of messages not addressed to them.

This technique uses the channel capacity of the bus very well, but some special situations should be taken into account:

- One participant does not pass on the token (the participant is asleep)
- One participant does not pass on the token because he has a great amount of information (the participant talks too much)

In a token passing bus system, a solution has to be found for both of these situations so that, in case of the failure of one participant, the bus system will still operate. Therefore, locally running time controls are necessary to find out how long a participant is active and how long he is not active. It must be locally and reliably checked if the transfer of information takes too long. A total breakdown of one participant will be noted through other participants. After reaching the time limit, initializing procedures will be activated from a predefined participant that will result in a new configuration of the whole logical ring.

Fig. 8 shows the timing characteristics. The basic waiting time is defined by the passing of the elementary token operations. The more information that has to be transferred over the bus, the longer the waiting time will be. There is an upper limit to the access time of each participant.

Carrier Sense Multiple Access with Collision Detect (CSMA/CD) (Fig. 7d)

In this contention control concept each participant can talk whenever she likes (Multiple Access), but she has to be certain that at that moment no other participant has the right to talk (Carrier Sense). In spite of this, collisions can still occur if two participants start to talk at the same time or nearly the same time. Since the participants listen whether a collision has occurred while talking (Collision Detect), each participant can back out as soon as she realizes the collision. She tries to gain the right to talk again after waiting some predetermined or random time. In the latter case the choice of this time depends on the number of collisions that have already occurred, as well as on other facts. Failures of the whole procedure occur when one participant with the right to talk, which she once acquired, does not return it. Each participant is locally responsible for herself to give back the right to talk after an upper time limit.

The waiting period for each participant depends on the probability of a collision. Fig. 8 shows the waiting period schedule. The basic waiting period is very low, because virtually no administration is necessary. Through increasing data exchange of the participants, the probability of a collision grows and therefore the waiting period increases. This process can end in the collapse of the whole bus system which happens when the number of collisions is so frequent that the collision situations must be dealt with permanently. Then no information can be transferred.

Timing Characteristics

The real-time requirement, namely, the guarantee of a maximum waiting time, will be ensured by a bus-allocation procedure according to the non-contention principles, namely the master process, the TDMA procedure, and the token passing procedure, which are deterministic. The CSMA/CD bus is not deterministic and thus access can only be guaranteed statistically for a given bus traffic load. If the bus collapses, however, no net information at all can be transferred.

For a real-time environment, the token passing procedure would therefore be preferable. For a fault-free operation, a guaranteed maximum waiting period is ensured. However, in case of malfunctions caused by bus participants (i.e., sleeping or too much information), certain initializing procedures are necessary, that consequently create incorrect timing behavior. The total initializing procedure is very complex, as shown by the necessary hardware required and the resulting down times are relatively long.

In comparison to the more established CSMA/CD procedure, the token passing procedure only recently became supported by highly integrated modules suitable for real-time processing, and so, in spite of its inherently non-deterministic property, many applications in a real-time or in a real-time-like environments make use of the CSMA/CD principle (Ethernet) the capacity of which can be enhanced by using high-speed access and high-bandwidths channels and by keeping the number of participants at a minimum.

FIELD BUSES

The basic properties of field buses are not different from those of other bus systems. However, a few requirements play a major role, and as a result, specialized bus systems are on the market and field-bus systems are heavily discussed in standardization committees.

The characteristics of field buses are as follows:

- low cost
- low power consumption
- low reaction time

- low failure rate, i.e., high reliability
- low sensitivity against electromagnetic interferences

The requirement for low cost arises from those fields of application in which a large number of signals must be connected by means of a distributed technical process. If a given signal point is, for example, a temperature sensor, the investment for the connection of that point to the field bus must be comparable to the costs of the sensor proper. The requirement for a low power consumption follows from the fact that the power distribution in the distributed technical process is often realized via the bus transmission lines. Low reaction times are necessary because applications like machine tools require a reaction time of milliseconds or less. Low failure rates are important to ensure the same or better reliably than that which is achieved with conventional analog techniques. Low sensitivity against electromagnetic interference is required because many technical processes in factories are strong sources of electromagnetic interferences.

Field-bus systems are of special significance in security-relevant control applications. Here it is important that all participants are continuously supervised. If the configuration changes, i.e., if it stations fail or are added, all participants must be informed before the next piece of information is sent. In the case of a token passing bus, this supervision function can be fullfilled if the message protocol includes the sender, i.e., the source address. This information is used to update the station configuration tables - the so-called *"live list"*. The live list facilitates the initialization procedures and the fault handling. With this mechanism, a real-time self-configuration will be supported. Self-configuration means the adaptation of the communication to the valid station configuration. This feature is used, for example, in the PROWAY B standard proposal (IEC SC65). Compared to the IEEE 802.4 Standard, every station holds only the predecessor and the successor for the token rotation. This leads to more complex procedures for initialization and self-configuration, a very important fact for real-time applications. Fault handling has as a consequence a non-deterministic timing characteristic or a very large basic time, if fault handling is considered in the waiting period.

The PROFI bus defined in the standard DIN 19245 is based on the ISO/OSI reference model. Three of the seven defined layers are explicitly described:

- Layer 7: Application Layer
- Layer 2: Data-Link Layer
- Layer 1: Physical Layer

Layers 3 to 6 are not defined, because only a few functions are used in field-bus applications.

The physical layer is determined by the following characteristics:

- Twisted-pair wire
- Transmission rates between 9.6 Kbit/sec and 500 Kbit/sec
- Line lengths of 1200 m maximum at 9.6 Kbit/sec down to 200 m maximum at 500 Kbit/sec
- Maximum of 122 participants

On the data-link layer, four services are defined:

- Send Data with Acknowledge (SDA)
- Send Data with no Acknowledge (SDN)
- Send and Receive Data (SRD)
- Cyclic Send and Receive Data (CSRD)

The application layer is subdivided into two parts:

- Field-Bus Message Specification (FMS)
- Lower-Layer Interface (LLI)

FMS describes the communication objects, services, and resulting models from the viewpoint of the communication partner. The LLI transforms the FMS services (layer 7) to layer 2 and additionally defines the parts of layers 3 to 6 that are necessary for the functionality of the PROFI bus. For further information see [2]

STANDARDIZATION

The necessity of standards is obvious when components and systems of different manufacturers must work together. In the area of parallel bus systems, international standards and some company standards exist that have gained general acceptance. The standardization for field-bus systems has not been completed yet. National and international committees are working on a solution.

Figure 9 shows standardization committees and the field-bus candidates. As an example, the PROFI bus, as a candidate for a field-bus standard, is supported by DIN (Deutsche Industrie Norm), the standard of the German standardization committee.

Standards-Gremium	Candidate
DIN 19244	UART, FT 1.1
DIN 19245	PROFI-Bus
DKE K 933.3.2	Fieldbus (Mirrorgremium to IEC SC 65 C Proway)
IEEE	Bitbus, MiniMAP, MAP-EPA
FIP France	Fieldbus, Upper Layers
INTEL/Bosch	(Inter) Controller Area Network (I) CAN
VALVO/Philips	Multimaster Serial Bus M^2 S
Motorola/Ford	UART

Figure 9: Field bus standardization

Bus systems that were designed in cooperation of a semiconductor manufacturer with a car manufacturer are of special interest since the first possess knowhow for mass production and the latter have enormous marketing potential. This applies especially to the so-called *(I)CAN bus*, which was developed jointly by INTEL and Bosch.

Concerning parallel and serial bus systems, a number of standards have already been accepted internationally. As parallel bus systems, the IEC bus and the SCSI bus have been standardized. The VME bus and the multibus are represented as standardization candidates in IEEE committees (top of Fig. 10).

Project 802 of the IEEE organization has made a breakthrough with respect to serial bus systems [3]: This applies to the basis of serial bus systems, defined in IEEE 802.1 and 802.2, the Ethernet bus as defined in IEEE 802.3 and the token passing bus, laid out in IEEE 802.4. The Tokenring and the Metropolitan Net are standardized und 802.05 and 802.06. Fig. 10 summarizes these important standards in the area of serial and parallel buses.

IEC 625-1,2	IEC-Bus
ANSI X379.2/84-40	SCSI-Bus
IEEE P1014	VME-Bus
IEEE P796	Multibus
IEEE 802.1	Basis
IEEE 802.2	serial busses
IEEE 802.3	CSMA/CD (Ethernet)
IEEE 802.4	Tokenpassingbus
IEEE 802.5	Tokenring
IEEE 802.6	Metropolitan Net
IEC SC 65	Proway
ANSI X3T9,5	FDDI (TAXI)
OSI/ISO Reference-model	Communication of heterogene systems

Figure 10: Standardization of parallel and serial bus systems

The OSI/ISO reference model used for communication in heterogeneous systems is accepted by all major manufacturers and therefore has gained general recognition. It is now the basis for the digital communication of serial systems.

For all bus systems, standardization is necessary; this is particularly true for field buses which have an enormous marketing potential. Fig. 9 shows an subset of committee members that are presently involved in the standardization of field buses. As of this writing, this process has not been completed yet, since company interests play a dominant role in the setting of standardization rules.

REFERENCES

[1] G. Färber, "Bussysteme", München-Wien, R. Oldenburg Verl
lag, 1987.

[2] Handbook Training-college "Serielle Bussysteme",
VDI Bildungswerk, 18./19. Sept. 1990, Düsseldorf, Germany.

[3] IEEE 802 Standards Office, IEEE Standards Office, 345 East 47th
Street, New York, N.Y. 1071.

Further reading:
F. Halsall, "Data Communications, Computer Networks and OSI",
2nd ed., Workingham, Addison-Wesley, 1988.

11

DISTRIBUTED SYSTEMS FOR REAL TIME APPLICATIONS
USING MANUFACTURING AUTOMATION AS AN EXAMPLE

Helmut Rzehak
Universität der Bundeswehr, München
Fakultät für Informatik
Werner-Heisenberg-Weg 39
D-8014 Neubiberg

THE TERM "REAL TIME DATA PROCESSING" IN DISTRIBUTED SYSTEMS

With the term "real-time data processing," we denote applications that have special emphasis on the following:

- Algorithms that can be executed within a guaranteed response time window, and
- Processes (that is, the performance of an algorithm) with overlapping time windows which must nevertheless be kept in logical and timely order.

The first item is also called the *requirement for guaranteed response time*. This requirement is the source of the term "real-time data processing." For manufacturing automation, the time window is typically within the region of 10 to 1000 ms. Guaranteed response times are not synonymous with fast computing.

Sources of temporal and logical dependencies are intrinsic dependencies of the application tasks. This situation is typical of manufacturing automation as well as of many other application classes. Other sources of dependencies are limited resources of the computing system (e.g., processors and memory), and communication needs between the computing processes in the performance of their tasks.

Fig. 1 provides an example. Track-keeping vehicles run between two parts of a factory on a single track (one direction at a time). Consequently, coordination among tasks, that control switches and movements of vehicles is necessary. We will use this example to illustrate concurrency control later on.

278

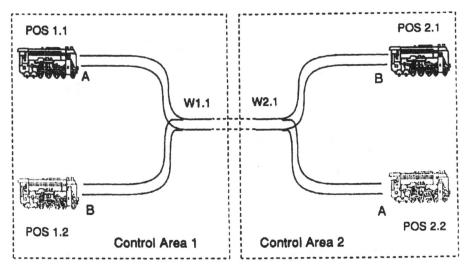

Figure 1: An example for the need of concurrency control

SPECIAL ASPECTS OF DISTRIBUTED SYSTEMS FOR REAL-TIME APPLICATIONS

Distributed systems have some peculiarities that in part simplify real-time applications but that also have repercussions on programming and therefore need to be known. Some of the problems raised have to be solved by the application programmer.

Scalable Processing Power

A decision in favor of a distributed system has some advantages if we follow some general rules (e.g., standards for communication and well-structured application programs). Additional nodes are easy to install, and with the additional computing power, time constraints are easier to meet. For example, highly time-critical tasks can be assigned to nodes of their own, thereby reducing the overhead for context switching.

Requirements for the Communication System

For manufacturing automation, the use of standards for communications is strongly recommended. The reasons as are follows:

- Some nodes, e.g., numeric-controlled machine tools or robots, are not programmable in the common sense, i.e., by software people. Nevertheless, control information ("programs") has to be transferred and state information has to be sensed. The interface problems of concern need standards for economic solutions.
- No supplier has the full range of equipment (e.g., computers and manufacturing devices with embedded intelligence) cooperating together in a medium-sized factory. Furthermore, dependence on only one supplier should be avoided. A proprietary communication system is no ideal solution.
- For application programs, we need standard interfaces to the communication system, for reasons of reusability and portability.

Hence, distributed systems for manufacturing automation are heterogeneous networks. Some kind of transformation by the communication system is necessary to ensure proper cooperation of the heterogeneous nodes.

Time constraints in a distributed system include the demand on limited message delays for communication. Bounds for message delays have to be guaranteed and must be sufficiently small for a certain application. In other words, we need a deterministic behavior of the communication system. In industrial applications, it is often required that message delays between two application processes should not exceed some 20 ms. On the basis of today's technology this is hard to obtain without compromises concerning universal interfaces to the application (see the section "Time Spent on Communication", below). In this context, the transfer rate on the medium plays only a minor role.

No Timely View over the Global State

Message exchange needs time that generally must not be omitted. The state of a process sensed in a remote node may have changed before we receive the state information. The time need may vary considerably, and therefore the time order of events rising in different nodes does not always appear to be the same from node to node. **Fig. 2** illustrates this effect.

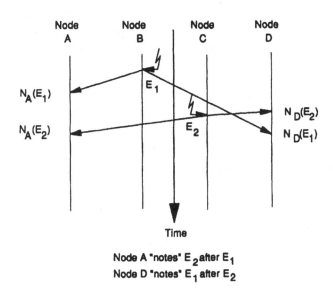

Node A "notes" E_2 after E_1
Node D "notes" E_1 after E_2

Figure 2: Relativistic view of the time order of events

The local clock of an arbitrary node is a special case of a local state variable. With respect to varying message delays, we cannot determine the current local clock on a remote node. Furthermore, we cannot suppose that all local clocks coincide. Synchronizing local clocks is a difficult and expensive task to perform if the tolerance is in the range of 1 ms or even less . Solutions with a global clock (available by each node) are possible on the restricted area of a job shop floor, but need additional measures and precautions with respect to a breakdown of this central clock. A special case of a global clock is the use of a radio timing signal as proposed in Chapter 2.1 (by Halang). To receiving such a signal on a job shop floor may be a hard task or even impossible. Generally, for clock resolutions of one second or more it may work quite well. But for resolutions of one millisecond or less the uncertainties in signal delays of the radio wave and within the receiver are of the same magnitude or more and therefore clocks will run asynchronously. Generally, we are forced to solve problems with local clocks, and in special cases the application programmer is faced with the above-mentioned relativistic event ordering.

Conflicts and Concurrency Control

Conflicts may arise from dependencies among processes running in parallel and may lead to incorrect results or disturbances and faults during the manufacturing process. Fig. 3 gives a hypothetical example concerning banking. Two different bookings on the same account should be done in parallel. The result shown is incorrect even if each of the transactions is itself perfectly organized.

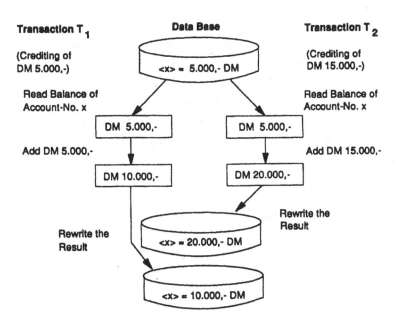

Figure 3: Ambiguity during parallel updating of files

A second example concerns the track-keeping vehicles shown in Fig. 1. An application process that moves engine A from position POS 1.1 to POS 2.2 has to perform the following sequence of actions:

Scan if target position is empty;
If not, wait until it is empty;
Make reservations for target position;
Scan if track to target position is empty;
If not, wait until it is empty;
Make reservations for the track to the target position;
Set switches;
Start drive;
Control velocity;
Stop drive if engine is in final position.

Now we are looking what happens if simultaneously another application process has to move engine B from position POS 2.1 to POS 1.2. We can simply copy the sequence of actions (changing the variables "engine" and "position"). The result is shown in **Fig 4**. In a distributed system, messages among some nodes are necessary to scan a truck section and to make reservations. We have a good chance that both asking processes will receive the state "free," because none of the processes has finished its reservation and both processes are trying to make reservations for the same resources. This conflict can be resolved if status sensing and reservation making are both performed in an indivisible (atomic) action.

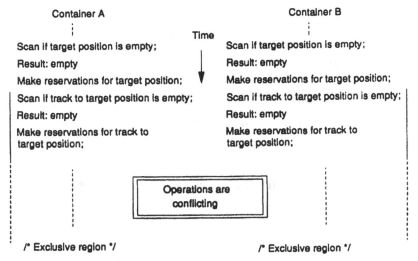

Figure 4: Conflict in the case of a partial breakdown

These conflicts arise if resources inside or outside the computer must be used exclusively by a process during work on it. Such a situation may be unavoidable for physical reasons or because the status of the resource may be changed during further usage. Furthermore, it may be necessary to reset the resource to an earlier status if concurrent processes cannot deliver a required partial result. An example is to put back a workpiece that cannot be tooled further.

Operation in the Case of a Partial Breakdown

We have characterized real-time data processing by the need for timely results. Response times are certainly missed in the case of a system breakdown. If a node fails we are happy to be using a distributed system and therefore to be able to transfer essential tasks to some other nodes. Continued operation is possible, eventually with reduced functionality (graceful degradation), but guaranteed remaining functionality solutions for the following problems are required:

- Since stored data are not available after the breakdown of a node, essential data, necessary for continuing operation e.g., for executing tasks transferred from the failing node must be stored several times on different nodes (replicated data). For write-operations on these data, all copies must be updated.
- Aborted processes on the failing node must not block resources on other nodes.
- For transferring tasks that communicate with the manufacturing devices, the corresponding variables must be available for the new host. This is possible if we have redundant sensors and actuators, or if we have multiple access to the original ones, e.g., through a field-bus.
- The relocation of application entities requires a remapping of addresses to get the new communication partner (e.g., with the PORT concept)
- A recovery plan is necessary to reset the entire network to a new consistent status after a partial breakdown. Resetting to an earlier status (backward recovery), as widely used in database management systems, is not always possible in real-time systems because

time is not reversible. Instead, one can try to define a new consistent status as an extrapolation of an earlier known status, taking into account the amount of the partial breakdown (forward recovery).

- Generally, status variables must be saved for recovery. However, in manufacturing automation, some of these variables represent the status of external devices. In this case, storage is not necessary because these variables can be replaced by status sensing.

STANDARDIZATION IN DATA COMMUNICATION

The ISO Reference Model

The ISO standards for Open Systems Interconnection (ISO-OSI) use a reference model that is presented in **Fig. 5** in the common form for local area networks.

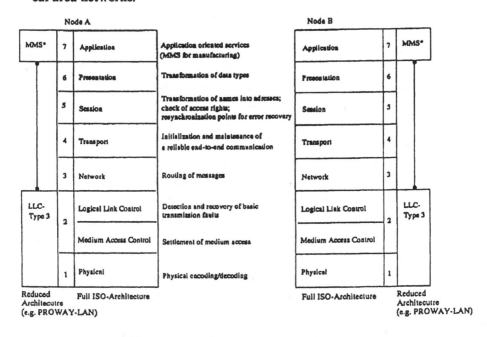

Figure 5: The ISO reference model

A request for communication arising in an application entity (on top of layer 7) runs through the layers from the top toward the bottom, where layer (i-1) provides some services for layer i (protocol stack). A service request must satisfy certain interface conventions. At the target node, the message runs through all the layers from the bottom to the top to the target application entity. For the tasks at each layer to be performed, agreements between the corresponding layers have to be settled. We call these agreements a *protocol*.

We have to distinguish between the transmitter and the receiver of a message. At the transmitter a message runs from the top to the bottom of the protocol stack and at the receiver from the bottom to the top. A certain layer gives an order to the adjacent layer by calling a service primitive. With this order, data of a well-defined structure called a *protocol data unit* (pdu) are delivered. This forms the *sevice data unit* (sdu) of the lower layer. An order given from the top to the bottom (transmitter) is called a *request*, and an order from the bottom to the top is called an *indication*. For some kinds of requests, a confirmation is sent back to the requesting layer. Some services need an answer from the receiving node (response); in this case, the roles of the receiving and transmitting node are inverted. For protocol handling, some control information is necessary. This information is called the *protocol header* and is placed in front of the pdu. The rest of the pdu normally contains user data supported by the calling layer. The corresponding layer of the receiver evaluates the header and discards it. The rest, if any, will be transferred as an indication to the next upper layer. As a result, we have the characteristic structure of pdu's shown in Fig. 6. One can see that the information from the application entity may be increased substantially by the protocol headers of all of the layers.

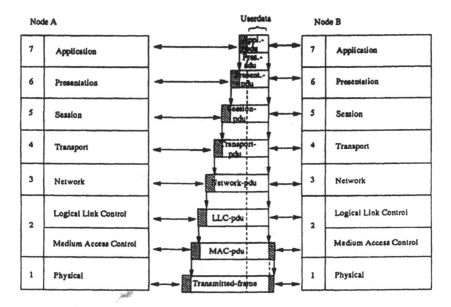

Figure 6: Hierarchy of the protocol data units

Several application entities of a node can independently maintain associations for communication. For example, for this reason the transport layer can maintain more than one transport connection. To identify an association, we need the number of the corresponding *service access point* (sap) in addition to the node address.

Standards for open systems interconnection are subdivided into two kinds of documents: One kind describes the services offered by a layer, of the communicating nodes, and the other kind defines the protocol of the same layer. For some layers, there are more than one version of protocols, partly due to different boundary conditions and partly due to historical reasons. For communicating with nodes that use different protocols, we need protocol conversion (e.g., gateways and bridges).

Necessary Transformation Processes

The communication system should basically support a reliable transport service based on standards. Therefore, we have to consider that

- several application entities of a node can independently maintain associations for communication;

- the sequence of messages is essential and therefore must not be changed;
- communication lines are never free of interferences; transient interferences should be recoverable by the transport system;
- the user should receive an error message in standardized form, if a requested service cannot be performed successfully;
- messages should be put into the transport system only as quickly as it can treat them in order to avoid traffic jams.

The lower 4 layers of the ISO reference model form the transport system. Within a network, it is supposed to use identical protocols. A connection between two networks using the same ISO protocol on layer 4 is easy to establish, in spite of the use of different protocols on the lower layers (layers 1 to 3).

Layer 5 is responsible for establishing and maintaining communication attributes, e.g., duplex or semiduplex telecommunication or resynchronization points for error recovery. These attributes are negotiated at association establishment.

The task of layer 6 is to represent and transform data structures. In simple cases, this is necessary to convert internal representations of basic variables, e.g., floating-point formats. Generally, information on the data structure has to be transmitted, together with the data for valid interpretation by the receiving application entity. Usually this structure is defined by type declarations in the application program, which are evaluated by the compiler and then made available at program run-time. Since several application entities may use different programming languages and since no standards exist for internal representation of data, the Presentation Layer performs the necessary transformation using a standardized description of all used data types (Abstract Syntax Notation 1: ASN.1) prepared by the application entity.

Finally, layer 7 provides special services for the application entities. This layer is composed of an application-independent kernel and special services for an application class. The application programmer looks at the network as the amount of services of the application layer. For manufacturing automation, these services are defined by the Manufacturing Message Specification (MMS).

Time Spent on Communication

To describing real-time behavior we use the client-server model shown in Fig. 7. We are interested in the message delay T_D representing the time until the receiver gets the message and the response time T_A representing the time until the client gets an answer. The service time T_S is application dependent. For a communication system comparison, we assume T_S to be very small.

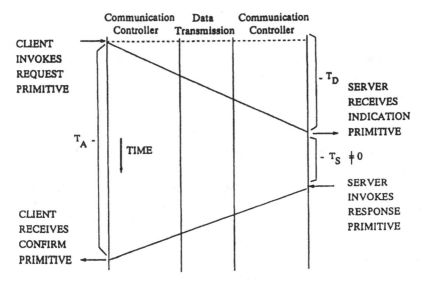

Figure 7: Server-client relation of two application entities

To show the order of magnitude, we present some results of measurements on a MAP network (Carrier Band Version with 5 MHz transmission speed). The following table give results for layer-to-layer communication in a network without further load.

Layer	T_D, Layer	T_A, Layer for $T_s = 0$
Medium Access (MAC)	3 ms	6 ms
Logical Link Control (LLC)	4.5 ms	9 ms
Network (NET)	6 ms	12 ms
Transport (TP)	9 ms	22 ms
Application Entity	20 ms	48 ms
(estimated; e.g. *read simple variable*)		

With load on the network, the measured time increases drastically. The results for the response time are shown in **Fig. 8.** Certainly, they depend on the communication controller. We use a state-of-the-art controller (commonly used VLSI-Chips for the medium access protocol; MC68020 with 12,5MHz for protocol handling of the other layers). However, it is not possible to speed up the network by some degrees of magnitude using today's technology. To reduce the time spent on communication for real-time applications, we propose a reduced architecture hooking the application layer on top of the LLC layer (cf. **Fig. 5**). This is in accordance with the PROWAY-LAN of the Instrument Society of America and the Mini-MAP concept. For reliable communication, we need acknowledged LLC services (LLC-Type 3) instead of the not acknowledged connection-less services (LLC-Type 1), and the missing network layer restricts communication to a single segment, because routing is not possible and the receiving node must be addressed by its LLC service access point. The transport protocol of layer four is dispensable because with the LLC-Type 3 services we have a reliable communication. With respect to the missing layers five and six, a predefined context (e.g. transfer syntax) has to be used. This situation

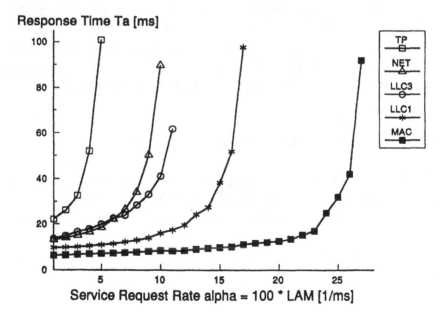

Figure 8: Response time for increasing load

may be simplified in the future if we have standards for common data types in all programming languages.

For predictable response times, we need a deterministic medium-access protocol. The frequently used CSMA/CD protocol (ISO 8802.3), for example, is not deterministic because it does not quarantee an upper bound of the waiting time and should not be used for most real-time applications. However, for predictable response times, deterministic protocols like the Token-Bus (ISO 8802.4) or the Token-Ring (ISO 8802.5) need the observance of some conditions and carefully selected protocol parameters. We especially need message priorities to separate time-critical and non-time-critical messages. For more details, the reader is referred to Chapter 3.5 ("Bus systems in real-time environments") and [12].

Communication Standards for Manufacturing

Computer Integrated Manufacturing (CIM) requires communication between all parts involved in manufacturing. A local area network with standard interfaces for all devices seems to be the best solution. This was the goal of a strategic initiative in 1980 by General Motors defining the *Manufacturing Automation Protocol* (MAP). The current version MAP 3.0 has been valid since 1988. It is based on ISO standards as far as possible. For interoperability, agreements on selecting protocol modes and options are necessary. The interface to the application called *Manufacturing Message Specification* (MMS) will be described in the next section.

THE APPLICATION-LAYER SERVICES

In this section, we give an overview of MMS describing basic principles and kinds of services. As an example, we describe the means for concurrency control offered by MMS to solve problems such as those shown in Fig. 4.

The Virtual Manufacturing Device

MMS is based on a *virtual manufacturing device* (VMD), which is thought to be some kind of abstraction of a real manufacturing device. A VMD may represent a manufacturing cell or a production line. It is part of the application, as well as part of the application layer. The operations on elements of the VMD have to be performed by operations on the real hardware of the target node (executive function). MMS can distinguish between several application entities, each covering a part of the VMD.

A VMD consists of the following elements:

- an executive function mapping the services of the VMD onto the local resources;
- at least one domain, which is a logically consistent part of the functionality of the VMD; e.g., a VMD "manufacturing_cell" may have a domain "robot" and a domain "milling_machine";
- if necessary, a virtual file store (using FTAM) for local files of the VMD;
- within the domains, several objects like programs, variables, events, semaphores, and journals accessible by MMS services.

Objects like variables and events may be defined globally for the VMD or locally for a domain.

MMS Services

MMS-services are composed of the following groups:

1. Context Management - *Initiate, Conclude, Abort, Cancel, Reject*
2. VMD Support - *Status, GetNameList, Identify, Rename, Unsolicited Status*
3. Program Invocation Management - *Create/Delete-Program Invocation, Start/Stop/Resume/Reset/Kill GetProgramInvocation Attribute*
4. Variable Access - *Read, Write, ScatteredAccess*
5. Semaphore Management - *Define/Delete-Semaphore,*

> *TakeControl, RelinquishControl, ReportSemaphoreStatus, Attach-
> ToSemaphore*
>
> 6. Operator Communication - *Input, Output*
> 7. Event Management - *EventCondition, EventAction, EventEnroll-
>   ment, EventNotification, ...*
> 8. Journal Management - *InitializeJournal, WriteJournal, ReadJournal,
>   ReportJournalStatus*
> 9. File Management - *Obtain File, File Get*
>   Using FTAM: - *File Open/Read/Close   File Rename, File Delete,
>   File Directory*

Nearly all MMS services are confirmed services, i.e., an acknowledgement is returned to the invoking application entity. Thus a sequentialization of events in different nodes is possible. Methods for resolving conflicts are described in the following section.

Communicating Distributed Applications

Some well known basic concepts exist for resolving conflicts like
those in the examples in the section above ("Conflicts and Concurrency
Control"). We consider these concepts to be logically equivalent because
one can simulate each of them by another. MMS provides the semaphore
concept, and therefore we will show how it works, recognizing that some
problems remain with the use of semaphores. The semaphore concept basically solves the problem of mutual exclusion. In the most simple form,
a (binary) semaphore is a synchronizing variable with two states: free and
blocked. It is accessible by two procedures:

- With P(sema) or REQUEST(sema), a process enters the critical
  region, and
- with V(sema) or RELEASE(sema), it leaves the critical region.

The structure of a program looks like this:

.
.

P(sema)

.

.(statements of the critical region)

.

V(sema)

.
.

A "critical region" of a process is the dynamic section of an application process during which it uses a resource (protected by the sema) exclu sively. All processes using the same resource must use the same sema-variable for entering (or leaving) their critical region.

The access procedures P(sema) and V(sema) are defined as follows:

```
procedure P(sema);
begin
      If sema = free                  atomic action
      then sema := blocked;
      else /* put calling process on a waiting-queue */;
end; /* procedure P */
```

It is important that checking the condition (if sema = free) and changing the value from *free* to *blocked* must be done in an indiv visible atomic action in order to prevent checking by more than one process prior to changing the value to *blocked*.

```
procedure V(sema);
begin
      if      /* waiting-queue is empty */
      then    sema := free;
      else /* continue a waiting process and discard it from
              the waiting-queue */;
end; /* procedure V */
```

Usually the implementation of a semaphore uses a special instruction (the *test-and-set* instruction) for testing the old value and setting the blocked status. This instruction is not interruptable and therefore performs the necessary atomic action. It works only if we have a common memory for all processes. Solutions by algorithms exist for implementing semaphores in a real distributed system, but these are time consuming and cause considerable communication overhead. In MMS the semaphore management is based on a semaphore manager established for each semaphore on a single node. Processes trying to use a semaphore have to direct their requests to the corresponding semaphore manager. If it is free, the requesting process gets back an *is_free* message as a special pattern (token). To release, the semaphore must be returned to the semaphore manager at the end of the critical region.

MMS provides semaphores in the more general form of counting semaphores. The semaphore manager can issue at most k tokens having in mind that the corresponding resource may have at most k users simultaneously. To granting further requests, tokens have to be returned by some using processes. If k=1, we have the above-mentioned binary semaphore. For convenience, named resources may be attached to a semaphore (pool semaphore) with automatic requesting and releasing if the resource is used.

The semaphore manager normally belongs to the application layer; however, a delegation to an application entity is possible. In the first case, the semaphore manager can answer requests without user interaction, but an application entity may be invoked for performing some supplying tasks. In our example for concurrency control (see section "Conflicts and Concurrency Control"), the belonging semaphore manager can give the privilege of handling a switch to that application entity to which the token has been sent.

Now we can give a sketch for the activity of moving the engine in our example (see **Fig. 1**):

Engine A:

.

.

/* Make reservations for track section;

        S_track_1, S_track_2; semaphore, nonautomatic */

TakeControl (S_track_1);

        /* Make reservations for track in control area 1;

           calling process gets handling privilege

           for switch W1.1 */

TakeControl (S_track_2);

        /* Make reservations for track in control area 2;

           calling process gets handling privilege

           for switch W2.1 */

Set switches;

Start drive;

Control velocity;

If engine has left control area 1:

        RelinquishControl (S_track_1);

        /* Handling privilege for switch W1.1 released */

If engine is in final position:

        Stop drive;

        RelinquishControl (S_track_2);

        /* Handling privilege for switch W2.1 released */

.

.

.

The reservations take place similarly for moving engine B. To avoid deadlocks, the requests for semaphores must be made in the same sequence, that is S_track_1 prior to S_track_2. The example should show how the means for the concurrency control provided by MMS work. For more details on concurrency control, we refer to [1] and [3].

Directories

Application entities in open systems need means to provide information on services offered by other application entities in the network. In this way, new applications can be added using the already available services. Additionally, an application can ask for other nodes offering the desired services in the case of a reconfiguration. For this purpose, directory services are provided by the ISO standards.; they are not part of MMS, but are mandatory for MAP networks. A directory is a list of known objects and information on how to access these (basically addresses). As shown in **Fig. 9**, central or decentral directories are possible. But there are no means defined in the standard to maintain consistency for a decentralized directory. The user must decide how to cope with this.

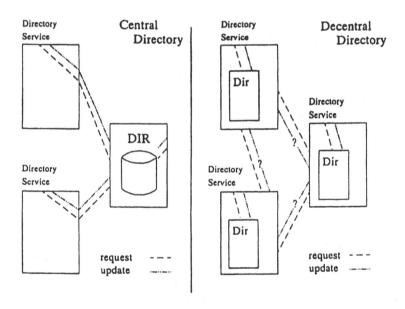

Figure 9: Directory structures

PREREQUISITES TO RECONFIGURATION

As pointed out in the section "Operation in the Case of a Partial Breakdown", a common motivation for using a distributed system is some kind of fault tolerance. A faulty node should not necessarily result in a total breakdown. In this section we discuss the tools and services required for reconfiguration. Basically, these are

- recognizing the applicable faulty status and determining how to recover (recovery plan);
- addressing the reconfigured application entities;
- management of replicated data.

The MAP specification (and MMS as a part of MAP) is not sufficient to cover the various needs. Some problems are left to the application, and some concepts do not seem to be the best-fitting solutions of the problems. The next sections give an overview.

Replicated Data

In addition to ensuring access to data locked in a faulty node, replicated data can increase performance, because the access time to local copies is shorter. However, to perform write operations, all the copies have to be changed identically. The problem has two major aspects:

- If changes cannot be performed on a node - probably only a tempo- rary problem - then copies should not remain inconsistent forever.

- Generally it is not possible to update all the copies exactly at the same time. The time during which reading of different copies may have different results must be minimized.

The state of the art is the use of a *two-phase commit protocol*. In a first phase a coordinator - normally associated with the application entity performing an update request - sends an update request to all local agents holding a copy. Each local agent creates a new version of the copy, where the *old* version still remains valid. Then it sends a ready-to-commit message back to the coordinator. If the coordinator receives a ready-to-commit message from all the agents it starts phase two, sending a commit message to the agents declaring the *new* version as valid. If the coordinator does not receive all the ready-to-commit-messages during a time-out condition, it sends an abort message and the update request fails. This protocol is not satisfying with respect to real-time applications because it may be meaningful to successfully committing an update request with an outstanding ready-to-commit message. This reduces the number of valid copies by one but may be necessary to meet real-time requirements.. MAP offers no services for management of replicated data. Therefore, the user must implement a two-phase commit protocol with the basic services offered.

Addressing Communicating Entities

In the case of reconfiguration, addresses of communicating entities and their objects change. Therefore, one cannot use direct addresses. We can look for a solution using unique global names and can dynamically redefine the corresponding addresses. Now we are faced with the problem that objects may exist twice (or more), but only one item is the active one. Conflicts in naming must be resolved in such a situation.

Another possible solution is the PORT concept. A PORT is an object maintained by a network operating system; it has a unique name together with some attributes. An application entity communicates with a PORT subsidiary for the target application entity. It has no knowledge of what is behind the PORT. An association is made with the target application entity by connecting PORTs with compatible attributes, which is a service of the network operating system. For reconfiguration, the connections of PORTs must be changeable dynamically.

For MAP networks, we can follow only the first concept. Using directory services, we can redefine addresses of named objects. Immediately this concerns only the directory entries. In the case of reconfiguration, an application entity must look for valid addresses of the objects to be accessed. Furthermore, decentralized directories may have invalid entries, which leads to a search through the directories.

Initiating Reconfiguration

Monitoring an unwanted status, we start a reconfiguration phase. We can provide a central monitor or decentralized monitoring. In the latter case, we have to commit a reconfiguration case taken for granted, because different nodes may have different information on the global status. MAP networks provide event-management services that are well suited and sufficient for monitoring.

For reconfiguration, all resources blocked by aborted processes have to be released. If necessary, they have to be requested again by the substitute entity. The semaphore concept as described above raises some problems. To abort a process during its critical region leaves the corresponding semaphore locked. On the other hand, the usage is anonymous, and one cannot check which process has blocked it. This situation is not resolvable without storing more information within a semaphore.

As shown above, we need a semaphore manager to implementing semaphores in distributed systems like MAP. This semaphore manager can maintain additional information on the entities using the semaphore. We can get information on the status and the calling entities by the MMS services *report semaphore status* and *report semaphore entry*. A *free-token* can be returned subsidiary to reconfiguration. Hence, failures of processes using semaphores are tolerable.

Not tolerable is a breakdown of the node containing the semaphore manager, because we cannot implement a subsidiary manager without accessing data linked to the semaphore. A copy of these data in another node may have become obsolete. If we suppose that the access for a resource and the corresponding semaphore manager are assigned to the same node, there is no difference if a resource is not accessible or if it is blocked by the semaphore. In either case, we need a new resource.

CONCLUSIONS

We have discussed the problems of distributed systems for real-time applications and have used the Manufacturing Automation Protocol to show how to cope with these. Generally, the lessons learned by studying MAP are

- The delay time caused by the communication system is far from negligible. Conditions for an upper bound of delay times are theoretically well understood, but the missing support of message priorities by the ISO protocols is a weakness undermining these conditions.
- The offered functionality is sufficient for most applications. Programs using semaphores tend to be error prone because the *Relinquish Control* operation may not be performed in unusual situations. Therefore, using semaphores for concurrency control may not be the best choice.
- The user can establish some kind of fault tolerance. Generally, only poor means are provided for this purpose.

REFERENCES

[1] Bic, L.; A.C. Shaw: "The Logic Design of Operating Systems"; 2nd Edition, Prentice Hall 1988, German Edition: Betriebssysteme, Hanser Verlag 1990

[2] Ciminiera, L.; A. Valenzano: "Acknowledgement and Priority Mechanisms in the 802.4 Token Bus"; *IEEE Trans. on Industrial Electronics,* Vol. 35 (1988), No.2

[3] Elmagarmid, A.; Y. Leu: "An optimistic concurrency control algorithm for heterogeneous distributed database systems"; *Proceedings of the 6th International Conference on Data Engineering,* Feb. 1990

[4] Gorur, R.M.; A.C. Weaver: "Setting Target Rotation Times in an IEEE Token Bus Network"; *IEEE Trans. on Industrial Electronics,* Vol. 35, No. 3, August 1988

[5] Hartlmüller, P.: "Wahrung der Konsistenz wesentlicher system-interner Daten in fehlertoleranten Realzeitsystemem"; *Dissertation,* Universität der Bundeswehr München, 1988

[6] Kaminski, M: "Protocols for Communication in the Factory"; *IEEE-Spectrum,* April 1986

[7] Lamport, L.: "The Mutual Problem: Part I - A Theory of Interprocess Communication;Part II - Statements and Solutions"; *Journal of the ACM,* Vol. 33 (1986), S. 313-348

[8] Levy, E.; A. Silberschatz: "Distributed File Systems: Concepts and Examples"; *ACM Computing Surveys,* Vol. 22, No. 4, Dec. 1990, pp. 321-374

[9] Manufacturing Automation Protocol Specification, Version 3.0;European MAP Users Group, 1989

[10] Millner, R.: "Communication and Concurrency";Prentice Hall, London 1989

[11] Process Communications Architecture; Draft Standard ISA-DS 72.03 - 1988; Instrument Society of America

[12] Rzehak, H.; A.E. Elnakhal; R. Jaeger: "Analysis of Real-Time Properties and Rules for Setting Protocol Parameters of MAP Networks"; *The Journal of Real-Time Systems*, 1 (1989), pp. 219-239

[13] Rzehak. H.: "A Three-Layer Communication Architecture for Real Time Applications"; *Proceedings of the EUROMICRO'90 Workshop on Real Time*, June 6-8, 1990; IEEE Computer Society Press, pp. 224-228

[14] Stankovic, J.A.: "The Spring Architecture"; *Proceedings of the EUROMICRO'90 Workshop on Real Time*, June 6-8, 1990; IEEE Computer Society Press, pp. 104-113

[15] Stoll, J.: "Anwendungsorientierte Techniken zur Fehlertoleranz in verteilten Realzeitsystemen"; *Informatik-Fachberichte*, Bd. 236, Springer-Verlag, 1990

12

ROBOT PROGRAMMING

K. Fischer, B. Glavina, E. Hagg, G. Schrott, J. Schweiger, H.-J. Siegert
Technische Universität München Institut für Informatik
D-8000 München

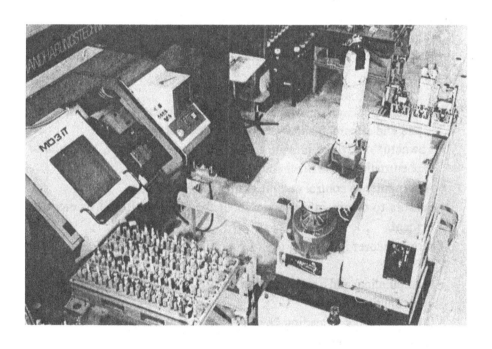

Autonomous mobile robot system in a flexible manufacturing environment (Source: iwb)

INTRODUCTION

The term robot first came into use about 1923, following the appearance of a play by the Czech author Karel Capek, called Rossum's Universal Robots. In this play artificial creatures strictly obeyed their master's order. These creatures were called 'robots', a word derived from the Czech robota, meaning 'to work'. In common usage a robot still means an apparently human automaton, intelligent and obedient but impersonal. A scientific definition is, for example, given by the Robot Institute of America (RIA):

A robot is a reprogrammable and multifunctional manipulator, devised for the transport of materials, parts, tools or specialized systems, with varied and programmed movements, with the aim of carrying out varied tasks.

To perform these tasks an operational robot consists of four interacting elements:

- A mechanical system which comprises the actual limbs (arms, joints and end effectors) and eventually devices for locomotion.
- The actuators which provide controlled power (electrical, hydraulic or pneumatic) to the mechanical system.
- A computer to control and program the robot.
- Sensors to obtain information about the robot itself and its environment.

There is an ever increasing use of robots in various fields:

- Manipulation and transportation in factory automation, e.g., in car production lines
- (Mobile) robots as loading devices for machine tools in flexible manufacturing systems
- Exploration of inaccessible areas, e.g. submersible robots, walking robots
- Special applications, like sheep-shearing robots

Robotics not only comprises the further improvement of robots but also the development of all domains associated with the use of robots [37]. The main interests of research concentrate on autonomous, intelligent and learning robots. The following sections give an overview over intelligent robot programming and the results on task-oriented robot programming achieved by us in the joint research project SFB331[1].

PROGRAMMING OF ROBOTS

Robots may be programmed by the user on several abstraction levels which differ significantly in their methods and expense.

Teach-in

The user manually drives the robot to a required position and then presses a button which causes all the joint positions or tool coordinates to be stored. The robot can afterwards repeat the motions taught, moving to the stored positions one by one. Driving the robot manually is either done by a hand-held teach-box with buttons or joysticks to change the position and orientation of the effector or by guiding the effector directly by hand; in this case the gravity of the robot arm is compensated by special control loops for the actuators. Teaching can be done without deep knowledge of the robot, and complex trajectories are possible without mathematical specifications. The disadvantage of teaching is the lacking flexibility of points taught when the situation changes. Moreover, the *move* instructions taught are only one part of a robot program in a manufacturing environment; synchronisation with other robots or machines and on-line reaction to sensor data has to be programmed separately.

---

[1] *This work was partially supported by the Sonderforschungsbereich 331 (Informationsverarbeitung in autonomen, mobilen Handhabungssystemen)*

Teleoperation

Teleoperation means that robot actions are remotely controlled by a human operator who executes manually the movements the robot effector should execute. For example, to control a multifingered gripper, the human operator can use a kind of glove which senses the positions of all fingers and transmits them to the gripper. In order to control the robot, the human operator in his remote environment must observe the scene and feel the reaction (forces, torques, etc.) of the effector. It is therefore important to transmit an image of the scene and a faithful simulation of the remote reactions to the human operator. Teleoperation is mostly used in dangerous environments (mining at large, deep sea mining, handling of radioactive parts of reactors).

Robot-level Programming

The alternative to teaching a robot by driving it through its cycle of operation is to type a program, which in the simplest case consists of a series of commands of the form 'move straight to point G', where G is given by its six-dimensional coordinates. These commands are expressed in some language designed for robot programming. Such robot programming languages are derived from a high-level language using its data structures, control structures, arithmetic and logic operations and subroutines or module concepts. Since robot control is a typical case of real-time control, the language and the operating system must be suitable for real-time purposes; this usually implies the ability to handle interrupts, multiple processes and the communication between processes. In addition to these general requirements, a robot programming language needs the following features:

- Geometric and kinematic calculation: Functions and data types to allow an efficient expression of coordinate systems and their transformation.
- World modelling: The ability to define objects and their relations (e.g., one object is attached to another).
- Motion specification: Different kinds of motion between points including interpolation and definition of speed.

- Effector control: Interfaces to control a variety of effectors.
- Communication: Input/Output signals to interact with the environment (e.g. switches, machine tools, sensors and computers).

Robot manufacturers often provide a language to go with their products; well-known examples are Unimation's VAL for the PUMA robot and Siemens' SRCL for Manutek and KUKA robots. Programming a robot in a robot programming language instead of teaching has many advantages. The robot proper is no longer needed for programming and if a robot simulation system is available, is not even needed for initial testing. Only for final tests the robot is used to verify that the simulated behavior and the real one coincide. Robot programs are more flexible and may be parameterized for changing applications or sensor-feedback.

Task-level Programming

Task-level programming is situated on a level higher than programming in a robot programming language and is usually synonymous to intelligent robot system (Case studies are given in [7,19,20,26,31,36]). At this level, how a task is solved is not programmed by a sequence of robot *move* instructions but by the specification of the task itself, e.g.,*'get tool T and put it into machine M'*, both *T* and *M* being symbolic names. This specification is analyzed and transformed. One or more robot-oriented programs and/or programs for machine tools such as Programmable Controlers (P/C) or Computerized Numerical Control (CNC) are created including the necessary synchronization between them. To solve this problem, the following complex modules are needed:

- A knowledge base to store the world model of the factory and all its parts.
- A knowledge base with rules defining how to decompose a task in subtasks and how to synchronize these tasks (task planning)
- Algorithms for assembly planning.
- Algorithms for motion planning, i.e. computing collision free moves of the robot.
- Algorithms for grip planning
- Integration of multiple sensor systems

A typical transformation of the task-level to the robot-level in a flexible manufacturing system is done at various abstraction levels (Fig.1). At the level of production planning and control system (PPC), manufacturing orders for products are planned according to their deadlines and to the actual load on all machines. The output of the PPC is sent to the task-oriented programming level via the active real-time knowledge base (cf. the following section). An active knowledge base not only stores data but also serves as the communication medium for all processes which are engaged in performing the transformation and are usually located on different computers in the local area network of the manufacturing system. The production planning system is a global layer which distributes single tasks among the various autonomous units (machine tools, (mobile) robots, etc.); all other layers are present for each autonomous unit. The local production control layer optimizes the incoming tasks relative to the current tasks. Both layers divide the original task into a set of subtasks of the kind 'get and put' for robots and 'load, mill and store' for machine tools. The synchronization of the tasks running on autonomous units is done in the task coordination layer. The task planning and execution layer finally generates a robot program in a robot-level programming language. An important part of this layer is the motion planning for robot movements; the geometric model of the environment which is stored in the knowledge base is used to plan collision-free motions for the robots. The robot-level program can be uploaded to the robot control system and executed. The lowest level of the system controls the joint actuators. The real-time constraints of this layer are about one millisecond, whereas the response times on task-level programming level are in the range of seconds and on PPC level in minutes. The program may also be tested using a simulation system for the robots in the factory environment. An interactive graphical robot programming workstation for task-level programming was developed in our group. In order to improve the off-line task-level programming as described above and to achieve on-line reactions to changing environments, to sensor information, and to errors, it is necessary to introduce a feedback between the task execution layer and the robots, machines, and the sensor systems (see Fig. 1). The task execution layer sends single commands directly to the robot control system and waits for their correct execution. Error replies and the changing world model in the knowledge base are used each time when the next action is planned by the task-oriented transformation.

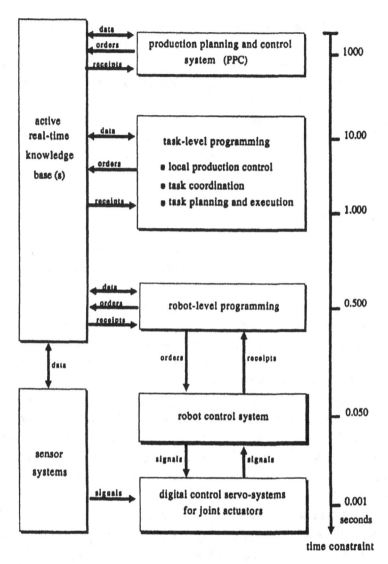

Figure 1: Levels of Abstraction in Task-level Programming

In the following sections, concepts, solutions, and implementations for task-level programming are explained in more detail.

ACTIVE REAL-TIME KNOWLEDGE BASE

To integrate all the single systems for running a factory autonomously, i.e., robot systems, planning systems, sensor systems, etc., a common medium is required which stores and handles their common information. As shown in [12] and Chapter 3.3, classical database systems are not sufficient for this purpose. Therefore, one research area in the joint research project SFB 331 is the development of a distributed, active real-time knowledge base which can be this medium. The basic concepts of the knowledge base currently in use are introduced in the following section.

Modelling Concepts

The modelling concepts of the knowledge base as shown in **Fig.2** depend on an object-oriented approach. Therefore, the basic structure in the knowledge base is the object. The objects can be divided into prototype objects called *classes* and individual objects called *instances*. Classes and instances describe properties of real-world entities. In **Fig.2** the classes *autonomous_sys, locomotion_sys, manipulation_sys* and *mobile_robots* are shown. The class *mobile_robots* has one instance named *mb1*. Classes and instances can be compared with records in classical programming languages like Pascal. A class is equivalent to the declaration of the record. An instance of a class correlates with the corresponding set of values for the items of the record of the class. The items are synonymous with the so-called *attributes* of a knowledge base class. An attribute stores information about one property of a real-world entity. Attributes are classified into *class, instance* and *administration attributes*.

Class attributes describe properties whose values are equal in all instances of the class. In **Fig.2**, the class *locomotion_sys* has one class attribute named *service_range*. Instance attributes express properties which are different in every instance of the class. So in **Fig.2**, the instance attributes *manip_loc* has in the instance *mb1* the value *approach*. Administration attributes contain information about which instances are an element of the class and about the inheritance of attributes (in **Fig.2** *instances, superclasses, subclasses*). A class inherits all the attributes of its superclasses, which means that the attributes defined in the superclasses are added to the one defined in the considered class. In the same way, the subclasses of a class inherit all attributes of that class.

The special case in which a class has only one superclass is called *single inheritance*, otherwise it is called *multiple inheritance*. If single inheritance is given, superclasses are generalizations, whereas subclasses are specializations of a considered class. An example for inheritance is shown in Fig. 2. The class *mobile-robots* inherits all attributes from the classes *autonomous-sys*, *locomotion-sys* and *manipulation-sys* (multiple inheritance). The class *locomotion-sys* is a specialization of the class *autonomous-sys* (single inheritance). For further explanation of classes, instances, attributes, inheritance see [3,5].

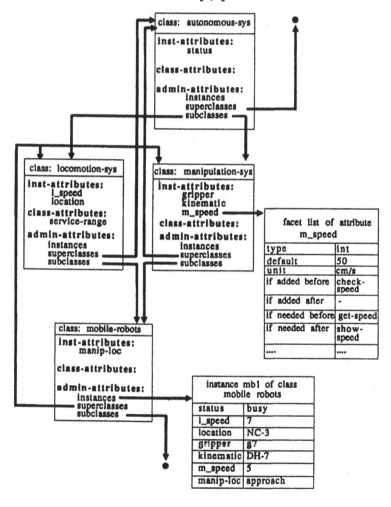

Figure 2: Modelling Concepts of the Knowledge Base

Furthermore, each attribute is described by a set of facets. Some facets are standard for all attributes and contain information like the type or the default value of an attribute. The user can define additional facets with arbitrary meaning. In **Fig. 2** the attribute *m_speed* of the class *manipulation_sys* has the facets *type, default, unit* and so on. Some special facets include the identifier of an attached procedure, called *demon*, which is executed if the attribute is used in a certain way. For instance, if the attribute is modified, the demon of the facet *if_added_before* is executed before the attribute value is modified. In the same way there are facets for *if_added_after, if_needed_before* and *if_needed_after* demons. Demons are written in classical programming languages by the user and allow the storage of procedural knowledge in the knowledge base. In **Fig. 2** the attribute *m_speed* of the class *manipulation_system* has the if_needed_before demon *heck_speed*. For a further explanation of demons see [4,5,34,35].

Demons are also the basis for the active component of the knowledge base [4]. The active component automatically distributes changing, specially marked data to interested application processes. This means that an interested application process needs no polling to get the new data immediately; a message with changed data is sent to it. This active component provides a comfortable, uniform communication facility for the connected application processes.

Interfaces and Implementation

The objects of the knowledge base can be handled by a set of services. To offer multiuser access, the knowledge base consists of a kernel process and a remote service interface linked to every application process. If a service of the remote service interface is called in an application procedure, the request is sent as a message over a communication interface to the knowledge base kernel process where it is executed. The result of the request is sent back to the corresponding application process.

The objects of the knowledge base are stored in a relational database system. As the mapping between the objects and the database tuples is fixed and the functionality of a query language for commercial applications cannot be used, the knowledge base kernel interfaces with the tuple layer of the database system. Therefore, the knowledge base kernel maps every knowledge base request into a sequence of tuple layer operations of the underlying relational database system. The mapping is explained in detail in [33].

The knowledge base kernel uses a well-defined tuple layer interface, so that it is possible to use the same kernel with different database systems. The actual implementation of the knowledge base kernel can be configured with the commercial database system TRANSBASE [46] and a main memory resident database system developed in the SFB 331 [41]. In contrast to the main memory resident database system which keeps the tuple as a AVL-tree resident in the working memory, the TRANS-BASE database system permanently stores every tuple in a B'-tree on the disk.

The knowledge base is implemented on a VAX-Station 3100 under VMS as well as on a DEC-Station 3100 under Ultrix. Depending on the underlying database system the knowledge bases differ in capacity and access time. The TRANSBASE knowledge base is able to manage huge amounts of data but with long access times. The main memory resident knowledge base provides short access times, however the amount of data is restricted by the available main memory. The approximate access times of the different knowledge base versions are shown in the following table. The times are measured on DEC-Station 3100 under Ultrix.

service	disk-based KB	main memory resident KB
read-service	150 ms	5 ms
write-service	500 ms	5 ms

On the basis of the remote service interface several tools have been realized to support the knowledge base administrator. A graphic and a VT100 shell allow calling the services of the knowledge base interactively. With these shells the knowledge base administrator can easily design the object structure of the knowledge base. With a flexible monitor the knowledge base administrator can watch relevant changes in the knowledge base. A browser is available which can display structures and contents of the knowledge base.

Applications and Future Work

Our knowledge base is already integrated into some applications in the SFB 331. In one application, the knowledge base is used to store and handle the topologically structured geometrical data for the path-planning system of an autonomous transport system [30]. In a second application, the environment of a small factory is modeled in the knowledge base to enable a comfortable access for a shop floor control system. Thereby, the transmission of tasks and acknowledgements between the shop floor control system and the machines executing the tasks is managed by the knowledge base [28]. A third application is described in the following section.

To get a knowledge base which provides a large data capacity as well as fast data access, future research is concentrated on the connection of the disk-based and the main memory resident knowledge base versions to a distributed knowledge base. The distributed knowledge base consists of several local knowledge bases which are extended by a mechanism for distributing data between them. Since the applications do not require permanent consistency between different local knowledge bases, the goal mentioned can be achieved. For example, a main memory based local knowledge base is used for fast data access and a disk based global knowledge base for permanent data storage. The global knowledge base is updated only when significant changes have occurred in the local knowledge base.

Since the applications of the knowledge base manufacturing environments are subject to real-time constraints, the integration of time constrained-relations into the knowledge base is necessary. For instance, the data of a sensor system refer to an instant in time which corresponds to a time-related status of the real world. Applications using the data of several sensor systems should not be allowed to mix data of different real world status. A first approach to integrate time-constrained relations is found in [47]. There the instances of the classes are marked with a time stamp which shows the time range of validity of the instance. Taking the time stamp into account when executing knowledge base services, a mixing of old and new data is avoided. In future work, this concept has to be integrated into the distributed knowledge base.

KNOWLEDGE-BASED TASK PLANNING

The principal item of task oriented robot programming is the task planner which has to transform higher level tasks to a set of primitive actions. To get a powerful and flexible interface, it uses the knowledge base which contains all information about the manufacturing environment. A knowledge-based task planner can be defined as a function

$$f : K^* \times T \rightarrow P^*$$

where K is the set of states of the knowledge base, T the set of tasks and P the set of primitive actions [17]. K^* expresses that the changes in the knowledge base are immediately combined into the act of planning and keep the planning running. The function f is (for a specific task) given by a set of rules which we call behaviour pattern, as it describes the behaviour of an autonomous unit when it solves a given task. This concept is comparable to the views of Schoppers [40], who indicates planning as selection of appropriate reactions for expected situations. Our concept offers the advantage that the behaviour patterns can be executed by a standard rule interpreter - we use OPS5 [9]. Another even more interesting advantage is, that in our concept planning in the normal case and planning of exceptions can both be described by the same mechanism, i.e., the behaviour patterns.

The single behaviour rules have a uniform structure

{ ‹ s › (state ˆphase p ˆcurrent c) }

(condition$_1$)

...

(condition$_n$)

$\longrightarrow$

(action$_1$)

...

(action$_m$)

They consist of a condition and an action part. If all conditions of the condition part are satisfied, the rule is ready to "fire"; that means, the actions of the action part are executed. The first condition is especially significant and will be outlined further. It also shows the structure of OPS5 facts. Such facts are plain tuples, where the first component is a classifying symbol. The other components can be addressed by names and can take up symbols or numbers. In a test (match), if a certain fact satisfies a certain condition, corresponding components are compared. In the following description, a state fact means a fact of the OPS5 working memory which can satisfy the first condition of a behaviour rule.

The Architecture of the Task Planner

The task planner is divided into three parts. These are

- the Task Interpreter (TI)
- the Interpreter of the Behaviour Patterns (IBP)
- the primitive actions of the domain

Fig. 3 shows how the single components cooperate. Each of them runs in a separate process. The primitive actions can be structured into more than one process.

The Task Interpreter

The Task Interpreter (TI) establishes the task-oriented interface for the upper application layer. It is implemented as an OPS5 process and is structured in two parts. The first part is domain-independent and establishes the asynchronous fact entry as described below. The second part is domain-dependent and consists mainly of the rule set which analyses the incoming tasks the TI is able to handle. The actual interpretation of the tasks is done by looking for the appropriate behaviour patterns in the knowledge base. These behaviour patterns are passed to the IBP, which performs the actual transformation of the task into a sequence of primitive actions. The combination of existing behaviour patterns to more complex ones is a planning step which is comparable to the concept of planning by scripts [14] and for this reason must not be underrated. Furthermore, the TI handles the finish and error messages of the IBP. The implementation of the TI as a separate process was done first of all for technical reasons. Besides a clean separation of competence, this brings the advantage that the TI can always receive new tasks even if planning is active. By this means, it is possible for the TI to do tasks of high priority (for example, abortion of a running task planning) immediately after they are present.

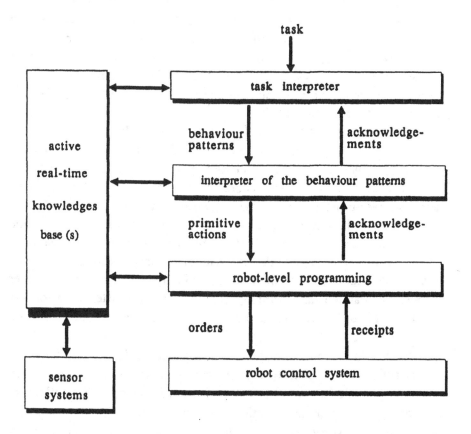

Figure 3: Architecture of the Task Planner

The Interpreter of the Behavior Patterns

The Interpreter of the Behavior Patterns (IBP) is the essential part of the task planner. It serves the robot-level programming layer, i.e., the interface to the robot control system, where the actions are actually performed. The behaviour patterns describe how the TI should split up a given task into primitive actions. Like the TI the IBP consists of two parts. The first part which is domain-independent provides the asynchronous fact entry as described in the following section, a watch mechanism for knowledge base elements and a finite automaton-like model with states and transitions (called state concept) for behaviour patterns. Furthermore, it is easily possible to set up timeouts to restrict the period of occurring wait states. The second part, which is domain-dependent, provides the error and exception handling which is given by fixed behaviour patterns built into the IBP.

The state concept takes into account that the rules of the behaviour patterns fire mostly sequentially in a relatively strict order. In all our applications, only a few meaningful cases occurred in which rules could be fired concurrently. Our experiences till now have shown that the state concept gives a better idea of the sequence in which the rules will fire. Furthermore, the location of an error is found more easily because this location is given by the actual state of the task planning when the error occurred. It is also possible that the state of the task planning is the only precondition for some primitive actions. Finally, the state concept makes it possible that the task planning can be reinvoked after an error has occurred. Here the starting state does not necessarily have to be the initial state of the whole task planning.

Even though the rules of the behaviour patterns fire sequentially in a rather rigid way, the primitive actions can be carried out concurrently. In fact it is possible to control any desired number of primitive actions simultaneously. The primitive actions are called on the right-hand side of a behaviour rule. Mostly, calling a primitive action means in fact sending a message to a separate process where it actually is performed. The primitive actions then run parallel to the IBP process on the same processor (e.g., the path-finding algorithm for the robot effector [21]) or on a separate processor (e.g., in the control of the robot). When a primitive action is finished, it is indicated by writing an entry into the knowledge base. Nevertheless, there is also the possibility to wait immediately after a call of a primitive action for its end. The call of the primitive action, in this case, has the semantics of a remote procedure call [1], as the primitive actions are always done in a separate process. This possibility was introduced because in those cases in which primitive actions have to be carried out in a strict sequential order, only one primitive action could be called on the right-hand side of a behaviour rule and for this reason the number of behaviour rules would become unnecessarily high. By the possibility of remote procedure calls, the behaviour patterns will be more compact and therefore intuitively easier to understand.

The watch mechanism for knowledge base elements was introduced to support exception handling. When this mechanism is used, exception handling is invoked if the value of a specified knowledge base element differs from a certain value or just alters its value. The conditions in the left-hand side of a behaviour rule normally describe the circumstances which have to occur in order to allow the execution of certain actions. The execution of rules is a logical derivation. Some of the conditions have to remain true for a sequence of derivation steps, i.e., a sequence of single rule executions. If all error situations are to be caught, an extra rule has to be formulated for each combination of truth and falsity of these conditions to specify the necessary actions for the error situation. Using the watch mechanism, one can formulate general strategies for exceptions raised when certain knowledge base elements have a wrong value. Such a strategy might be, to slow down or stop the motion of a robot in order to win time to analyse the error situation. The watch mechanism can be adapted to certain situations by using the state concept. This means that the watch mechanism can be enabled and disabled for a knowledge base element in any state of a behaviour pattern.

Processing Asynchronous Data with OPS5

As the knowledge is written into the knowledge base asynchronously in real-time, a way had to be found to get this knowledge into the OPS5 working memory asynchronously. This means that asynchronously arriving messages at an OPS5 process have to be inserted as facts into the OPS5 working memory under real-time constraints . The OPS5 rule interpreter executes the rules in a fixed recognize-act cycle. Only at a certain point of this cycle (immediately after the act phase) is it possible for the interpreter to take over asynchronously incoming messages as facts into the working memory. For this purpose, two procedures have to be formulated in the PASCAL interface of OPS5. One of them ('interrupt') is called asynchronously by the operating system while the other ('completion') is called synchronously by the rule interpreter.

```
procedure interrupt;
begin
    put the incoming message into the list lif of incoming
    facts;
    remark that the procedure completion should be invoked
    during the recognize-act cycle;
end; interrupt

procedure completion
begin
    forbid interrupts;
    for all facts f in the list lif of incoming facts:
        put the fact f into the OPS5 working memory;
    remove the remark that the procedure completion must be
    invoked during the recognize-act cycle;
    allow interrupts;
end; completion
```

In the interrupt handling procedure 'interrupt' asynchronously incoming messages are taken into the PASCAL interface. By the procedure 'completion', which is invoked by the rule interpreter itself, immediately after completing the act part (right-hand side) of a rule [17], the facts are put into the OPS5 working memory. In this manner it is possible to deal with asynchronously incoming facts in an OPS5 system. If no more rules can be fired and the computation of the program is not finished, the system has to wait for more asynchronously incoming facts. The simplest solution to manage this would be, to invoke an endless loop, which always will be active if no other rules can be fired. In order not to waste CPU time, we use another solution in which no dummy rules are fired. The main idea is to formulate a 'rien-ne-va-plus-rule' (r-rule) as described in [29]. In this r-rule a PASCAL procedure which handles the wait state is invoked. Using this procedure, no superfluous rules are fired.

More important, all asynchronously incoming facts are checked at least once in the match and the conflict-resolution phase, before a possibly unbounded wait state is entered. We call this scheme *asynchronous fact entry* of the OPS5 system. Because the whole flow of information comes from the knowledge base, all asynchronously arriving messages are interpreted as knowledge base elements. That means, they are mapped on facts of an OPS5 fact class *wb-element*. The effect is that exceptions caused by asynchronously incoming knowledge base elements can be handled easily. For this purpose it is sufficient to write down a rule which matches a knowledge base element in its first condition [17]. This is due to the following three arguments:

- the point in the recognize-act cycle where the facts are passed into the OPS5 working memory ensures that these facts are the most recent ones at the next match and conflict-resolution phase.
- by the MEA conflict-resolution strategy, those rules which match in their first condition the most recent facts of the OPS5 working memory are fired first.
- the first condition of a behaviour rule cannot be satisfied by an asynchronously incoming knowledge base element, because of the convention that behaviour rules match in their first condition those state facts which cannot arise asynchronously.

Behavior patterns for both regular and exceptional handling of a given task can be formulated by this mechanism. When multiple exceptions occur, it is possible that a hierarchy of behaviour patterns is activated. In it the active behaviour pattern can, after its own completion, always decide which behaviour pattern has to be reactivated.

MOTION PLANNING

Introduction

The motion planning problem consists in finding a sequence of movements for a given manipulator from a certain start configuration to a certain goal configuration so that the manipulator does not collide with any of the environmental obstacles. Every possible arrangement of the n manipulator links is uniquely described by the robot's configuration vektor. This vector has n scalar components, one for each joint of the n-link manipulator. So, the solution of a motion planning task can be represented by a sequence of configuration vectors, or by a one-dimensional curve in the n-dimensional configuration space (C-space).

Due to its importance, the findpath problem has been worked on by many researchers. For a concise review of previous motion planning work, see the article of Sharir [43]. Most of these treated only a two-dimensional workspace, which is of minor interest in real, three-dimensional applications.

Some approaches dealing with three-dimensional environments are, in principle, capable of handling any number of degrees of freedom, but their inherent exponential space and time complexity limits their practical usage to three or four degrees of freedom (DOF) [24,32,44]. Others restrict the manipulator's kinematics to a simpler, 4-DOF model, thus using only a fraction of the robot's moving capabilities [10,21].

In spite of the necessity of dealing with more degrees of freedom in many practical tasks — like moving bulky loads or controlling redundant manipulators — there are only a few implementations that cope with three-dimensional object models and six (or more) degrees of freedom.

Donald's search by local expert operators [13] has run-times of several hours for a flying polyhedral object with three translational and three rotational degrees of freedom. The algorithm is not directly applicable to a revolute-joint robot manipulator, because it needs decoupling of translations and rotations.

Barraquand and Latombe [2] report a challenging motion planning example for a 31-DOF stick-limb robot, which is beyond the capability of other existing planners (the run-time is 15 minutes on a DECstation 3100). The performance, however, is hardly comparable because common robots have only six degrees of freedom but voluminous limbs.

The mixed approach of Faverjon and Tournassoud [16] combines the advantages of local and global search methods. To restrain the exponential resource consumption, they use a coarse-resolution free-C-space representation of only the first three robot links for the global search. The remaining degrees of freedom are considered only in the local low-complexity search procedure, which is very fast and efficient, but limited in its applications by possible deadlocks. It is left to the global part to provide appropriate subtasks for the local planner. Realistic examples --- a PUMA-like robot with and without bulky load --- are reported with run-times of a few minutes. However, the heuristic assumption for the bulky-load solution results in a search which is not complete but 'solves many practical cases' [16].

The approach proposed here is a heuristic one, with emphasis on short run-times for practical motion planning problems with six-axes manipulators and three-dimensional environments. Its aim is to find quickly paths for commonplace tasks, and not to solve the *general piano movers problem* --- though, in principle, this would be possible too with this algorithm.

CGR Search in n Dimensions

The main intention of the proposed motion planner is to provide a fast generation of collision-free motions for a general three-dimensional manipulator with six degrees of freedom.

One way to avoid large run-times is to restrict the construction of the free C-space to the first three or four degrees of freedom [16,24,32,44]. However, this reduction is only appropriate in cases where the remaining degrees of freedom have only minor effects on the space occupancy of the manipulator. It is not recommended with bulky loads or with manipulators which cannot be decoupled in mainly position-effective and mainly orientation-effective links.

The algorithm proposed here, the combination of goal-directed and randomized search (CGR search) tries a different approach to treat general 6-DOF manipulators with arbitrary load: no explicit construction of the free C-space at all. Any information about C-space obstacles is computed only on demand and only at single points along one-dimensional subspaces of the arbitrarily high-dimensional C-space. This strategy allows a fast computation of collision-free motions for complex manipulators in three-dimensional workspaces.

CGR search, the combination of goal-directed and randomized search, is a synthesis of an efficient, local collision-avoidance method (goal-directed search) and a simple but effective global decomposition method (random subgoal generation) [22]. The CGR search has shown its superior efficiency in the results of various three-dimensional motion planning experiments.

Goal-Directed Search

The goal-directed search tries to reach the goal in a straight line in configuration space. Tests for collisions with obstacles are performed in small periods, called *steplength*, beginning at the starting point. The steplength depends on the geometric and kinematic properties of the manipulator and of the tolerances and the exactness of the environment description, and it must be small enough to ensure a safe path between two adjacent collision-free test points.

If an obstacle is met, the search tries one or more side steps, i.e., orthogonal to the current search direction in C-space which is treated like an Euclidean vector space. Supposing the C-space obstacle's surface is locally a linear manifold [13], we find a collision-free side step in at most $2*(n-1)$ tests (with n being the number of dimensions of the C-space). After this slide step, the direct search towards the goal is continued from the new position. Subsequent alteration of steps towards the goal and steps to the side result in sliding along the C-space obstacle surface, and finally, in circumnavigation of this obstacle (see Fig. 4).

Clearly, in some cases with concavities or dead ends, no way out can be found by this strategy, and the goal-directed search is stuck. Therefore, in these cases the goal-directed search gives up, pretending that there is no direct way between start and goal, because trying to escape from such local minima of the goal-distance function by complete search would imply the unwanted exponential expenditure.

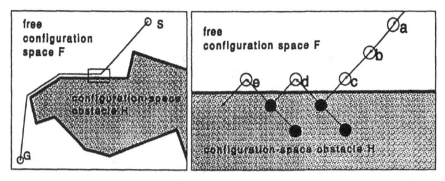

Figure 4:

Left: Goal-directed straight-and-slide search. This strategy is strictly local; therefore, the complexity of the search does not depend on the size of search space, which grows exponentially with the number of degrees of freedom. The rectangle marked in the center of the figure is enlarged on the right side.

Right: Collision avoidance by straight-and-slide search. All circles are test points on the way towards the goal. The light points *a*, *b*, and *c* are collision-free configurations on the direct path to the goal. The black circle after *c* marks an unsafe configuration and is the base for a side step to *d*. Starting now from d, the straight search again runs into a colli-sion, which produces a side step to *e*. The final collision-free path for the robot to move in is through points *a*, *b*, *c*, *d*, and *e*.

Random Subgoal Generation

If no collision-free path could be found by the goal-directed search, the random part of CGR search generates a new subgoal. This subgoal is constructed by choosing random values for all components of the ma-nipulator configuration vector, and testing this configuration for colli-sions. If necessary, this process is repeated until a collision-free configu-ration is obtained. The initial problem is now decomposed into finding a path from the original start to the new subgoal and finding a path from this subgoal to the original goal. The basic assumption of this approach is that there are many possible solutions of the findpath task, and that the probability of hitting a good subgoal in a few random attempts is sufficiently high. The experimental results strongly confirm this as-sumption.

Configuration Graph

In more complex environments it may not suffice to decompose the original task into only two subtasks. Then more than one subgoal is necessary to guide the goal-directed search around the obstacles. So, all previously generated subgoals and all partial paths found are stored in the *configuration graph* as nodes and as edges between nodes respectively. At the beginning of the search this graph consists of the start node and the goal node and an empty set of edges. Subgoals created subsequently and paths found are added in progress. The CGR search stops when the start node and the goal node lie in the same connectivity component of the configuration graph.

The configuration graph is a very abstract, approximate model of the topology of the free C-space. In contrast to the vast complexity of the geometry of free C-space, the topology of even high-dimensional C-spaces is extremely simple. Accordingly, the experimental results show that even in six-dimensional C-spaces already a few random subgoals (producing only a small configuration graph) lead to a solution of the motion planning task. So, after completion of the CGR search, a simple graph search technique suffices to extract the final path from the start configuration node to the goal configuration node.

Computational Complexity

It is well known that there is a polynomial-time motion planning algorithm for any particular manipulator [42]. The very large exponents of its computational complexity, however, make its implementation impractical [10]. Furthermore, findpath worst-case complexity is exponential in the number of degrees of freedom of the manipulator [42].

Because exact solutions of realistic motion planning problems seem to be computationally intractable, heuristic shortcuts might be necessary [43]. The motion planner proposed here uses the heuristic CGR search to tackle the problem of vast-complexity search spaces. Its worst-case behavior remains — according to theory — exponential, but the average-case task is solved much faster by this heuristic strategy (some experiments suggest that the influence of the number of degrees of freedom is not significant in practical cases). The basic assumption of CGR search is that the motion planning problem allows many different solutions, and that it suffices to find quickly any one of these paths. Thus, no optimal path will be found by this method.

Due to its random nature, the algorithm is probabilistically resolution-complete [2]. That means, our planner will solve any findpath task if a collision-free path exists (at a given resolution). However, the statement of an upper limit for the run-time is impossible. As a consequence, the algorithm is unable to decide that a problem is unsolvable. In practice, this is no drawback because, first, realistic tasks are designed to have many solutions, and second, with generally short run-times, one can assume that tasks requiring more than a certain limit of computation time are (likely to be) unsolvable.

Object Representation

The robot's workspace with its objects is considered by the CGR search only through the collision check, which is applied to single configuration vectors. The tests for collisions between the manipulator and the environment obstacles have to be carried out sufficiently often on a path to ensure safety of the whole path. Thus, a fast computation of the test procedure is essential to gain small overall search times.

Because the goal-directed search does not need more information on a given configuration than simply *collision* or *no collision*, there are certain possible optimizations. First, if one object colliding with the robot is found, the rest of the objects need not be tested. Second, if an object is so far away from a robot limb that its bounding box (or some other hull) does not intersect the hull of the robot limb, no further treatment of the internal structure of these objects is necessary.

The best way to represent the environment objects is a hierarchical model [15]. The workspace is a tree whose leaves are the polyhedral surfaces of the real objects. The internal nodes represent virtual objects which are composed of the virtual or real objects of their subtrees. Every virtual object is a geometric hull of its subobjects. Choosing a simple hull (like bounding box or sphere) results in a fast collision test between virtual objects.

Skilful grouping of real and virtual objects makes the collision test of a particular robot configuration very efficient. The expenditure is basically only dependent on the number of obstacles in the immediate reach of the current manipulator configuration, and nearly independent of the number of objects in the whole workspace.

Experimental Results

The CGR search for six degrees of freedom and the collision check for three-dimensional objects were implemented at our Institute in the programming language *Pascal*. All run-time measurements reported here were made on a VAXstation 3100 with the operating system VMS. The speed of this machine is quite comparable to that of a SUN 3 workstation.

The collision check works with a three-level tree: object hulls (at the most abstract level), facet hulls, and real facets (at the leaves). Thus, the most primitive objects are the surfaces of the environment solids. An additional composition of several neighbouring objects to super-objects was not necessary in the examples. But with a larger number of environment 'objects' this grouping will be advantageous.

With motion planning by CGR search no preprocessing is necessary, like computation of C-space obstacles [16,24,32,44], free C-space [10], artificial potential functions [2], or similar supporting structures. The run-time includes all necessary computations. Several motion planning experiments were carried out with the CGR search strategy. They are chosen to show the performance of this method in a range of different applications.

Figure 5 depicts three motion planning tasks in three different environments. The first two tasks, on the left and in the middle, are comparable in their difficulty to examples given previously by authors of classical C-space approaches [10,16,24,32]. The run-times, of about one minute, are of the same order of magnitude as those of the other methods.

The right part of Fig.5 shows a snake-like 6-DOF manipulator, which is more versatile than a PUMA-type robot and cannot be decomposed into mainly position-effective and mainly orientation-effective links. Therefore, a reduction of the degrees of freedom is not possible, and classical C-space methods are not applicable. The average run-time of the CGR search for this problem is 241 seconds.

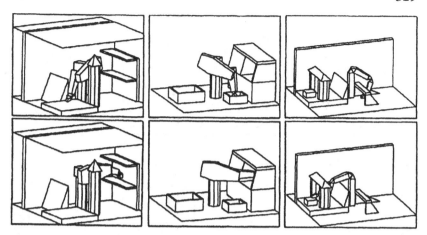

Figure 5: Example of motion planning tasks. All six degrees of freedom are taken into account.

Left: No subgoal is needed for this task, the run-time is 70 seconds.

Middle: The PUMA robot in a fabrication environment loads the machining center with a piece of raw material or with a tool. Average run time is 60 seconds. A solution needs 1.9 generations of subgoals, on the average.

Right: This 6-DOF snake-like manipulator cannot be decoupled into a 3-DOF arm and a 3-DOF hand like a PUMA-type robot (with the arm mainly responsible for movement). So all other methods which rely on reduction to three or four degrees of freedom must fail to plan motions for this construction. The average run-time with CGR is 241 seconds, with 6.1 generations of subgoals, on the average.

As with the previous example, no reduction to three or four degrees of freedom is possible in the task of **Fig. 6** because all six links are necessary to reorientate the bulky load. The average run-time is 224 seconds. The solution shown is arbitralily chosen out of 1000 experiments. To our knowledge, only Faverjon and Tournassoud [16] give a comparable example. Their planner uses "a few minutes" for the generation of a trajectory (plus an unspecified amount for preprocessing), but the method is not complete.

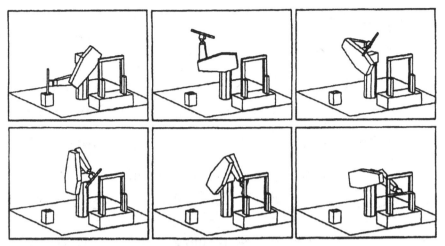

Figure 6: Example of a very difficult motion planning task because the load is bulky and requires reorientation. So all six degrees of freedom have to be considered in order to get a solution. The average run-time is 224 seconds, with 5.3 generations of subgoals, on the average. The motion shown is a typical path, and it has been computed in 226 seconds, using 6 random subgoals. The whole path consists of 234 discrete configurations.

Open Issues

As the algorithm searches without consideration of global aspects, the path found is not optimal. If desired, path shortening, path smoothing, or other optimizations (time, energy) are possible by further processing steps, which are not implemented yet. Of course, only suboptimal solutions will be produced by this postprocessing, because absolute optimizations are not traceable in high-dimensional complex-structured C-spaces.

The random nature of the search results in a variation of the run-times for a certain motion planning task. A parallel treatment of the same task on several workstations or coprocessors will further reduce the average run-time for the first solution found. For example, parallel search on two machines in the case of the task of **Fig. 6** would reduce the average run-time from 224 seconds to 129 seconds. Four and eight parallel processes would result in 80 and 54 seconds, respectively. The generation of subgoals, the goal directed search between subgoals and the collision test show promising opportunities for further parallelism. With a few parallel working RISC processors — which are five times faster than our current machine — real-time motion planning applications seem to be feasible.

SENSOR INTEGRATION

Sensor Systems in Industry

This section describes an approach to the construction of a flexible manufacturing system's execution layer. This layer, which may be found in Fig. 1 and which is specific for each autonomous system, bridges the gap between an interface to devices (these are sensor devices and active devices like robots, transport systems and machines) and an interface of complex sensory results or sensor-controlled operations, respectively. The services included in the latter are activated by the coordination layer and also contain non-linear plans. These plans describe not only a sequence of actions to control the flow of material and the steps of production, but also include the supervision of actions and error recovery.

Because we think of autonomous systems equipped with many sensors, we conceptually assume sensor information being involved in the behavior of every autonomous unit. Therefore, sensor modules control the interfaces to active devices. Sensor signal processing turns out to be the main task of the execution layer. In this context our experiences gain importance which show that control systems of active devices like robots are not capable of performing complex sensorial computations. As a consequence, we decided to spread sensor data processing over a net of processors (see below).

In an industrial automation environment the geometric and physical properties of objects and their structure are known quite well. But due to the rough manufacturing environment and the limited precision of active components, there remain some uncertainties, for example in the position of objects or in the correct execution of actions. Therefore, sensors are needed to verify and define precisely the environment model and to supervise constraints which guarantee correct performance passively or actively.

In spite of that, sensor systems in industry, especially multisensor systems, have not gained the common acceptance they should be expected to find. One reason for this can be found in the cost of building sensor applications. Nowadays, sensor hardware components are becoming more powerful and less expensive, but a strong need remains for a cheeper and more efficient way of producing software for sensor applications. To overcome the cost of software developement, we have to abandon the common procedure of developing special sensor systems for each application. Instead, an approach has to be found to create sensor applications in a structured and modular way in order to reuse software components and modules.

Functional Decomposition of Multi-Sensor Applications

Henderson [25] and later Rowland [48] tried not only to characterize sensor devices by the type of data they measure, but treated program modules in the same way. Program modules become virtual or logical sensors. These produce, for example, as a result an object's centre of gravity which they compute from a camera's data. But an analysis of sensorial functions which the execution layer in an industrial environment has to provide shows that these Logical sensors need to be supplemented. Other units, Logical actors, transform, for example, sensorial information into instructions for active devices or access and manage a model of the industrial environment.

Through this concentration on the results of sensor units, a program system for sensor signal processing changes to a hierarchical system consisting of Logical sensors and actors. These logical units evaluate sensor data, filter and transform them to the higher (or more symbolic) notation of the environment's model. Logical actors receive the results of this hierarchical process and control active devices, robots, transport systems etc. All sensor units receive virtual or physical sensor data from below, that means from sensor units which are nearer to the sensor device level. They change them into the calling of an active device's services or to a more abstract item which is given to other interested units. For each pair of communicating logical units, the type of transmission (e.g., wait, no wait) is specified.

In this process there are three possibilities to trigger a sensor unit: a master unit which is interested in another unit's result sends an activation message to a logical unit below; a sensor unit which after computing a result finds several constraints (data constraints or time constraints) fulfilled, transmits the item to an interested logical unit above. Thus, the value of a result and, more general, sensorial knowledge may control the way of signal processing (data-driven activation). In the case the activation is driven by time constraints, sensor units receive data, for example, once every clock cycle. Sensor units which implement control loops or supervise certain conditions in the environment, need this kind of activation.

Logical Sensors and Actors

The idea of Logical sensors, developed and implemented by [25] for image data processing, has been adapted to the needs of a multi-sensor environment in industry automation. We define Logical sensors as program units characterized by the result they produce. This result is described syntactically as a (nested) record structure which is known from Pascal or other high-level languages and which is composed of a given number of basic types. Logical sensors, thus, describe one transformational step in obtaining a world model from other (raw) sensor data. They integrate information from multiple sensor devices or from other Logical sensors, use their own view of an environment description and are linked together to form a hierarchical net structure.

In sensor applications, Logical sensors provide the interface to sensor devices, filter sensor data on the various levels of abstraction, transform sensor signals to these levels, and supervise the manufacturing process by evaluating constraints on sensorial results. Logical sensors are also used to combine sensorial information from different sources and to identify or localize objects by matching sensorial knowledge about the environment and model knowledge.

Logical actors do not produce any virtual result, but they manipulate the real manufacturing environment, its computer-based model and the flow of control in a sensor application. We differentiate three types of Logical actors according to their function:

Acting: Logical actors are our construct to transform sensorial results into any kind of action sequences out of a given set of services executed by manipulators, machines and other active devices. By introducing this type of Logical actors, we achieve our objective of an homogenous interface to the device layer.

Adjusting an environment model: Logical actors use sensorial information which describes the real environment to update the computer model. The latter in turn may be accessed by all sensor units.

Plan execution: We distinguish two main types of plans. Planning systems generate in their task transformation the first type which describes the manufacturing process, its supervision and, if needed, correcting functions for errors discovered by Logical sensors. The second type of plans incorporated in Logical actors describes how to proceed in performing a sensorial function. By this, we support concepts both for active sensing and for sensor strategies. As a notational form for the second type of Logical actors, which we consider to be constructed off-line by a human user of our sytem, we use transition graphs stated in [8]. Logical actors for plan execution, in contrast to all the other sensor units, connect and cut dynamically their communication links to servers.

LSA Net and LSA Function Graph

Logical sensors and actors work in parallel. Therefore, we can consider a sensor application as a hierarchical graph - we call it LSA net - with sensor units as the set of nodes and the edges resulting from the flow of data and control. LSA nets, which are generated off-line, supply all the sensor services. The leaves in an LSA net control physical sensor devices; the various roots of the network (sensors and actors) form the interface to the planning system and supply to task planers a set of elementary functions.

Figure 7 shows a simple example from [39]: An LSA net to process the signals of two tactile arrays which are placed in both gripper fingers of a manipulator. There are two Logical sensors which access the physical sensor devices. Logical sensors implement three stages of processing the tactile information: after being read from both devices the data pass through noise filters and the filtered results (given by Logical sensors) indicate a single or double contact. During gripping, a Logical actor (labeled

'self-centring grip') transforms a single contact into a correcting movement of the robot to achieve double contact, which is returned to an interested master by a designated Logical sensor ('double contact'). In this
example the tactile matrix from the right gripper finger should show a
characteristic drilling hole. The tactile image of this is used after gripping to determine the workpiece's exact gripping position.

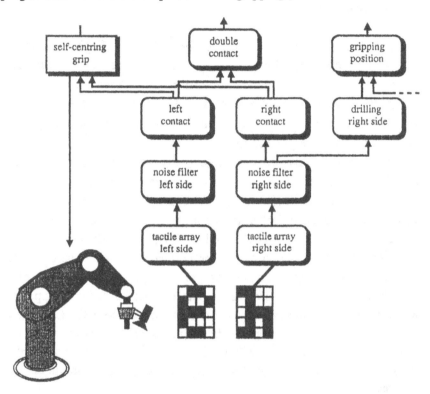

Figure 7: Tactile Arrays for Gripping - LSA Net

Sensor systems in a manufacturing environment are active only at
few, discrete points of time. The tactile arrays from above, for example,
are needed just in the first phase of gripping a workpiece. We, therefore,
have to consider not only the static links in a LSA net but also the dynamic changes in these connections. A LSA net's static edges define all
possible interconnections; the current communication links in a LSA net
form at each given period of time an additional structure, the LSA
function graph, which shows the flow of data in a specific part of the

sensor application. In addition, a LSA function graph focuses the point of view upon this part of the application and contains all the Logical sensors and actors together with their links which are necessary to achieve one sensorial function.

Implementation

As one of the main advantages, the concept of Logical Sensors and Actors (LSA concept) allows us to construct multi-sensor applications in a modular way. This modularity not only holds true for the algorithms incorporated in an application but also forms the basis to separate algorithms, flow of data, flow of control, time management and modeling of the manufacturing environment.

Several of these aspects in building a sensory application can be classified and handled uniformly. This holds true for communications, error recovery or for the access to an environment model. We, therefore, are able to describe a Logical unit's behavior and communication needs in terms of its classified aspects. Additionally, we developed a framework (toolbox) of standard components for Logical sensors and actors which is configured according to a description of an LSA net and supplemented by the algorithms for the pure sensor data processing. Sensory applications are generated automatically from the toolbox and from the LSA net's description. The toolbox supports distributed sensor units and uses procedural languages to implement low-level sensor nodes; a first attempt with distributed, rule-based modules for the derivation of symbolic properties showed good (performance) results.

In executing sensor functions we always have to take care of time constraints. Time constraints arise from Logical actors which control active devices and spread over all sensor units which provide sensorial information to them. In addition we have to estimate an upper limit for the time the whole task of supervising actions and taking corrections needs. Control loops have to be executed periodically, with an exact clock frequency being kept.

In order to separate algorithms for sensor signal processing from those for time constraints, we introduce special Logical sensors, time sensors, which produce the current system time as a result. Time sensors trigger periodically other Logical units or the execution of special branches in a plan. Managing time constraints not implicitly, as part of a Logical unit, but explicitly, in terms of an LSA net's components, gives one necessary prerequisite to quantify off-line the activation frequency of sensor units and the times needed to react on sensory events. Time management becomes part of building an LSA net's description. In developing descriptions of LSA nets, we support users by analyzing the communication needs of Logical units, thus determining the flow of sensorial information and activations for sensor units. This also helps in distributing a sensory application over the underlying hardware basis in such a way that time constraints can be fulfilled.

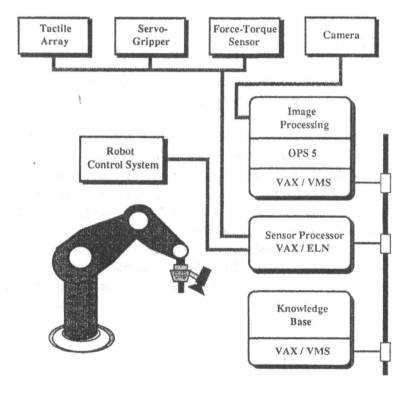

Figure 8: Hardware Structure of the Execution Layer

A PUMA 560 robot equipped with various sensors - see Fig. 8 - gives us the basis for our implementations. All of the robot's sensors (a camera near by the robot's wrist, a force/torque-sensor carrying an elektronic servo-gripper and tactile arrays enclosed in the gripper's fingers) are connected to sensor processors in a computer network. Sensor processors under the real-time operating system VAXELN perform direct control loops and other time-critical sensor functions. A so-called *cartesian real-time interface*, developed at our Institute, gives one of these accesses to the robot system and allows us to use the robot as an active device. VMS opens to us a rule-based approach to sensor signal processing. VMS machines contain image processing boards and host our knowledge base. The sensor processors form the hardware basis for LSA nets which process the robot's sensor data and which are generated using the LSA toolbox.

CONCLUSIONS

Task-level programming of autonomous robot systems requires solutions in different areas of software engineering. Methods of artificial intelligence are integrated for rule based transformation of tasks and for knowledge representation. Numerical problems have to be solved for motion planning and sensor data processing. At the lower layers (e.g., sensors, robot control) hard real-time constraints have to be met; real-time features also appear in robot programming languages and in real-time operating systems at the robot control level. Our investigations yield valuable ideas and solutions to these complex problems. The future work will concentrate on distributed real-time knowledge bases and will improve the task transformation facilities. Sensor feedback loops including image processing, complex motion and grip planning and sophisticated error detection and recovery mechanisms will be integrated to achieve highly autonomous robot systems.

REFERENCES

[1] Andrews G. R., Schneider F. B.: "Concepts and Notations for Concurrent Programming". *Computing Surveys*, Vol. 15, No. 1, March 1983.

[2] Barraquand J., Latombe J.-C.: "A Monte-Carlo Algorithm for Path Planning With Many Degrees of Freedom" *Proc. IEEE Int. Conf. on Robotics and Automation*, Cincinnati, USA, May 1990, pp. 1712--1717

[3] Bocionek S.: "Dynamic Flavors", Technical Report, Technische Universität München, *TUM I8708*, June 1987

[4] Bocionek S., Meyfarth R.: "Aktive Wissensbasen und Dämonenkonzepte", Technical Report, Technische Universität München, *TUM I8811*, September 1988

[5] Bocionek S., Meyfarth R., Schweiger J.: "Handbuch zur A4-Wissensbasis-Shell". *Benutzeranleitung Version 2.1*. Technische Universität München, February 1990

[6] Bocionek S., Fischer K.: "Task-Oriented Programming with Cooperating Rule-Based Modules", To appear in: *Int. Journal for Engineering Applications of Art. Intelligence*, Pineridge Press Periodicals, 1989

[7] Bocionek S.: "Task-Level Programming of Manipulators: A Case Study", TU Munich, Institut für Informatik, *Report TUM I9001*, January 1990, 52 pages.

[8] Bocionek S.: Modulare Regelprogrammierung, Vieweg-Verlag, 1990.

[9] Brownston L. et al.: "Programming Expert Systems in OPS5", Addison- Wesley 1985.

[10] Brooks, R. A.: "Planning Collision-Free Motions for Pick-and-Place Operations", *Int. J. Robotics Research 2*, No. 4, 19-44 (Winter 1983)

[11] Coiffet Ph. and Chirouze M.: "An Introduction to Robot Technology", *Kogan Page Ltd.* London, 1983.

[12] Dadam P., R. Dillmann R., Kemper A., Lockemann P. C.: "Objektorientierte Datenhaltung für die Roboterprogrammierung", *Informatik Forschung und Entwicklung*, 2/87, 151-170, 1987

[13] Donald, B. R.: "A Search Algorithm for Motion Planning with Six Degrees of Freedom", *Artificial Intelligence 31*, 295-353 (1987)

[14] Dorn J., Hommel G., Knoll A.: "Skripte als ereignisorientierte Repräsentationsmechanismen in der Robotik", *Proceedings GI-Jahrestagung 1986*.

[15] Faverjon, B.: "Hierarchical Object Models for Efficient Anti-Collision Algorithms. *Proc. IEEE Int. Conf. on Robotics and Automation,* Scottsdale, USA, May 1989, pp. 333-340

[16] Faverjon, B.; Tournassoud, P.: "Motion Planning for Manipulators in Complex Environments", *Proc. Workshop on Geometry and Robotics,* Toulouse, France, May 1988, pp. 87-115

[17] Fischer, K.: "Regelbasierte Synchronisation von Roboter und Maschinen in der Fertigung", TU München, Institut für Informatik, *Report Nr. TUM I8816,* December 1988.

[18] Fischer K.: "Knowledge-Based Task Planning for Autonomous Robot Systems", *Proc. of the Int. Conf. on Intelligent Autonomous Systems,* Amsterdam (Netherlands), Dec. 1989 , p. 761-771.

[19] Freund, E.; et. al.: "OSIRIS - Ein objektorientiertes System zur impliziten Roboterprogrammierung und Simulation". *Robotersysteme 6,* 1990, p. 185-192.

[20] Frommherz, B. J.: "Ein Roboteraktionsplanungssystem", *Informatik Fachberichte 260,* Springer, Berlin, 1990.

[21] Glavina B.: "Trajektorienplanung bei Handhabungssystemen", TU München, Institut für Informatik, *Report Nr. TUM I8818,* December 1988.

[22] Glavina B.: "Solving Findpath by Combination of Goal-Directed and Randomized Search", *IEEE Int. Conf. on Robotics and Automation,* Cincinatti (Ohio), 1990 , p. 1718-1723.

[23] Hagg E., Fischer K.: "Off-line Programming Environment for Robotic Applications", *Proc. of the INCOM '89,* Madrid, Sept. 1989

[24] Hasegawa, T.; Terasaki, H.: "Collision Avoidance: Divide-and-Conquer Approach by Determining Intermediate Goals", *Proc. Int. Conf. on Advanced Robotics,* Versailles, France, October 1987, pp. 295-306

[25] Henderson T. C.: "The Specification of Logical Sensors", In *A. Saridis, A. Meystel (Eds.): Workshop on Intelligent Control,* Troy, New York, 1985.

[26] Hörmann A.: "Petri-Netze zur Darstellung von Aktionsplänen für Mehrrobotersysteme", In: *Autonome Mobile Systeme, Beiträge zum 4. Fachgespräch,* Ed. P.Levi und U. Rembold, Karlsruhe, 57 71, November 1988.

[27] Kampmann P.; Schmidt G. K.: "Multilevel Motion Planning for Mobile Robots Based on a Topologically Structured World Model", *Proc. of the Int. Conf. on Intelligent Autonomous Systems,* Amsterdam (Netherlands), Dec. 1989.

[28] Kogler M.: "Modellierung der IWB-Fabrikumgebung in einer objektorientierten Wissensbasis", TU München, Institut für Informatik 6, *Diplomarbeit*, October 1990

[29] Krickhahn R., Radig B.: "Die Wissensrepräsentationssprache OPS5", Vieweg 1987.

[30] Lainer W.: "Einsatz einer aktiven, objektorientierten Wissensbasis zum Speichern von topologisch strukturierten Geometriedaten", TU München, Institut für Informatik 6, *Diplomarbeit*, January 1990

[31] Levi P.: "Verteilte Aktionsplanung für autonome mobile Agenten", *KI Fachberichte 181*, Springerverlag, 1988, p. 27-40.

[32] Lozano-Pérez T.; Jones J. L.; Mazer E.; O'Donnell P. A.: "Task-Level Planning of Pick-and-Place Robot Motions", *Computer 22*, No. 3, 21-29 (March 1989)

[33] Meyfarth R.: "Objektorientierte Datenhaltung für Wissensbasen unter Verwendung von B- Bäumen", *Technical Report, TUM I8815*, Technische Universität München, November 1988

[34] Meyfarth R.: "ACTROB: An Active Robotic Knowledge Base", *Proceedings of the 2nd Int. Symposium on Database Systems for Advanced Applications (DASFAA 91)*, Tokyo, to appear in April 1991

[35] Meyfarth R.: "Demon Concepts in CIM Knowledge Bases", *IEEE Int. Conf. on Robotics and Automation*, Cincinatti (Ohio) 1990, p. 902-907

[36] Milberg J.; Lutz P.: "Integration of Autonomous Mobile Robots into the Industrial Production Environment", *IEEE Conference on Robotics and Automation*, Raleigh, USA, 1987.

[37] Rembold U. et. al.: CAM-Handbuch. Springer, Berlin, 1990.

[38] Rowland J. J., Nicholls H. R.: "A Modular Approach to Sensor Integration in Robotic Assembly", *6th Symposium on Information Control Problems in Manufacturing Technology (INCOM'89)*, Madrid, 1989.

[39] Scherer C.: "Einsatz eines taktilen Arrays zur wissensbasierten Objekterkennung beim Greifvorgang", *Diplomarbeit*, Institut für Informatik, TU München, 1990.

[40] Schoppers M. J.: "Universal Plans for Reactive Robots in Unpredictable Environments", *Proc. of the 10th IJCAI*, 1039-1046, August 1987.

[41] Schuster H.-D.: "Entwurf und Realisierung einer Hauptspeicher-Version des MERKUR Tuple- Layer", TU München, Lehrstuhl für Prozessrechentechnik, *Diplomarbeit*, June 1989

342

[42] Schwartz J. T.; Sharir M.: "On the Piano Movers Problem. II: General Techniques for Computing Topological Properties of Real Algebraic Manifolds", New York University, Department of Computer Science, Courant Institute of Mathematical Sciences, Report 41 (1982)

[43] Sharir M.: "Algorithmic Motion Planning", *Computer 22*, No. 3, 9-20 (March 1989)

[44] Siméon T.: "Planning Collision-Free Trajectories by a Configuration Space Approach", *Proc. Workshop on Geometry and Robotics*, Toulouse, France, May 1988, pp. 116-132

[45] Todd D. J.: "Fundamentals of Robot Technology", *Kogan Page Ltd*, London, 1986.

[46] TransAction Sorftware GmbH: Transbase Relational Database System - Manuals. Munich, 1989

[47] Wirth M.: "Analytischer Vergleich wesentlicher Aspekte von Versionierungsverfahren und Systementwurf zur Integration von Versionen in eine objektorientierte Wissensbasis", TU München, Institut für Informatik 6, *Diplomarbeit*, November 1990

[48] Rowland J. J., Nicholls H. R.: "A Modular Approach to Sensor Integration in Robotic Assembly", *6th Symposium on Information Control Problems in Manufacturing Technology (INCOM'89)*, Madrid, 1989.

13

REAL-TIME DATA PROCESSING OF THE SENSORY DATA OF A MULTI- FINGERED DEXTROUS ROBOT HAND

Alois Knoll
TU Berlin
Fachbereich 20
D-1000 Berlin-10

INTRODUCTION

In recent years, multi-fingered dextrous end-effectors for robots have attracted a steadily increasing amount of attention in the robotics community (cf. also Chapter 4.1). Their design is modelled on the physical structure of the human hand thus allowing for a limited imitation of human grasping. The Belgrade-USC artificial hand (see Fig. 1) is the most recent result of research efforts aimed at the development of robot end-effectors, capable of reproducing most of the functionality of the human hand. The development goals for this hand included minimal weight, the ability to handle a large class of grasping tasks, and mechanical dimensions similar to the human model. Rather than stressing maximum flexibility and dexterity, the design emphasizes the use of synergies between the motion of finger joints and fingers. Its versatility in grasping makes this hand suitable for robotic applications such as the handling of toxic or nuclear waste, but it may also be utilized in the field of prosthetics.

The hand is equipped with sensory feedback for finger positions and finger pressure. An ultrasonic sensor generating high resolution images of the objects to be grasped may be used for pre-shaping the hand prior to grasping. Theoretically, this sensor may generate a new range image within very short time intervals, typically within the order of a second.

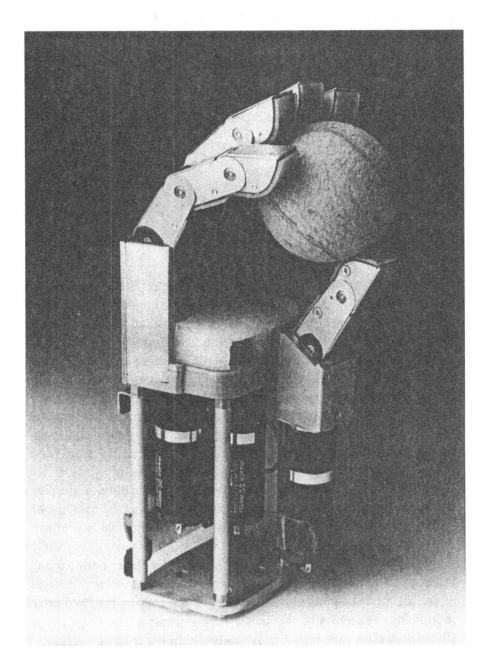

Figure 1: The current model of the Belgrade-USC hand

The time required for the complete closure of the Belgrade II hand is presently 2 s. Future models will offer a reduced closing time to mimic the behavior of the human hand, which requires about 100 ms for closing. Even though the time constants of the current model seem to be rather long at the first glance, they impose very hard restrictions on the execution time of the software controlling the whole setup, i.e. hand and attached sensors. Three areas are particularly crucial if a steady, uninterrupted motion of the hand is to be obtained:

- The control program of the hand must be able to keep track of all external forces exerted on the hand or one of its fingers and it must also be able to react accordingly. When the fingertip is brought in touch with the object (at a nondetermined point in time), the controller must immediately reduce the voltage applied to the motor driving the finger. Moreover, if the hand collides with an unexpected obstacle, the motion should be interrupted to avoid any damages. In both cases, the controller has to react immediately (within a few milliseconds).
- The setup for acquiring ultrasonic data (a stepper motor, an ultrasonic transmitter and a transient recorder) are to be controlled in sync with the motion of the hand and the grasping. Data acquisition and processing time should be as low as possible and can be minimized only by acquiring and processing data simultaneously.
- The processing of ultrasonic sensory data is particularly hard to complete within the time frame of even 2 s. The image generation algorithm currently realized, requires 16 x 32 kBytes = 512 kBytes of image data to be subjected to a Fourier transform, a manipulation of the resulting spectra and to an inverse Fourier transform. To do all the necessary computations on a minicomputer within the 2 s available is impossible. Instead, a digital signal processor must be used, but even then the requirements are hard to meet.

The structure of the software controlling both the sensor and the hand is shown in Fig. 2. A process called *Control* directs the mechanical section of the hand and the entities of the ultrasonic sensor. The software system exhibits only static parallelity, therefore the creation and termination of processes during runtime is unnecessary. Runtime efficiency, on the other hand, is of paramount importance and is the major reason why the whole system was programmed in Occam-2. This language perfectly meets all our real-time requirements.

We will return to the software in more detail, after explaining the mechanical section of the hand and the tactile and ultrasonic sensors.

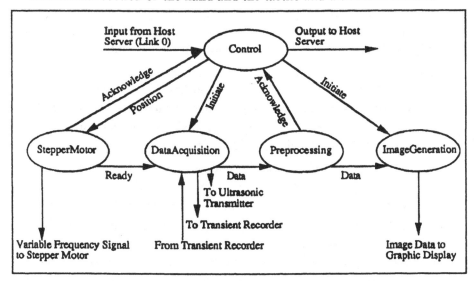

Figure 2: Structure of the software system controlling the ultrasonic system. Nodes of the graph denote parallel processes, arcs denote communication paths.

THE MECHANICAL SECTION OF THE BELGRADE-USC DEXTROUS HAND

The current model of the Belgrade-USC hand is the latest descendant in a line of several hands for prosthetic and robotic applications. The first hand was developed and built in the late 1950's [1]. It had five fingers driven by a single servomotor. The basic principle of grasping was simple: When one of the fingers touched the object, the others continued to move until all finger tips exerted approximately the same pressure on the target object. The current hand [2] was designed on similar principles; in particular, it supports autonomous shape adaptation.

The hand possesses five fingers whose joints cannot be articulated independently of one another. Instead, finger joints are moved in a coordinated fashion similar to the human hand when grasping a target object. The index finger (finger 2) and the middle finger (3) (as well as the ring finger (4) and the little finger (5)) are combined as a pair driven by a single actuator; therefore, the fingers within a pair are not independently controllable. As a result, they perform a *synergistic motion* during the closure of the hand. Two actuators control the thumb: The first actuator rotates the thumb about an axis parallel to the wrist. The maximum rotation angle is approximately 120 degrees; the thumb may thus be rotated from its fully extended position into on that is opposite to all other fingers except the little finger. The second actuator controls the flexion of the thumb.

The coupling between the two fingers of a pair is compliant, i.e. when the motion of one finger is inhibited, the other finger continues to move. This way a dynamic and sensitive adaptation of the finger configuration to the shape of the object is possible. This adaptation is performed autonomously without any interference on the part of the controller.

A five-fingered hand with three joints per finger would require 15 actuators for complete controlability [3]. The current state of technology, however, makes it impossible to design multi-fingered hands that offer adequate power and have more than five directly coupled actuators. With the Belgrade-USC hand, the kinematic complexity was reduced by incorporating joint coordinating mechanisms into the fingers such that the fingers follow the pattern of human finger joint activation when enclosing a target object. When the knuckle joint of a finger is moved, the other two joints will move as well (see Fig. 3).

The four d.c. servomotors used as actuators are located inside the wrist of the hand. The speed of the motors (max. voltage 36 V, max. power 6 W) is controlled by varying the voltage applied. A four channel D/A converter is loaded with an appropriate bit combination through the parallel port of a PC. The motors are directly connected to the fingers via small levers; there are no cables or tendons which would inevitably introduce mechanical hysteresis, dead zones, etc. A rocker arm applies the force generated by the servomotor to the finger and provides the compliant coupling between the two fingers of a pair (see Fig. 4). When one of the fingers touches an object, its motion is inhibited and the other finger continues to move at twice the original speed until it touches the object as well (Fig. 5).

348

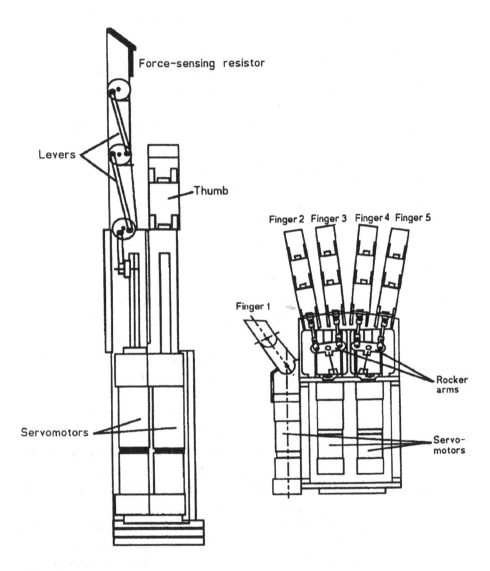

Figure 3: Coupling of finger segments

Figure 4: Mechanical structure and coupling of finger pair

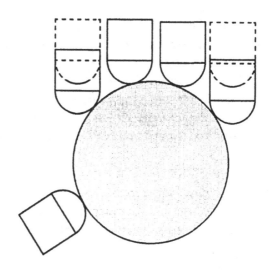

aption of the finger configuration to the target object's
pe

s that this mechanical structure is well suited to most grasp-
'ith target object sizes of the order of centimeters. There are,
large number of other tasks (e.g. turning a screw) that require
gree of dexterity than this hand can provide. The maximum
nd can carry is about 20 N; complete closure of the fingers
fully extended positions takes about 2 s.

Force Sensing

eshaping of the hand necessary for approaching the object is
erived from a two-dimensional image supplied by a compu-
ystem. During the grasping operation, control of the hand re-
sensory feedback and the autonomous shape-adaptation fea-
is purpose, the hand has been equipped with a set of internal
nsors for each finger pair and the thumb a set of external
s is mounted on the fingertips.
w finger speed does not require precise position control and
oviates the need for precision optical shaft-encoders. Instead

of shaft-encoders, high-resolution conductive plastic potentiometers are used. These potentiometers have a high life expectancy and they are small enough to be mounted inside the finger structure. Their resistance is proportional to the position of the knuckle joint and their output voltage may be fed directly into the position controller. The use of potentiometers offers additional advantages over shaft encoders: No switches are necessary to mark the initial position and the position of the fingers is always available even directly after an emergency shutdown.

The force-sensors are made of force-sensing resistors [4]. These resistors, which can be constructed in various shapes, consist of a compound of conducting and non-conducting particles suspended in a polymer matrix. The particle composition provides for a low temperature dependence. The force resistor changes its resistance in proportion to compression, i.e. in proportion to the force exerted on it. The range of forces that can be measured lies between 200 mN and 50 N.

A block diagram of the controller is shown in **Fig. 6**. It essentially consists of P-controllers for the position realized in software. The entire controller hardware fits on a standard PC plug-in board. The driver for the hardware was written in the "C" programming language.

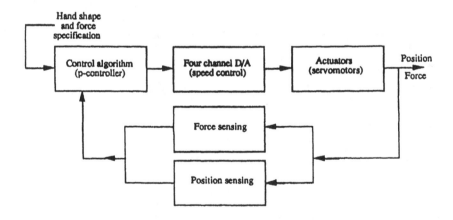

Figure 6: **Block diagram of the position controller**

Ultrasonic Sensing

To support the vision system used for pre-shaping the hand, an ultrasonic holographic sensor system has been developed [5]. This system - which is quite unlike laser-holographic systems - provides high resolution images of the object contour in two dimensions. Acoustical holography holds the potential for exploiting all information in both the axial and lateral directions, thereby making the generation of comparatively high resolution images of the object contour (in two or three dimensions) possible. This method has been used successfully in the area of nondestructive evaluation (NDE) as well as medical and marine applications for detecting and localizing objects. As will be shown below, it is also suitable for robotic applications.

The purpose of holography is the reconstruction of sound wavefronts based on the knowledge of the value of the sound field on a prescribed surface. Object recognition follows the reconstruction in the following manner: At all points where a sound source or a reflector is located, the reconstructed wave field takes on high values; at other points its value is low. Suitable thresholding therefore isolates the contours of the objects within the reconstruction region.

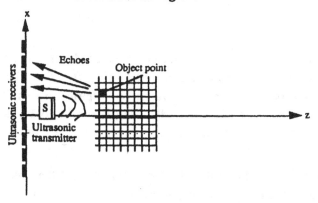

Figure 7: Basic setup for the acquisition of utrasonic echo data

The sensor consists of a single transmitter and several receivers along an aperture line (Fig. 7). At the beginning of an image generation, the transmitter emits a short sound impulse. The impulse is reflected by solid objects in the reconstruction area and travels back to the receivers

352

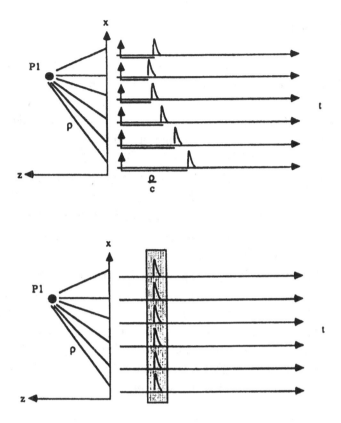

Figure 8: Principle of reconstruction. Measured delay (as denoted by impulses on t-axes) and delay calculated for P1 (denoted by bar below the t-axes) match. If impulses are shifted back by a time proportional to the length of the underbar and are summed up, a high value will result (grey box)

where it is recorded and stored. It can be shown [5] that the sound field at the time of recording can be computed for every point inside the reconstruction region if the time-signals arriving at the receivers are delayed by an appropriate propagation-time and added. Propagation time is the time it takes the signal to propagate from the object point to the receiving point.

The principle of the method is illustrated in Fig. 8: Suppose, a reflector at point P1 was "illuminated" by the ultrasonic transmitter and the reflection at P1 takes place at time t = 0. Then a pattern of impulse arrivals results on the aperture that uniquely identifies the echo as having come from point P1. To determine the spatial origin of the impulse as being P1, all recorded signals are shifted by a delay, proportional to the distance between P1 and the points of the aperture. Then they are summed up for t = 0 (see the grey box in **Fig. 8**). A high resulting amplitude indicates that there had been a reflection at P1. If, on the other hand, the impulse came from a different point, the shifts made for P1 will not match the actual delays, and the sum will be low.

In practice, the time that elapses between the emission of the impulse and its arrival at the point for which the sound field is computed must also be taken into account. It is obvious that the computation is not possible for all points in space; reconstruction must instead be limited to points within a suitably discretized grid. The granularity of this grid depends on the processing power available and on the ability of tranducers to discriminate between pulses arriving within very short time intervals, i.e. the bandwidth of the transducers. It is important to note that the evaluation of the sound field may be carried out for all points in parallel, a necessary precondition for very fast recognition. Given a powerful parallel processor, recognition time will only depend on the propagation delay of sound in air.

The Ultrasonic Transducer. We will now briefly describe one of the key elements of the transmission channel: The broadband ultrasonic transducer. Apart from a wide-band transfer function, development-goals for these transducers focused on providing a field of view that is sufficiently large to be used for the illumination of medium sized objects, and on making them rugged. The transducers realized (named L^2QZ) are composed of several alternating layers of electrically active piezoceramic material and elastomers (**Fig. 9**). The former provides for good electromechanical properties while the latter lowers the acoustic impedance of the entire element. The parameters of the plastic material were chosen so as to come close to the acoustic impedance of air, resulting in a high coupling factor. The active element is surrounded by damping material that, in combination with the plastic foils, lowers the Q-factor of the oscillating system and extends its bandwidth by flattening the transfer function. The low quality- (Q-)factor reduces efficiency, but the good coupling between oscillator and air due to the low acoustic impedance compensates for this reduction.

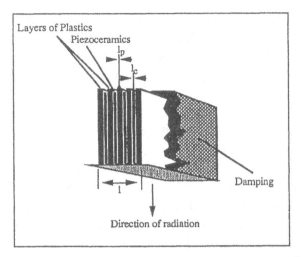

Figure 9: Schematic representation and dimensions of L^2QZ transducer. l_c is the thickness of piezoceramic layer, l, the thickness of plastics layer. Element width l is 5 mm

The operating frequency of the transducer is $f_0 = 200$ kHz. The bandwidth obtained is as large as 200 kHz , which must be compared to standard transducers for air, consisting of a monolithic block of piezoceramics, whose bandwidth is well below 10 kHz. An operating frequency of 200 kHz was selected as a compromise between attenuation in air and the minimum size of detectable objects. The frequency can be increased if the mechanical dimensions of the active element are reduced. **Fig. 10** shows a real output signal of the receiving transducer and the result of signal preprocessing. Two reflectors spaced 2 mm apart in the axial direction served as the object, which reflected a sound impulse emitted by a L^2QZ transmitter.

Since the transducers have a very low efficiency, their output signals are very small and must therefore be filtered before being processed further. Signal preprocessing starts with a correlation of the receiver output with a calibration echo obtained from a calibration reflector. The correlation is realized by convolving the receiver output and the calibration echo. This operation frees the output signal from noise and, most importantly, detects faint echos. In a second step, the signal is demodulated by taking absolute values of the convolved signal. Filtering away all frequencies by a low pass filter yields an output signal as shown in **Fig. 10**. Finally, thresholding cuts away residual noise. The level of the threshold is

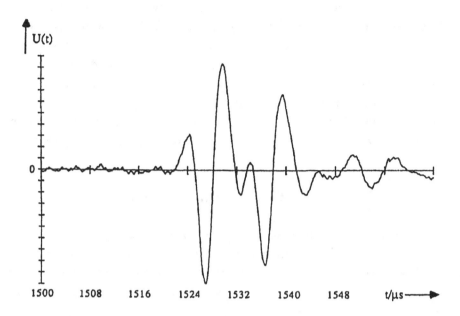

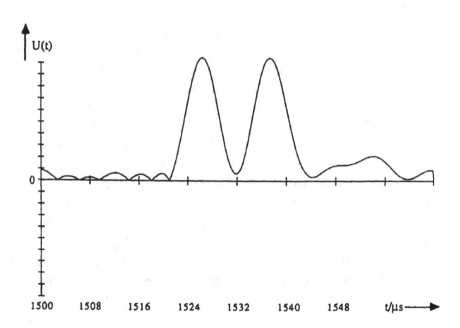

Figure 10: Top: Output signal of L²QZ transducer. Signal returned
from two slabs 2 mm apart in depth direction. Bottom:
Signal after pre-processing

generated automatically by observing the receiver output at times where the absence of echoes can be guaranteed. This dynamic adaptation makes a correct recognition of echos possible even in the presence of ambient noise. In this case, the threshold level is increased automatically which results in a reduced sensitivity to noise. It was shown experimentally that axial resolutions of 2 mm and absolute precisions of 0.5 mm are attainable. A necessary presupposition for measurement precision is, of course, that the propagation medium air is completely undisturbed.

Setup for Image Generation - Hardware and Software

The setup for holographic imaging is comparatively simple: A fixed ultrasonic transmitter continuously irradiates the objects. The transmitter is excited by a power oscillator generating an impulse peak of a duration $t_p \langle 10$ ms and an amplitude $U_p \rangle 200$ V. The set of transducers was of the aforementioned L^2QZ type. Since an array of transducers has not yet been available, it was simulated by a fixed transmitter and a moveable receiver.

The receiver was mounted on a moveable carriage; its height z_0 was adjustable. The carriage was actuated by a spindle drive under the control of a micro computer. The spindle drive required a variable frequency signal which was generated by a software process running on the control computer (see below). The variable frequency signal was fed into a power amplifier. The frequency range was 10...400 Hz; in order to move the receiving transducer between its positions, the frequency was increased continuously so as to obtain a constant acceleration until the receiver was halfway between the old and the new receiving position. After passing this point, the carriage was decelerated continuously and finally stopped.

Since the L^2QZ receiver provides output voltages of only a fraction of one millivolt, a special low noise amplifier operating at frequencies up to 1 MHz and matched to the transducer parameters had to be developed. The amplifier is connected to a one-shot transient recorder sampling the input signal at 8 MHz and digitizing it to 8 bits. The total size of memory is 32 kByte, thus the maximum time for filling the entire storage with data is 4 ms. Within this time interval ultrasound travels about 1.35 m, hence the round trip distance between transmitter, object point and receiver must not exceed this limit.

The structure of the software controlling the transducer drive, the transient recorder and the image generation was shown in Fig. 2. The software package is divided into five processes running in parallel. The programming language used was Occam-2, a language well suited to a medium-sized real-time software system such as ours. We will now briefly describe how our software was laid out, because to this day, the mapping of technical processes to software processes, i.e. the derivation of the structure of a software package from the real-world problem to be solved, is still very much dependent on the intuition and the experience of the designer:

For lack of a good theory, there are a number of "rules of thumb" to design the structure of a software system; one of which is to map real world entities in a one-to-one fashion to software processes. In our case this implies

- the realization of a process for the spindle drive (*StepperMotor*),
- a process controlling the A/D-converter and the transient recorder (*DataAcquisition*),
- a process storing and preprocessing the data of a single shot (*Preprocessing*) and
- a process that compiles the output image from the preprocessed data after enough data have been acquired for a complete image.

These four processes are triggered and controlled by a master process (*Control*). Let us look at the interaction between the processes in our experimental setup: The user interacts with the process *Control*. *Control* prompts the user for the necessary parameters. These parameters are the aperture width, the receiving positions spacing, the amplifier gain, etc. From these input parameters *Control* determines the absolute positions of the receiver carriage and transfers them to the process *StepperMotor*, which acknowledges the reception of the data. Based on these parameters, *StepperMotor* calculates the number of steps as well as the motion speed of the stepper motor and starts the motor. Once the receiver is in its first position, *StepperMotor* sends a *ready* signal to the process *DataAcquisition*. This process, if previously initiated by *Control*, triggers the transmitter to produce a short sound impulse. It waits some time and then triggers the transient recorder. As soon as the transient recorder has all the data, they are transferred to an internal memory section of *DataAcquisition* and from there to the process *Preprocessing*. After *Preprocessing* has completed its operations on the data, i.e., filtering and converting to a format suitable for the next step of compiling the image, it sends the data to the process *ImageGeneration* and transmits an acknowledgement to *Control* that the next shot is to be taken. The whole procedure repeats until all shots are recorded, i.e. the receiver is on its last position on the aperture. Then, the process *ImageGeneration* is initiated. Based on the preprocessed data, it generates the ultrasonic image to be rendered by the graphic display. On completion, all processes return to their idle state.

Experimental results

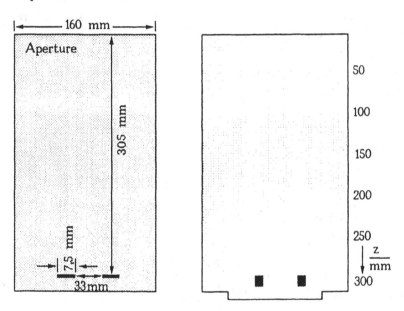

Figure 11: Geometry (left) and reconstruction result (right) for dual reflector scene

To explore the potential of the holographic approach, several objects of different surface structure were imaged. Reconstruction was carried out on a grid consisting of rectangular elements whose dimensions were chosen differently, depending on the desired maximum resolution. The sound field was computed for each element, and a suitable threshold was applied to it. If the sound field exceeded the threshold, the raster element was marked as belonging to the object contour. Figs. 11 and 12 show the results for two different scenes consisting of simple objects. In **Fig. 11** two plane objects 7.5 mm wide were imaged using a quadratic element size of 10 x 10 mm^2 for reconstruction. The resulting picture shows that the object position is recognized correctly within the chosen resolution. limit both in axial and the lateral direction. The reconstruction of the object of **Fig. 12**, which was turned into an oblique position with respect to the direction of sound propagation, also yields a correct image. The raster element size was reduced to 5 x 5 mm^2 here. Given the small aperture of **Figs. 11** and **12**, an increase of the inclination of the object will direct most of the reflected signal energy past the aperture and produce a blurred image. If greater inclinations are to be recognized as well, the aperture width must be increased.

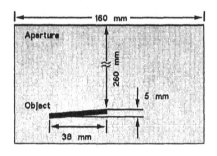

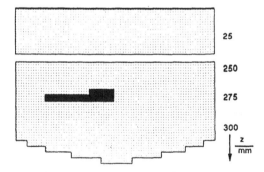

Figure 12: Geometry (top) and reconstruction result (bottom) for oblique reflector scene.

360

In both cases echo data were acquired using only one transmitter and by recording at 17 locations where the distance between the location was 10 mm. Other examples of results are shown in Figs. 13 and 14. Here, a segmented approach was used, i.e. several images were generated based on echo data acquired from different transmitter positions. The resulting reconstructions were superposed. The spacing of transmitter positions was d = 10 mm. The solid line indicates true object contours. Apart from some overshots in the lateral direction, the surfaces that are visible to the system are recognized accurately. Note, the varying resolution scale in

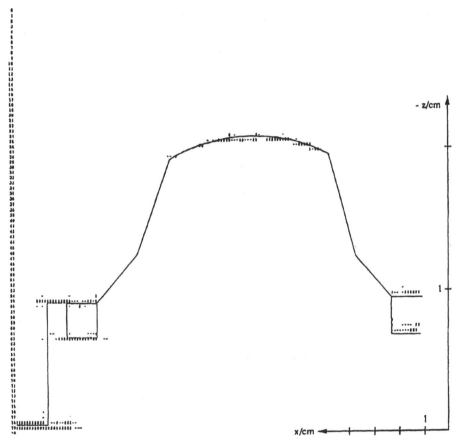

Figure 13: Reconstruction result of multifrequency holography, segmented data acquisition. Distance between transmitter positions is 10 mm. Solid line indicates true object dimensions. Character distance marks a lateral distance of 1 mm. Numbers on the left side indicate rows of reconstruction raster. The distance between rows is 0.375 mm

each of the figures. The lateral overshots can be reduced at the expense of data acquisition time if segment width is reduced. The images in Figs. 13 and 14 result from "real-world" objects used in automotive industry.

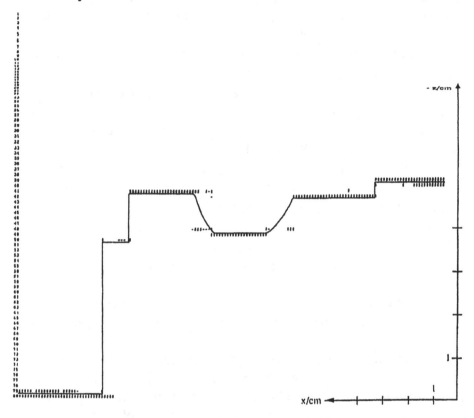

Figure 14: Reconstruction result of multifrequency holography, segmented data acquisition. Segment width is 10 mm. Solid line indicates true object dimensions. Character distance marks a lateral distance of 1 mm. Numbers on the left side indicate rows of reconstruction raster. The distance between rows is 1.25 mm

Data Processing on a Digital Signal Processor

The reconstruction process takes about 500 s on a standard PC. To reduce reconstruction time, a special PC board has been developed that incorporates a transient recorder and a digital signal processor (see **Fig. 15**). The sampling rate of the transient recorder is 20 MHz, its storage capacity is 128 kBytes. The clock frequency of the signal processor is currently 30 MHz, it may be increased to 40 MHz, if necessary. Using this PC board, reconstruction time can be reduced to approximately 2 s per image (which is the time necessary for the complete closure of the hand) but only at a reduced resolution. For high resolution images, the time needed by the board is well above 10 s.

To alleviate the problem of writing efficient algorithms for digital signal processors (DSP) without having to go down to the assembler level, a special language called ImDisp has been designed [6]. We will use this language in the future for specifying our algorithms for filtering, image generation, etc. It belongs to the family of imperative languages (such as Pascal or C) and has been designed with two goals in mind:

- The special functionality of widespread digital signal processor architectures should be accessible to the high-level language programmer.
- The language should provide constructs that permit the specification of typical DSP algorithms without sacrificing potential inherent parallelism of calculations, i.e. it should not force the programmer to sequentialize computations only because of constraints imposed by the language (as is usually the case with programs written in C or Pascal).

Like every modern imperative language, ImDiSP is strongly typed and offers all the control constructs necessary for structured programming (sequence, iteration, selection) as well as procedures and blocks for structuring the program text. The basic data types are boolean, integer, real and complex. These types may be used to form arrays and records. The expressive power of the language is based on its set of operators manipulating array structures, the predominant data type in DSP applications. All standard arithmetic operators are overloaded: They do not only operate on integers and reals but also on complex numbers, arrays, slices (sub-arrays) and matrices composed of the base types. A simple example is the multiplication of two vectors: Let a and b be vectors of the same dimension. Then, c := a*b multiplies these vectors element by element creating a new vector which is assigned to the variable c (also of type vector).

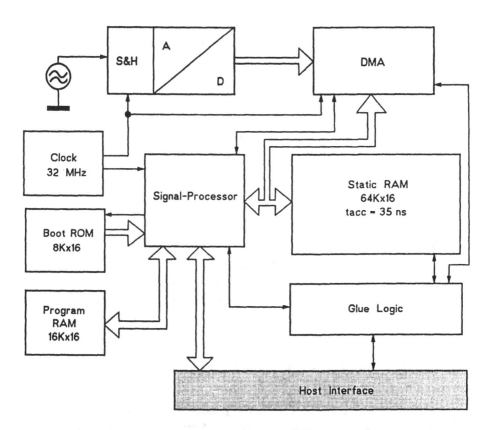

Figure 15: Block diagram of a DSP-based acquisition and processing unit

Like in APL, it is possible to combine "inner" and "outer" operations using a dot notation. For example, the dyadic operator ! . *, applied to two vectors (as in c := a +.* b), will first multiply the vectors element by element and subsequently sum up the products to yield a result of the base type of the vectors. Using this technique of combining operators, an expression like

$$y = \sum_{i=0}^{N} h_i x_{N-i}$$

is easily translated into the following code segment:

```
y := x[^ .. 0] +.* h
```

where " ^ ", used as an array subscript, denotes the upper bound of the array. The difference equation of a recursive filter

$$y_n = \sum_{\upsilon=0}^{N} a_\upsilon x_{n-\upsilon} - \sum_{\mu=1}^{M} b_\mu y_{n-\mu}$$

with coefficients vectors a (dimension N+1) and b (dimension M) can be realized in just a few lines of code:

```
loop

    x[^] := input;    -- get input value
    y[^] := a +.* x[^.. 0] - b +.* y[^-1 .. 0];-- do the filter
    output := y[^];   -- output result
    delay (x, 1);     -- shift contents of array x by one element
    delay (y, 1);     -- shift contents of array y by one element

end loop;
```

where input and output are I/O port addresses.

The standard procedure delay pushes the contents of an array down by an arbitrary number of elements leaving the top elements (high index) undefined. This procedure is normally applied to objects of a special one-dimensional array type circular array that implements circular buffers of any base type. If certain conditions are met, the compiler makes use of the modulo-n addressing mode (a feature available on most signal processors) when delaying the array contents. This results in very low overhead code. Other standard procedures of the language are biquad, used for realizing cascade filters and the procedure butterfly that implements DIT/DIF butterflies and the necessary address calculations.

Processor features directly accessible to the ImDiSP programmer are saturation arithmetic (which may be turned on and off before and after any instruction by means of a pragma (metacommand)) and hardware do-loops. Bit-reversed addressing of array elements is possible. Operators are provided that permit the uniform application of a certain operation to all elements of an array, the extraction of matrix diagonals and the determination of the smallest or greatest element of an array. To allow procedures to have state, all variables declared local to a procedure are static by default, i.e. their value does not change between calls. A pragma exists that can make them volatile which means that they become undefined upon exit of the procedure. Volatile variables can be held in registers; this removes the need for memory fetches and thus makes execution much faster.

A future extension of the language will include representation clauses of data types that direct the compiler to internally represent data types according to the specification of the programmer. If it turns out to be necessary, pointer types and associated operators will be added to the language. It is not intended to introduce language constructs supporting programming in the large (module, import, export statements, etc.) because ImDiSP is primarily intended for defining algorithms whose program size always remains on a relatively small scale. Therefore, procedures are sufficient for structuring the program.

Using ImDiSP, it is possible to specify algorithms involving structured data in a very compact form leaving it to the compiler to serialize the code as far as necessary. This way it is obviously much easier to generate efficient code than by usage of a notation that forces the programmer to write down sequential steps.

CURRENT STATE OF THE PROJECT AND SUMMARY

Both, the mechanical section of the hand and the ultrasonic sensor have been tested successfully. However, even though a high-speed DSP was used, the time required for the generation of a single picture is still above 10 s. This adds to the data acquisition time, which, due to the slow stepper motor, lies well above 30 s. The resolution of the ultrasonic sensor, however, is more than adequate for common grasping tasks. In addition, the position and force sensors also work well.

The ultrasonic holography approach offers several advantages over optical vision systems including its use of simple recording devices, low hardware cost, and its inherent ability to determine range. The latter obviates the need to extract the depth dimension from its two- dimensional projections. Since the image is focused synthetically in the computer memory, resolution is uniform over the entire depth range. However, due to the specularity of ultrasonic reflections, parts of an object visible to an optical system may not be detectable by the ultrasonic sensor (and vice versa). Therefore, a combination of an ultrasonic holographic sensor and a camera vision system lends itself to all tasks requiring high accuracy in the lateral direction.

To speed up the operation of the sensors and to further exploit the potential of the holographic approach, more work needs to be directed along the following lines:

- Development and construction of an array of efficient broadband transducer elements.
- Dedicated hardware designed to make use of parallelism inherent in reconstruction formulae.
- Fusion with sensors based on principles other than sound wave propagation.
- Development of a new floating-point DSP board to drastically reduce image generation time.

Finally, the whole circuitry should be miniaturized. This will not present a problem as the power of DSPs increases and the packaging volume of high speed static RAM decreases.

REFERENCES

[1] R. Tomovicz, G. Boni, "An Adaptive Artificial Hand", *Trans. IRE*, AC-7, 1962

[2] G. Bekey, R. Tomovic, I. Zelkovicz, "Control Architecture for the Belgrade-USC Hand", *Dextrous Robots Hands* (Eds.: S. Ventakataraman, T. Iberall), Berlin-Heidelberg-New-York, Springer-Verlag, 1989

[3] S. Jacobson, E. Iversen, D. Knutti, R. Johnson, K. Biggers, "Design of the Utah/MIT Dextrous Hand" *Int. Conf. on Robotics and Autom.*, 1986

[4] Anonimus, "The Force Sensing Resistor: A New Tool in Sensor" Technology", *Interlink Inc., Data Sheet*

[5] A. Knoll, "Akustische Holographie - Ein Hilfsmittel zur Bestimmung der räumlichen Position von Objekten in der Robotik" *Robotersysteme* 4, pp. 193 - 204, Heidelberg, Springer-Verlag, 1988

[6] V. Kruckemeyer, A. Knoll, "Eine imperative Sprache zur Programmierung digitaler Signalprozessoren" *Technischer Bericht 90/10 des Fachbereichs Informatik der Technischen Universiät Berlin*

14

FLY-BY-WIRE SYSTEMS FOR MILITARY HIGH PERFORMANCE AIRCRAFT

D. Langer, J. Rauch, and M. Rößler
MESSERSCHMITT-BÖLKOW-BLOHM GmbH
Aircraft Division
D-8000 München 80

INTRODUCTION

In the early times of aviation, aircraft flight controls were purely mechanical systems comprised of pulleys and cables deflecting the aircraft's control surfaces about their hinges. The control stick/pedal inputs applied by the pilot counteracted the aerodynamic forces acting upon the control surfaces.

During the years until 1960, the capabilities and handling qualities of flight control systems were enhanced by the introduction of hydraulic actuation systems and by including electronic processing of gyro-rate signals into the control loop as a stability augmentation system.

In the early 1980's, the first aircrafts incorporating Fly-by-Wire (FBW) systems were taken into service. With a FBW flight-control system, the pilot is completely mechanically decoupled from the control surfaces, which are responsible for guidance and control of the aircraft. The pilot's inputs to his control elements (stick, pedals) as well as the signals of the aircraft's sensors (e.g. linear and angular movements/accelerations, static and dynamic pressure, airstream directions) are converted into electrical signals and sent to the central flight control computer. All inputs are processed by the control law algorithms, resulting in electrical output signals to the actuators.

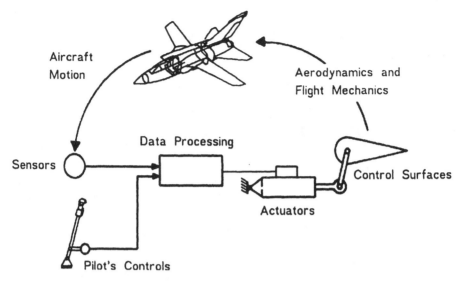

Figure 1: Schematic of a Fly-by-Wire flight control system

Figure 1 shows a schematic view of a FBW flight control system.

For conventional aerodynamically stable aircrafts, usually a mechanical backup system is incorporated to provide a flying home capability in the event of fatal failures in the FBW system. In this case, the pilot has the possibility to control the essential surfaces via mechanical links. During the last decade, in the development of modern high performance fighter aircrafts the fundamental principle of flight mechanical stability was left towards a dynamically unstable design offering the benefit of improved performance and high manoeuvrability. The required levels of aircraft stability must by achieved by artificial stabilization. This results in the ACT (Active Control Technology), i.e. integration of aerodynamic configuration and flight control system in the very beginning of the design.

Full-time FBW flight control systems are mandatory for artificially stabilized aircrafts. As these aircrafts are not flyable without the artificial stabilization, no mechanical backup system can be provided. Examples of modern aircrafts controlled by full-time FBW systems without mechanical backup are the European Fighter Aircraft EFA (European Fighter Aircraft) presently being developed by a British/German/ Italian/ Spanish industrial collaboration, and the Airbus A320.

The basic requirement, the fulfilment of which must be proven to the aviation approval authorities, is an aircraft loss rate due to flight control system failures of less than

- approx. 10^{-7} per flight hour for military aircrafts, and
- approx. 10^{-10} per flight hour for civilian aircrafts.

Consequently, the design and development of the FBW systems are dominated by these very stringent safety requirements. The use of three to four redundant processing channels including appropriate voting and monitoring algorithms is common to all modern aircrafts.

A key role in the development phase belongs to design, implementation and testing of the software for the flight control computer. Due to the complexity of the flight control software, which is being processed by a multiprocessor system in real time, it is mandatory, to

- use structured design methods,
- limit the size of software modules,
- limit the allowed software language constructs

in order to enable a thorough testing during the aircraft development phase. Widely used in flight control system applications are dissimilarity methods. These range from dissimilar testing, i.e. testing of the actual system implementation versus a dissimilar model, to the use of different hardware and software implementations for the redundant channels of the flight control system.

FUNDAMENTALS OF FLIGHT CONTROL SYSTEMS

For an understanding of the structure and complexity of the control algorithms the theory on flight control systems will be briefly summarized.

Consider **Fig. 2** which shows an aircraft with associated coordinate systems, components of the state vector and control inputs to the control system. It should be noted that throughout this paper, variable names and indices are used according to air vehicle specification LN 9300 (Luftfahrtnorm = Air Vehicle Specification) [1].

Let a body coordinate system (BCS) with axes X, Y, Z be aligned with the principal axes and originating in the center of gravity (c.o.g.) of the aircraft. Furthermore, let an inertial coordinate system (ICS) with axes X_g, Y_g, Z_g be located with its origin in the c.o.g. of the aircraft. Let the state vector of the system consist of components of translational and rotational motion of the aircraft. Let the components of translational motion of the state vector be the vector $\vec{r}$ with components (x, y, z) in the BCS and vector $\vec{v}$ with components (u, v, z) in the BCS of the aircraft. Similarly, let the components of rotational motion of the state vector be the vectors $\vec{\Theta}$ and $\vec{\Omega}$.

The vector $\vec{\Theta}$ has as components the Euler angles which describe the orientation of the BCS relative to the ICS (Inertial Coordinate System). The vector $\vec{\Omega}$ has as components the angular velocities in the BCS (Body Coordinate System) (p, q, r).

δ_A: Angle of deflection of aileron
δ_B: Angle of deflection of dive brakes
δ_E: Angle of deflection of elevator
δ_F: Angle of deflection of flaps
δ_R: Angle of deflection of rudder
δ_{RPM}: Change of power plant revolutions per minute

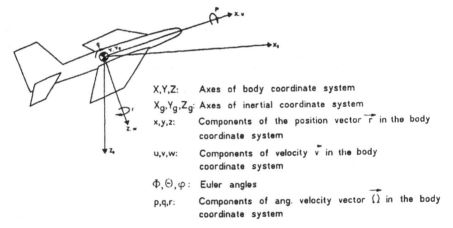

X,Y,Z: Axes of body coordinate system
X_g,Y_g,Z_g: Axes of inertial coordinate system
x,y,z: Components of the position vector $\vec{r}$ in the body coordinate system
u,v,w: Components of velocity $\vec{v}$ in the body coordinate system
Φ,Θ,φ: Euler angles
p,q,r: Components of ang. velocity vector $\vec{\Omega}$ in the body coordinate system

Figure 2: Coordinate systems and components of velocity and angular velocity vectors of an aircraft

The controls of the aircraft consist of the surface deflections $\delta_A, \delta_B, \delta_E, \delta_F, \delta_{RPM}$, which represent the angle of deflection of aileron, air brakes, elevators, flaps, rudder and the engine input δ_{RPM} representing the change of engine's revolutions per minute.

The assumptions for the derivation of the dynamics of the aircraft are that:

- the aircraft is a rigid body,
- the air flow along the aircraft is quasi-steady, and
- the mass of the aircraft remains constant for the duration of the dynamic analysis.

It is shown elsewhere [2] that the equations of dynamics of the aircraft can be written in vector matrix form as:

$$\dot{\underline{X}} = f_1(\underline{X}, t) + f_2(\underline{X}, U, t) \qquad (1)$$

where $\underline{X} = [xyzuvw\phi\Theta\Psi pqr]^T$ is the state vector

and $\underline{U} = [\delta_A\delta_B\delta_E\delta_F\delta_{RPM}]^T$ is the control input vector of the control system, and
$f_1(\underline{X}, t), f_2(\underline{X}, \underline{U}, t)$, are matrices of appropriate dimensions [2].

By means of a linearization process [2], the nonlinear equations of motion of the aircraft are linearized about an operating point, thus, producing the linearized equations of motion:

$$\delta\underline{X} = f_3(\underline{X}, t)\delta\underline{X} + f_4(\underline{X}, \underline{U}, t)\delta\underline{U} \qquad (2)$$

where
$f_3(\underline{X}, t), f_4(\underline{X}, \underline{U}, t,)$ are matrices of appropriate dimensions [2]. For the state and input vector of the nonlinear, Equ. (1), and the linear equations of motion, Equ. (2), the following relationships hold:

$$\underline{X} = \underline{X}_0 + \delta\underline{X}$$
$$\underline{U} = \underline{U}_0 + \delta\underline{U}$$

The quantities $\underline{X}_0, \underline{U}_0$ are constants which satisfy Equ. (1) at the operating point. Therefore, they are referred to as steady state values. The quantities δX and δU are mathematically infinitesimally small deviations from the operating point and are sometimes referred to as variational quantities.

Feedback Control

Control systems design methods are applied to the linearized equations of motion, Equ. (2), to generate a control algorithm that satisfies prespecified goals which are derived from the requirements outlined in the following section.

Furthermore, stability checks are applied to Equ. (2) to ensure that the linearized system remains stable when controls are applied to it. This implies that the corresponding deviations δX, δU from the respective steady state values are driven to zero in finite time.

If the control algorithm meets the design requirement for the linearized system it is, subsequently, applied to the nonlinear system to maintain it on a defined operating point and to drive disturbances about this point to zero in finite time, as well.

As the controller is designed for the linearized system but applied to the nonlinear system, it is mandatory to verify the envelope within which this assumption holds. In order to do this computer simulations, rig tests and, finally, flight tests are performed to assure a correct performance and local stability of the controlled system under test.

The choice of an operating point discussed above is such that every mode of the flight control system is associated with one or more operating points. The control laws corresponding to a mode of operation are, in general, independent of each other and can, therefore, be processed in parallel. Switching from one mode of operation to another is a phase of concern to the control engineer and is tested as rigorously as the performance of the controlled system mentioned above.

REQUIREMENTS ON FLIGHT CONTROL SYSTEMS

The design of the flight control computing system of a modern FBW high performance aircraft is driven by three factors:

- number and complexity of functions to be provided
- dynamics of the process (i.e. aircraft) to be controlled
- safety of flight

The number and complexity of functions and speed with which they have to be processed define the total amount of computing power (i.e. throughput) required. The dynamics of the process determines the sampling rates and the allowable computational delay. Given a certain type of processor with a known performance the number of parallel processes can be derived. To achieve the required safety of flight redundant computing channels have to be applied. The related redundancy management (data exchange between redundant channels, voting and monitoring), which has to run at the same speed as the functions above, is a considerable additional burden for the flight control computing system.

Functions

Modern FBW Flight Control Systems have to provide the following functions:

- **automatic modes** without manual control from the pilot, which keep the aircraft on a trajectory or in a certain attitude defined by the commands of sensors in the avionic system (e.g. terrain following).
- **basic manual mode:** In this mode the aircraft follows the pilot's commands via the stick, pedal and switch inputs. The pilot is not burdened with the task to monitor the structural and aerodynamic limits of the aircraft the exceedance of which can either damage or break the aircraft's structure or lead to departure, which would make the aircraft uncontrollable. Both cases are flight safety critical. Without any external interference the flight control system automatically optimizes the aircraft's performance (e.g. establishes trim conditions with minimum drag) and enables the pilot to fly the aircraft over the full speed and altitude range (flight envelope). Associated with this mode is the highest level of handling qualities, a measure of how well the aircraft can be controlled by the pilot.
- **degraded modes:** In these modes the aircraft is also controlled manually by the pilot, but some of the functions above are lost leading to an increased pilot's workload, reduction of the flight en-

velope, non optimal aircraft performance and reduced handling qualities. However, all the functions have to be available to safely recover the aircraft from any manoeuvre and to land it.

The requirement that an aircraft must not be lost with a probability greater than 10^{-7}/Fh (Flight hours) due to any failure in the flight control system has not been changed when turning from mechanical to fly by wire systems. However, the number and complexity of the functions and the quality with which they have to be provided have steadily risen. Furthermore, the environmental conditions became more stringent.

As not all the functions need to be available through all the phases of a flight or a mission, the application of a principle called *graceful degradation* is allowed. This means that functions which are not required to safely recover and land an aircraft may be lost with a certain probability which is determined by mission requirements rather than safety requirements. However, the system has to detect that such a failure condition exists and has to turn the affected function into a fail safe state where it cannot adversely influence other functions.

To mechanize the above functions FBW Flight Control Systems have to perform the following tasks

- **Measurement of input data** (sensors' and pilot's control elements) including
  - inertial data i.e. angular rates and linear accelerations along the three axes and the Euler angles (pitch and bank)
  - airdata i.e. static and dynamic pressure
  - airstream direction angles
  - pilot's commands i.e. stick deflection in pitch and roll, pedal position, switch states.

    Although no stringent technical requirement, it is a fact that all the above sensor data are digitally processed near to the location of measurement and transmitted to the central computer via a standardized serial digital interface.

- Sensor error correction: This applies especially to the airdata and airstream direction angles. The serial digital outputs represent the condition at the measurement location which are influenced by the aircraft itself and its engine. A complex compensation has to be performed to obtain the required undisturbed parameters.

- **Control law computation** closure of the feedback loops to stabilize the aircraft, calculation of the feedback gains, command filtering and limiting. The very high degree of non linearity leads to complex control laws.
- **Control of the aircraft surfaces:** The primary control surfaces of the aircraft which have to be available throughout all flight phases are driven by two-stage hydraulic actuators. A modern high performance military aircraft can have up to 7 primary actuators. Each stage is controlled by a digital control loop. Other than for the sensors, local processing is not possible due to environmental conditions.

  In addition, there is the requirement to control a similar number of secondary actuators, which may be less complex due to reduced performance and availability requirements.
- **Control of the data flow:** In one channel data from the sensors have to be routed to the control law computing element and from there to the actuators. Data between redundant channels have to be exchanged at least before being input into a control law computing element and before being output to the actuators. The delay between sensor measurement and actuator movement has to meet the timing requirements of section Timing Requirements.

  Data also have to be exchanged between the flight control system and other aircraft subsystems (avionics, utilities control).
- **Redundancy Management:** After data have been exchanged between redundant channels a non faulty parameter from one channel has to be selected for further processing. This is done in a voting and monitoring process which is illustrated for a triplex system in **Fig. 3**. The voter has to select one of several inputs while the monitor has to identify the faulty input by means of a majority decision. A midvalue select voter computes its output by forming the average value of two midvalues while an averaging voter computes its output as the average value of all its corresponding inputs, see **Fig. 3**. After a permanent fault has been detected the voting and monitoring stage has to be reconfigured such that the faulty input is no longer considered. However, if a fault of short duration, like a transient, is detected a reconfiguration must not take place.

378

Midvalue Select Voter

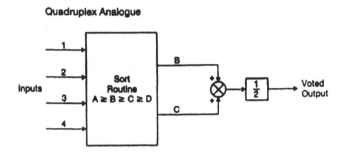

Monitor all VS Selected

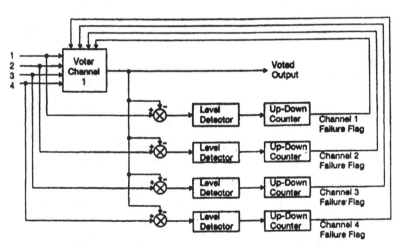

Figure 3: Example of a quadruplex voting and monitoring process

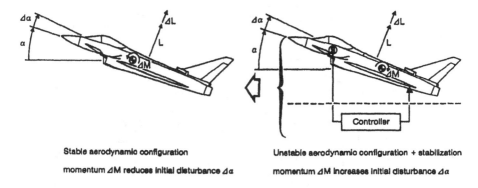

Stable aerodynamic configuration

momentum ⊿M reduces initial disturbance ⊿α

Unstable aerodynamic configuration + stabilization

momentum ⊿M increases initial disturbance ⊿α

Figure 4: Aerodynamic behavior of the airframe in the pitch axis and
the associated control system

Timing Requirements

The dynamic instability of the airframe which defines the timing
requirements of the flight control computing system is illustrated using
as an example the aerodynamic behaviour in the pitch axis, see **Fig. 4**.
If not stabilized by control loops, the angle between the aircraft's velocity vector and the fuselage (angle of attack) would increase in an exponential way. The related time constant, which also depends on the flight
condition, is for modern high performance aircraft of the order of several 10 msec up to several 100 msec. Using the design rule outlined in
section 4, the sampling frequency of the related control loop can be defined.

In order to meet the stability requirements, it is necessary that the
allowable phase loss from sensor measurement to aircraft control surface
movement is not exceeded. As the computing system contributes considerably to this phase loss (the other contributing factor is sensor and actuator bandwidth) due to transport and computing delays, an upper limit
must not be exceeded. Typical values for the basic stabilization loops are
listed in **Fig.5**.

380

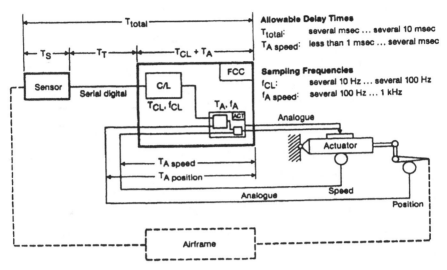

Figure 5: Typical values for basic stabilization loops

Usually it is required, that different processes in one computing channel are synchronized (intra lane synchronization). It is also required that the redundant computing channels are synchronized to each other (inter lane synchronization). Furthermore, in case that the redundancy management routes the output of a process in one channel to the input of the same process in another channel, the phase loss requirement above has to be met. Another reason for interlane synchronization is, that transients, caused by switching from one channel to another, have to be within defined limits. Interlane synchronization is not required if the output of one process is updated at a rate, which is considerably faster than the sampling frequency of the process downstream.

The timing requirements above apply to the basic stabilization loops of a fly by wire aircraft. Part of this loop are the processes sensor measurement and actuator movement which in itself contain control loops. The bandwidths of those loops are several times higher than those of the basic stabilization loops of the aircraft. The sensor processing is done outside the flight control computer and is not considered here. For the primary actuators the position and the speed control loop have to be closed in the flight control computer. The bandwidth of the position control loop is the same as that of the basic stabilization loop of the aircraft while the bandwidth of the speed control loop is several times higher. Furthermore, the secondary actuators, which usually have only one position loop, are controlled by the flight control computer.

In addition to the control loops above there are processes in a flight control computer which run at a lower iteration rate than the basic stabilization loops (e.g. gain scheduling with airdata). To perform these tasks and to ensure that the flight control computer has a deterministic behaviour it is required that the tasks are organized in a rigid time frame.

In order to provide the functions of the former section (Functions) and to meet the timing requirements above a structure has to be established which allows parallel processing. For modern flight control applications, the required throughput for each computing channel including a growth potential of 50% is equivalent to that of two to five MOTOROLA 68020 microprocessors.

ARCHITECTURE

In the design of digital flight control systems (DFCS) the sample rate of the overall system, i.e. the rate by which a control systems' output is computed as a function of its inputs, is a significant constraint. The sample rate influences the performance and, thus, the quality of the control applied to the aircraft. Furthermore, it constrains hardware and software architecture of the computer system which executes the control algorithms.

For a first, rough design of the DFCS, the sample rate of the overall system can be estimated based on the sampling theorem and on stability considerations. The sampling theorem states that the minimum sampling frequency (ω_s) of a digital control system should be at least twice as high as the highest frequency components (ω_c) within the system.

$$\omega_s \geq 2\omega_c$$

Another bound on the sampling rate is derived from results of stability checks applied to the closed loop control system.

Since the frequency components (ω_s) of the DFCS are related to the fastest eigenvalues of both the control system and the aircraft, lower bounds on computer speed and computer throughput can, thus, be provided.

A flight control system, which consists of several control loops, has loops which experience higher signal variations than other loops do. Consequently, control loops with high signal variations have to be sampled with a higher sampling frequency, while it suffices to sample the remaining control loops at lower rates. This introduces the concept of a multirate digital control system. Multirate DFCS are mathematically more difficult to treat than systems with a single sample rate. However, the computational load of the respective flight control computer is significantly reduced over that of an single sample rate system with identical control tasks.

Task Dedicated Processors

The hardware architecture and respective design considerations are discussed below. An analysis of the control effort, see the former section (Requirements on Flight Control Systems), reveals that the computational load is too high to be performed by a single processor at current technology level. Therefore, a multiprocessor system is required for control law computations.

Tasks are allocated to individual processors based on the following considerations. The structure of the control algorithm, see the former section (Requirements on Flight Control System), is such that translational, rotational, and air data computations are decoupled. Furthermore, the pitch and lateral components of the rotational control are loosely coupled. The control algorithms associated with the actuators, which act on the control surfaces, have to be sampled at a considerable higher frequency than the remaining controls. This suggests a dedicated processor for this type of computation. Finally, a separate processor is required for input/output operations and one or more processors for growth potential. The resulting multiprocessor system with dedicated processors is shown in Fig.6.

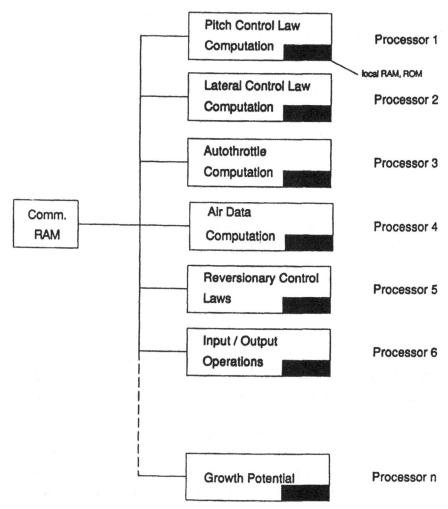

Figure 6: Multiprocessor system with dedicated processors

Task Scheduling

In the current software architecture a deterministic nonpreemptive scheduling mechanism was favoured over a stochastic one. In deterministic scheduling all tasks and subtasks of the flight control system are exactly timed and sequentially processed by a rigid controller executive until completion. This allows predictable execution sequences at all times and provides for ease of integration, testing, tracing, verification, and validation.

In stochastic scheduling, however scheduling algorithms have to be defined to ensure [3]

- a prioritization of tasks,
- a synchronization of dependent tasks,
- the prevention of deadlock conditions,
- the prevention of livelock conditions and
- the prevention of race conditions.

The above mentioned scheduling algorithms have to be tested in addition to performance and operational requirements of the flight control system. Hence, the overall complexity of a DFCS (Digital Flight Control System) with stochastic scheduling is much higher than one with deterministic scheduling mechanisms.

Timing and Synchronization

The timing of tasks is done by means of a timer module, Fig. 7. The timer comprises two cascaded binary counters, clocked from a precision frequency source. The first counter times a sub-frame or segment and then resets itself. The reset pulse is exported as a hardware timing reference, SMR (Sub Master Reset). The second counter counts the number of segments. On the completion of the last segment, this counter resets itself. This reset pulse is exported as a hardware timing reference MR (Master Reset). The segment count is driven by the frame timer and changes its value as a new segment commences.

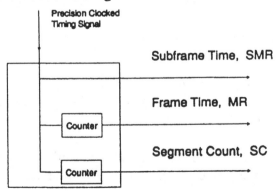

Figure 7: Illustration of a timer module

The frame time can be increased resp. decreased to synchronize itself with other, redundant flight control computers. The timing and segment count information is given simultaneously to all processors of the flight control computer to provide a unique timing reference.

Multiprocessor Structure

A multiprocessor system with task-dedicated processors is shown in Fig.6.

The architecture is characterized by a collection of cooperatively autonomous processors. These processors are based on the local processing principle, i.e. local RAM and ROM (Read Only Memory) areas are firmly allocated to one CPU (Central Processing Unit) and are not addressable by external processors. This implies, that software errors, which are generated in one processor, do not propagate to other processors within the system.

Interprocessor communication is done by means of a global communication RAM, which is accessed over a common bus.

This common bus typically consists of a 16 bit multiplexed address/data bus supported by control signals and a clock. This represents a more general and neutral design than a separate data and address bus.

The communication with other modules is software transparent to the processors, since the global communication RAM (Read and Write Memory) space is mapped into the processors' local address space. Although the use of a multiplexed bus requires slightly more control circuitry, this allows a reduced requirement for buffer devices, which in turn reduces the computers power consumption.

Only one module may drive the common bus at any one time, though one or more modules may sink the information (address or data) driven onto the bus.

Common bus activity is controlled by a bus controller. The bus controller ensures that each module requesting the bus is granted its use, in turn, according to a predefined order programmed into the bus controller.

Each common bus transaction consists of two bus cycles, an address cycle followed by a data cycle.

386

The general architecture of the internal bus is shown in Fig.8.

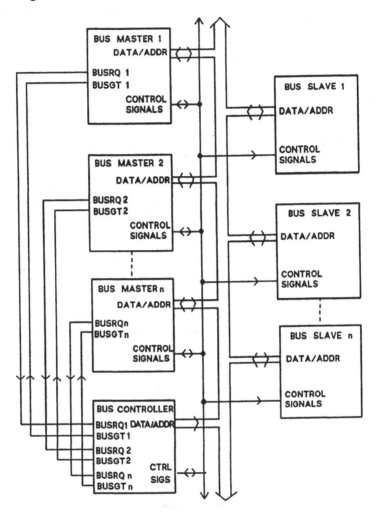

Figure 8: Common bus and associated bus masters and bus slave modules

Processor modules, identified as bus masters, are required to read as well as update (write) global communication memory. For either of these operations the processor, once granted the common bus, drives the address of the particular memory location onto the bus. The bus master also identifies the type of data transfer by driving the respective control signals to appropriate states throughout the address and subsequent data cycle.

The memory mapped card module, called bus slave, containing the addressed location is activated upon recognizing that the bus address lies within its assigned address range. The bus slave latches the bus address at the end of the address cycle allowing the memory location to be addressed throughout the data cycle. For a data write operation the bus master outputs data either byte-size or word-size to the common bus. The activated bus slave latches the data into its memory location, subsequently.

As soon as the two-cycle transaction is complete, the bus controller grants the bus to another requesting bus master and the common bus control process is repeated.

In order to optimize global communication, all write operations to global memory area are buffered, to prevent the bus master processor from being forced into a waiting state, until the data are transmitted. In case of a read operation, the processor must enter a wait condition, until it is allocated a time slot on the common bus.

The number of common bus accesses can be minimized by buffering all global data required by a processor into its local memory locations. All subsequent accesses to the data will use the copies in local memory and thus avoid incurring further access to the common bus. The processors data buffering tasks will be split into several small tasks throughout the frame. The data buffering tasks of each of the processors will be arranged at different times in a frame in order to avoid common bus clashes.

Software Considerations

Digital data processing, in general, is subject to quantization errors. These errors occur, if an analog signal is represented by a digital quantity with finite resolution. This is especially true for digital flight control systems, where a part of the physical states of the aircraft are converted from analogue to digital data format by means of an A/D converter (Analog to Digital Converter). Subsequently, these data are processed by a digital computer and, finally, signals are output, which cause a continuous deflection of the respective control surfaces of the aircraft. Quantization errors due to the number representation in digital computers can be reduced by using a floating point number representation over integer arithmetic. However, this is done at the expense of speed of computation which in turn is an important design constraint of real-time processing.

In the present system, Fig.6, those processors which perform control law computations are using floating point number representation and are augmented with a floating point processor, while the remaining ones use integer arithmetic to increase processing speed. An exception is the actuator processor. It is using integer arithmetic although it performs control law computation. This is done in order to meet the computing requirements at high sampling rates.

For present flight control applications, ADA is widely used as high order programming language. Generally, project baselined/validated compilers must be used. Assembly language is only used, where the capabilities of ADA do not allow specific performance requirements to be met.

The following constraints are placed on the use of ADA for flight critical applications. The use of ADA for the generation of target code will be subject to the projects' provisions and restrictions. Where translators, assemblers, compilers or any other support software are used in the preparation of high integrity software programs, the resulting flight code shall nevertheless remain fully verifiable at machine code level.

Language constructs, which are critical with respect to safety, must be avoided. These constructs are characterized by non-deterministic behaviour during program execution, e.g.

- constructs which are non-deterministic within tasking,
- constructs which provide dynamic storage allocation,
- certain access types.

Furthermore, for the generation of run-time code only libraries of proven integrity are used.

From the above mentioned constraints on language constructs a "safe-ADA" subset of constructs was defined for the purpose of coding the flight control software.

TESTING

Typically more than 50% of the total expenses of the development of a flight control system must be spent for testing. The important role of testing has two main reasons:

- For Fly-by-Wire systems there exists no possibility for an "uncritical" operation of the system in the target environment, i.e. from the first flight onwards the safety of both pilot and aircraft is of primary concern.
- The costs of late correction of failures are increasing tremendously with the progress of the project.

Recently, the effect of a failure in the FCS, which had not been detected before going into flight testing, caused major trouble for the Swedish JAS39 aircraft development program. An instability in the control laws during the landing phase led to the loss of a prototype aircraft (fortunately the pilot was not severely injured) and to a program delay of more than one year.

Throughout the FCS development process, testing activities are performed on all levels of system integration, beginning with unit tests of software modules and ending with aircraft flight tests. All testing must be brought in line with the final goal of reaching the flight clearance status. The clearance plan serves as a basic document, which lays down the required demonstrations of quality on all levels in a consistent way.

The major teststeps can be classified as follows [4,5]:

- Software module tests
- Equipment verification tests
- Performance validation tests
- Flight tests.

Software Module Tests

The software module tests can be classified as either "white box tests" analyzing internal paths and branches or "black box tests" analyzing interfaces and global data aspects. For compiled code, white box tests are an important teststep to avoid dormant errors. During these tests, all instructions must be executed and all decision paths be activated. Execution time aspects are investigated, as phase shift and transport delay are crucial issues for FBW applications.

The software module tests are performed against the requirements defined in the software specifications of the respective equipments which constitute the Flight Control System (e.g. Flight Control Computer, Inertial Measurement Unit, Air Data Computer). Most of the tests are performed in non-real time, part of it in a host computer environment.

Equipment Verification Tests

By means of the equipment verification tests it is demonstrated, that the complete software of an equipment is correctly implemented, i.e. the functions perform as specified and technical aspects of inputs, outputs and passage of data are correct. The integration of the software with the original equipment hardware is tested in real time.

A method often used for verification tests is the "Cross Software Testing". With this verification method, the equipment under test is run in parallel with a reference model by stimulating original equipment and model with identical inputs and comparing the outputs. All output disparities are automatically recorded for later analysis and correction where necessary.

The reference model is a dissimilar model of the equipment, which is coded, therefore, by an independent team and implemented in a different hardware, using the original requirement specification documents.

The Cross Software test method enables to carry out and automatically verify a very large number (of the order of 10^7) of test cases in short time. The tests are automatically documented and fully reprodukible. A block schematic of Cross Software Testing is shown in Fig.9.

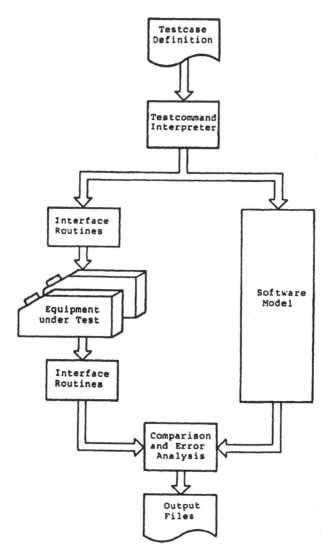

Figure 9: Cross-Software testing of FCS equipment

Validation of Performance

Validation of performance is the process of demonstrating, that the system performs according to the requirements and reacts favourably in all situations.

In this test stage, the different elements of the FCS, which contain embedded software and which have been verified on equipment level, are tested as integrated parts of the complete system. Parts of the system may be substituted by software models. To enable closed loop testing, a six-degrees-of-freedom model of the aircraft dynamics is executed on a simulation computer.

The performance test facility, called "rig", incorporates all essential control and display elements of the pilot. The tests are performed like flight manoeuvres, the pilot acting on the rig according to "flight charts" prepared by the test engineers. An important part of rig testing is the injection of all failure states, which have been identified to be safety relevant, into the system under test, and the verification of the specified failure recovery capabilities of the FCS.

The common goal of all teststeps described so far is to show that the item under test (software module, equipment with embedded software, integrated FCS) is behaving according to specified requirements. This fact explains one of the most critical problems of FCS development: a formal proof of correctness of the system is impossible [6], as most requirements are defined in specification documents using plain language and thus being necessarily imprecise. In order to overcome these shortcomings, different methods are presently being developed and already used for FCS applications, which make use of formal top-down specification practices.

Clearance for Flight Test

The final step towards the operational clearance of a flight control system is the flight test. When the FCS has successfully passed all laboratory tests described above, an initial flight clearance for development flight testing is issued by the aviation authorities.

During flight testing, the behaviour of the aircraft incorporating the newly developed FCS is investigated by test pilots. The judgement of the pilots is cross-checked against the customer requirements, which are laid down in the system specification. The analysis of identified problems is supported by traces of all important parameters, which are recorded in flight by a dedicated flight test instrumentation system.

Due to the complexity and safety criticality of a FBW (Fly-by-Wire) flight control system, the software development is structured into several packages. Each software package representing a certain FCS functionality must pass the complete testing process including flight tests, before the next set of functions can be introduced into the aircraft. The first software package must contain the basic control laws enabling a safe handling of the aircraft in all phases of the flight. Later packages will contain functions offering more handling comfort and mission effective-ness like autopilot modes.

A schematic showing the iteration loops during FCS development is shown in Fig. 10.

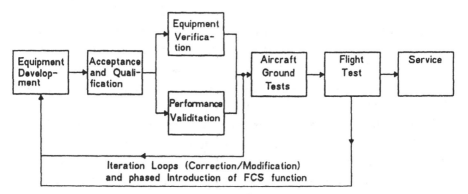

Figure 10: FCS Development iteration loop

CONCLUSIONS

As computer technology progresses, further increases in computer speed and computer performance over present computer capabilities will take place which, however, will not be as revolutionary as those happened in the past decades.

The overall speed of a computer system, for example, will increase mainly by optimizing the parallel processing architecture of a multipro-cessor system rather than enhancing the sequential processing speed of individual computing elements.

The impact of progress in computer technology on the architecture of real-time flight control systems of high performance aircraft is out-lined below.

The increase in computer speed will, nevertheless, be such that time critical software modules, which are coded for this very reason in assem-bler language, are going to be coded in a high level language, such as ADA. The language ADA contains more powerful language constructs than an assembler language does. Therefore, the lines of source code of a program in a high-level language over one in assembler language are significantly reduced at equal functionality, thus, providing for ease of design, testing, and verification of the software modules involved.

As the experience with operating systems of multiprocessor systems grows, a relaxation of requirements on non-deterministic behaviour in both task scheduling and high level language constructs will be allowed in the FCS design.

This is based on the assumption that more effective methods and testing tools will be developed in the future, which allow the detection or, even, the avoidance by structured design methods of problems associated with them. These are the detection of deadlock or livelock states and race conditions where not even a single occurrence of them can be tolerated in a real-time FCS. The additional design effort of a FCS with non-deterministic scheduling mechanisms over one with deterministic scheduling is justified. This is due to the fact that non-deterministic scheduling, in general, makes a more effective use of the available computer resource. This added benefit is bought, however, at the expense of speed of computation and extra administrative software overhead, as well.

With a general increase in computing power of future flight control systems more functions can be implemented within the FCS. On-line self checks of the FCS can be increased, for example, in both functionality and, hence, in complexity. Furthermore, the FCS can be burdened with more stringent control requirements, see the former section "Requirements on Flight Control Systems". Finally, an estimate of the airframe's response to given FCS's (Flight Control System) control signal can be provided in order to detect a degradation or damage of the airframe and to switch to appropriate degraded modes of operation. The application above is commonly associated with a real-time expert system which augments the flight control system of the aircraft.

REFERENCES

[1] Normenstelle Luftfahrt, (1970) LN 9300 (1970).

[2] McRuer, D. et al., "Aircraft Dynamics and Automatic Control ", Princeton University Press, Princteon, New Jersey, 1973

[3] Andrew S. Tannenbaum, "Operating Systems: Design and Implementation", Prentice-Hall, New York, 1987

[4] Gary L.Hartmann, Joseph E.Wall, Edward R.Rang, "Design Validation of Fly-By-Wire Flight Control Systems", *AGARD, Lecture Series Nr. 143*

[5] Michel Muenier, "Testing Embedded Software", *AGARD, Lecture Series Nr. 158*

[6] Karl N. Levitt, "Software Validation and Verification Techniques", *AGARD, Lecture Series Nr. 109*

15

ARTIFICIAL INTELLIGENCE TECHNIQUES IN REAL-TIME PROCESSING

Klaus Peter Kratzer
Fachhochschule Ulm
D-7900 Ulm

INTRODUCTION: A PARADIGM AND ITS IMPLICATIONS

Artificial Intelligence (AI) as an independent discipline has been conceived in the early fifties. Meanwhile, it has become a melting pot of diverse techniques aiming at research on and simulation of the human way of thinking. Originally, the field of AI could be considered an offspring of cognition science, but recently the application-oriented aspect has prevailed. AI has brought us, among many new concepts, a host of long-established techniques, veiled in a new terminology. The question is: Can these new techniques be applied to real-time problems?

The first problem arising when writing an essay on AI is, how to draw the borderline between AI products and the results of "conventional" programming. Winston [1] attempts a solution by coining a naïve, somewhat circular, but surely correct definition of the field:

> Artificial Intelligence is the study of ideas that enable computers to be intelligent.

This is followed by a description of the twin goals of AI to enhance the usefulness of computers and to find out, eventually, what intelligence is all about. To accept this definition for our purposes, one has to assume that all non-AI products and programs are non-intelligent. This assumption would be rather arrogant, so this definition is far too lopsided towards the AI position to be of further use. On a lighter vein, there are also citations which take the opposite point of view - for instance, Philippe Kahn's [2]

> "As long as you call it AI, it's not useful — as soon as
> it's useful, it's a program."

or Tesler's Theorem cited by Douglas R. Hofstadter [3]:

> "AI is whatever hasn't been done yet."

Alternatively, one might try to define AI as [4]:

> "the branch of computer science devoted to programming
> computers to carry out a task that if carried out by
> human beings would require intelligence."

Since endless arguments could be exchanged over what feats may or may not constitute intelligent behaviour on the part of a computer, it is probably better to avoid definitions altogether [5] and to consider AI as a collection of assorted programming techniques manipulating symbolic information and suited to serve specific fields of application.

Among those application areas are

- *logic and theorem proving,*

- *symbolic transformations,*

- *analysis of natural language,*

- *perception and pattern recognition*

- *man-machine interfaces,*

- *control of robots*

- *expert systems.*

Any successful program in those fields is a potential AI system; the discussion which properties to expect from such a program and the tools for programming are the domain of an AI expert.

TECHNICAL ASPECTS OF AI

A closer look at the true nature of AI reveals that an AI application program essentially consists of

- *standardized algorithms and methods (see e.g., [5,1], and usually observes*

- *one or several characteristic programming styles (see [6])*

if you strip off all components related to the application domain. A further approach attempts to do away with the need to model the application world using symbols, and to map microstructures of the human brain directly on computer structures and pathways. These

- *sub-symbolic models*

are not yet accepted as full-fledged AI methods. However, the resistance among the AI community against this approach is slowly diminishing.

Standardized Algorithms And Methods

Many AI algorithms are concerned with searching for optimal solutions in huge multi-dimensional spaces. In most cases, those problems and their solutions did not arise with the ascent of AI, but were incorporated from different fields and, in this process, considerably improved. The following paragraphs discuss some of the more important achievements in this regard; the list is by no means complete.

- *Search Strategies.* The search for any (or the optimal) solution of a given problem in a given domain of discourse is predominant in AI. Examples for such algorithms may be found in solutions for strategic games (e.g., chess) or for classical problems like the traveling salesman problem. Although the principal techniques to solve such problems like *back-tracking* or *graph-searches* have been common knowledge for a long time, AI succeeded in supplying meta-strategies, such as *hill-climbing, evaluation functions, minimax and alpha-beta procedures.* Their common goal is to produce optimal *bound conditions* to select existing partial solutions to a problem for further development and to indicate which partial solution will definitely or most probably not lead to an acceptable end.

- *Production Systems.* A major application of searching techniques are production systems as closed systems for symbolic transformation. A production system accepts a *grammar,* which in turn consists of an *alphabet* of symbols with a starting symbol and a body of *rules* to transform one or a sequence of symbols into one other or another sequence of symbols. Using a *control system* to select the rule to apply next to a chain of symbols and to determine when to consider the production finished, a production system can analyse strings of symbols whether they conform to a given syntax (e.g., parsing a high-level program), generate plans of action (e.g., in robotics or manufacturing), or assign meanings to received signals provided the signals are expected to behave according to some particular pattern. See **figure** 1 for an example of a simple production system producing symmetric patterns with an odd number of characters, which consist of the letters "a" and "b".

Alphabet $\alpha = \{a, b, S\}$

Rules $\rho = \{(S \to aSa), (S \to bSb), (S \to a), (S \to b)\}$

Starting Symbols $\sigma = \{S\}$

Starting from "S", all symmetric odd-sized symbol patterns containing "a" and "b" using rules from ρ may be constructed, e.g.

$S \to aSa \to aaSaa \to aabSbaa \to aababaa$

Vice versa, a given symbol chain may be analysed insofar as to determine whether those symbols conform to a given grammar (α, ρ, σ).

$bbabb \to bbSbb \to bSb \to S$

Figure 1: A simple production system

Initially, production systems were used only in a literal sense, i.e., to parse expressions conforming to formally defined languages. In an AI application, however, this notion is generalized to sequences of symbols which somehow map to real-world properties. Application of rules thus may show ways to reach target states and to avoid undesirable situations.

- *Deduction.* A further step is to leave pure symbol manipulation alone and to build a model of the real world based on a notation which in turn is derived from mathematical logic. In this case, the model consists of facts and rules, which may be transformed to new facts and, in some cases, even new rules. Though pure logic does not encompass an execution model to control such transformations, mathematical theory provides us with several approaches like *predicate calculus* to lay down a foundation for the application of rules to facts. Typically, they are based on a heavily restricted subset of logical clauses. Most notable among them is *Horn clause logic* which governs the power and expressiveness of the programming language *PROLOG* [7]. This language implements an execution model where clause manipulation is generalized to two kinds of algorithms, *unification* and *resolution,* to cover binding of variables and application of rules, respectively. The underlying virtual machine resembles a primitive kind of theorem prover which is able to answer simple questions relating to facts and their derivates, but not to analyse facts and rules in terms of completeness and consistency (*closed world assumption*).
The basic stategies to process rules and facts are either to start deriving new facts until such new facts satisfy some predefined goal condition (*forward chaining,* see **Fig. 2**), or to formalize alternative goal states (*hypotheses*) and to try and find the hypothesis which can be shown as a derivate of known and accepted facts (*backward chaining,* see **Fig. 3**). Further refinements do not only accept mere facts, but process statements in conjunction with some metrics for the truthfulness, probability, and the degree of reliability. In many cases, however, the results of such "confidence juggling" will be highly questionable.

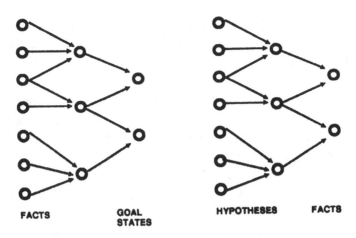

FACTS   GOAL
                STATES   HYPOTHESES   FACTS

Figure 2: Forward chaining   **Figure 3: Backward chaining**

Programming Styles

"Each AI system is a program" - thus, the principal impact of AI has been achieved by promoting and reviving a host of *programming styles* [6]. These, in turn, are coupled to new and progressive ways of *representing data.*

Basically, each AI problem is tackled along the lines of the following methodology:

- First, a cognitive theory is developed (e.g., by conducting a study of human problem solving or the characteristics of human memory).
- Next, one has to ponder how to represent and manipulate data. The products of this step are a *programming language* and a *programming style* as model of information processing.
- Finally, this apparatus is applied to some fictitious or real-world problem.

Manifold repetitions of this process by scores of communicating R&D personnel has produced a number of commonly accepted styles of programming in AI. The next paragraphs will discuss some of the more prominent ones.

- *Functional Programming*. The foundations for this style have been laid by Joe McCarthy in the late 50s [8] as alternative to conventional, statement-oriented programming, at that time mostly in FORTRAN. He created a *LISP* interpreter to implement a somewhat warped representation of Church's lambda-calculus [9]. The basic mechanisms of functional programming are *encapsulated terms* and *conditional expressions*. Functions in this context are depicted as mappings of structured objects to different objects; one can hardly discern a borderline between function declarations and data as required in conventional von-Neumann programming, since functions are objects and may be part of any data structure. LISP is one of the few programming languages capable of interpreting runtime generated code, and as such allows pretty flexible ways of dynamic access to data. Nevertheless, use of LISP in AI applications is nowadays waning due to the heavy penalties to be paid in terms of overall system performance.

- *Logic-oriented Programming*. The most important contribution of this programming style is its emphasis on the strict separation between all application-dependent information (*knowledge base*) and the execution model implemented as *inference machine* (see Fig. 4); the inference machine may be applied to knowledge bases from totally different domains. The knowledge base consists of axioms which resemble real-world facts and transformation rules. An application-independent rule processor uses the rules to generate new facts, in case of a knowledge base on temporal aspects of the real world, the rule processor would produce hypothetical new facts (*prognosis*).

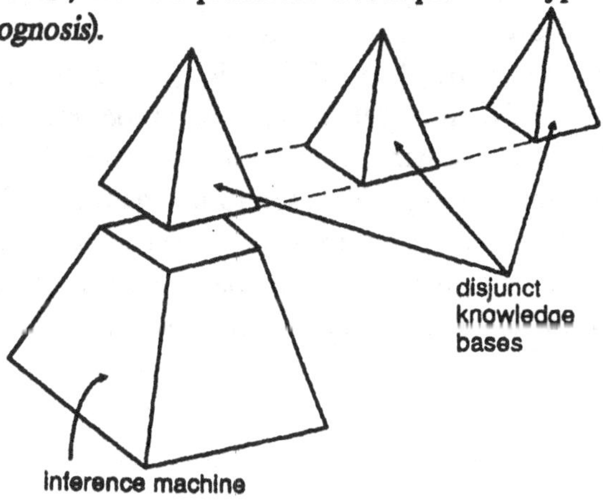

disjunct
knowledge
bases

inference machine

Figure 4: Separation of knowledge bases and inference machine

However, such a nice situation as depicted in **Fig. 4** is entirely fictitious. It seems to be impossible to cover all aspects of human expertise in one unified execution model. Thus, logic-oriented programming may well be used as the principal technique in an AI project, but sure enough, it will never be the *only* one.

- *Goal-oriented Programming.* This style is merely a refinement of logic-oriented programming and primarily suited for planning purposes. The execution model attepts to find a way from the current situation to a target situation by backward chaining.

- *Object-oriented Programming.* The most important programming style up till now did not arise from a classical AI environment. Object-oriented programming has a number of diverse origins; the most important ones are described in the next paragraphs.

    · *Data representation in "frames".* The term *"frame"* as coined by Marvin Minsky [10] roughly corresponds to the notion of an object in object-oriented programming. The idea of frames is derived from an all-purpose data representation method in AI, called *semantic networks* (see **Fig. 5**). A semantic network consists of nodes, which symbolize single phenomena, and of relationships between such phenomena. Semantic network are mainly used to represent common sense knowledge, e.g., for understanding natural speech in a given environment. The example shows how to declare factual knowledge (*"The SF Giants will win the World Series '92"*) as well as common sense knowledge (*"A sporting event is some kind of event"*).
    The most important relationship turn out to be the *is-a*-relationship, which by abstraction assigns an individual to a group of other individuals sharing some common traits, and the *kind-of*-relationship, which connects an abstraction to a more general abstraction. Minsky's idea was to normalize the information contained in a semantic network insofar as to separate abstractions (now called *frames*) from simple objects, which are connected to a frame by an *is-a*-relationship; these are now called *instances*.

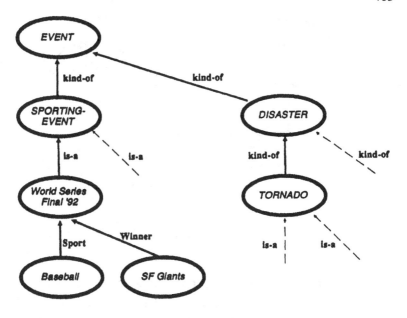

Figure 5: A semantic network

·*Programming Languages.* Semantic networks are static structures; there is no inherent execution model to cover the dynamic aspect. A object-oriented programming system now uses frames as representation model (while renaming them to *classes*) and adds a *message passing paradigm* as execution model. Each class defines to which messages the instances, which are its *members*, can respond. How to respond, is up to an instruction sequence (*method*) which is associated with one of the permissible message patterns and executed upon receipt of a message. In addition, each class passes to its members the information which bits of individual knowledge (*slots*) are to be associated with each member, optionally with default values or evaluation procedures. As shown in Fig. 6 the *kind-of*-relationship now exists between classes; one of them is the more specific (*subclass*), the other the more general one (*superclass*). All slots and methods associated to a superclass are passed to subclasses and, finally, to their instances (*inheritance*). However, all such inherited properties may be overruled at lower levels of the abstraction hierarchy.

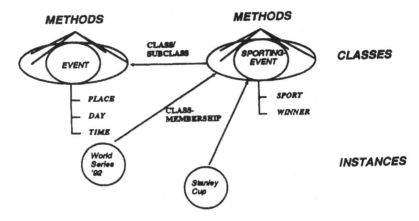

Figure 6: An aspect of an object environment

The world of objects can be imagined as a simulation of the objects in some segment of the real-world. Each object can perform all manipulations pertaining to itself and its neighboring objects *autonomously*. Thus, an object-oriented program basically consists of class and object definition, its execution is started by sending to one of the objects a message from an external source (e.g., picking a menu item on a terminal). The target object engages other objects by sending further messages, and eventually all relevant methods are being executed.

The first non-AI researchers, attracted by that simulative approach, created a new programming language called *SIMULA*, a derivative of good old *ALGOL 60*. Some other languages of no further importance followed. Currently, an extension of the C programming language, called C++ [26], attracts many software engineers due to its powerful data representation mechanisms along the lines of the frame model. However, some of the flexibility of the original concept got lost on the way, because C++ is strictly compilative, wheras only interpretation can preserve such useful features as dynamic creation of class hierarchies and methods.

· *Programming for novice users.* A further path of research led to object-oriented programming via user interface design. In particular at Xerox Corp.'s Palo Alto Research Center work was focused on computers accessible to computer illiterates (see e.g., [11]). The pinnacle was reached with a complete programming environment, called *Smalltalk* [12], allowing interactive, fully object-oriented software design. By the way, the claim that C++ is a direct offspring of *Smalltalk* is only true to a limited degree. C++ is strict on type checking, whereas *Smalltalk* could not be more liberal in this regard - and this fundamental difference in design pervades all aspects of programming.

Many of these programming styles by now have found their way into the commonly accepted toolbox of software engineering. Using C++ and object-oriented programming - as compared to pure C - for instance, is nowadays considered a wise decision to improve project management and software re-use rather than an indication that an application program contains traits of AI.

The Sub-Symbolic Approach

All approaches discussed above were based on symbols and symbol manipulation. There still remains a possible trouble spot: the mapping of real-world phenomena to symbols which is part of the design of an AI system. A way to avoid an explicit mapping is to avoid symbols at all. Instead, the system is configured as *simulator for cerebral microstructures* governed by the laws of physiology and neurobiology. Hofstadter elaborates on the concept of a *"grandmother cell"* structure, developing an intuitive notion of the term *"grandmother"*, instead of giving awkward rule-based definitions, which may resolve this term to simpler concepts but never leave the universe of symbols [3].

This idea has been conceived in cognitive science and neurobiology a long time ago. John von Neumann toyed with it before reverting to his well-known reference architecture. However, serious application prospects did not emerge until the mid-80s, due to paradigmatic new developments in this field (refer to [13]) for an overview and of. [14,15] for further study).

Transferring the physiological notion of a brain cell (*neuron*) to computer science, one arrives at a primitive processing unit with rather limited capability. Naturally, more recent research on the properties of central synapses casts some doubts on this simplistic view. However, as a *complex* called *neural network* these units can perform astonishing feats, although such systems merely work along the lines of a human brain as a gross abstraction.

A further noteworthy property of this class of models is their emphasis on *communication*. The state of a processing unit is defined by its *state of activation*, usually a scalar value. Thus, neural information processing amounts to switching the activations within a neural complex. To allow collectivism, the processing units are interconnected according to some predefined topology; to simulate synaptic information transfer, each connection is associated with a *weight* as measure for the influence of one processing unit to another. The configuration of weights is the sole means of knowledge representation.

Figure 7 shows an example of a neural network, in this case a so-called *feed-forward network*. A problem or question has to be encoded to be fed into input units as activation pattern (*stimulus*). The information is passed on via intermediate units until a reaction can be decoded using an activation vector taken from different units. It is evident that this procedure can be parallelized to a high degree; thus, investment in specialized hardware returns a particularly high dividend with such systems.

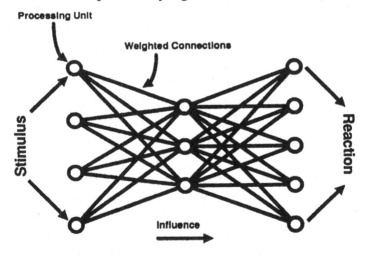

Figure 7: A neural network

A further incentive to "go neural" is the stunning *adaptability* of neural configurations. There is no need to define a weight structure from scratch; rather, it is generated step by step by feeding the network a representative set of data together with desired reactions. If a neural network stimulated with a given test pattern produces an erroneous result, the error measure is distributed among all processing units according to their "responsibility".

The consequences are

- You do not have to define a model of your application world as concisely as with symbolic systems.
- The definition of the desired function of the neural network is only descriptive in an extensional way; there is no need for complicated formalism apart from your encoding/decoding scheme.

If coding and adaptation are conducted properly, the resulting weight configuration is usually pretty robust against noise and slight variations of idealized test cases. If application requirements change, a neural network adapts quite dynamically.

However, the major drawbacks of neural networks up till now preclude large-scale application in a critical environment.

- The dimensions of neural networks are limited by available storage and processing power. Thus, configurations examined so far did not even get close to the 10^{11} processing units and 10^{15} connections in the human brain. One might argue that such a simulation is not needed for operation in a specific application domain, but as soon as the outside view of the network has to be curtailed due to size restrictions, input and output coding have to conform to an abstraction of the outside world, which means that modelling is again imperative.

- A running neural network may well be validated by testing, but yet there are no usable tools for tracing, debugging, or verification. Analysis mechanisms available draw nice pictures, which are informative with toyish examples, but useless for serious work. Consequently, the behavior of a neural complex in non-stereotyped situations is hopefully correct, but essentially unpredictable. A neural network in control of the nuclear power plant next door might be an unpleasant experience.

The state of the art in information processing all but precludes autonomous neural networks. Therefore, current development activities favour integration of neural components into conventional systems; by this means, improved control of behavior can be achieved.

AI APPLICATIONS

The previous chapters demonstrated that AI itself is only distinguished from other fields by some applications that have simply been declared AI, and by characteristic methods. In this regard, a good part of programs nowadays are subject to this characterization. The particular appeal of AI methods, in particular data representation techniques, are

- unprecedented flexibility and dynamics of data structures,
- useful data acquisition tools, and
- relaxation of the strict distinction between procedural and static aspects of application knowledge as enforced by conventional programming.

However, a good number of programs using such techniques would never be considered AI applications, since many of those techniques have found their way into the standard repertoire of a competent software engineer, e.g., data encapsulation, design of virtual machines.

Running An AI Application

Now, what are the typical traits of a true AI application? If the system is not run as stand alone environment, the distinction is made by the way of integration with the conventionally programmed application segment.

- *Loose coupling.* The AI segment is clearly separated from the conventional part, e.g., by running in a different process or sharing data via the file system only.
- *Programming environment.* The AI segment is designed in an environment considered typical for AI. For quite a time the program-

ming systems of LISP, PROLOG and their derivates as well as some specialized languages (e.g., OPS-5) were branded as such. Their use often entailed considerable investment in specialized hardware (LISP stations), cumbersome interfaces to file systems databases and all other peripherals, and much trouble hiring competent engineers.

Meanwhile, typical AI applications are not necessarily confined to a separate environment, therefore much easier to integrate, and less easy to distinguish. But anyway, as shown in the second section, it is the style and methods that count rather than circumstantial properties.

Expert Systems

The most famed class of standalone AI systems are expert systems (cf. [24,25]. An expert system is designed to simulate the behavior of a human expert, in particular

- to answer questions,
- to explain answers,
- to explain the application,
- to provide guidance in dialogs.

Success in developing and operating an expert system depends on three prerequisites:

- Human experts must be few and expensive.
- The task must be tedious and highly repetitive.
- The expert should be willing to cooperate.

Figure 8 shows a simplified overview of the architecture of an expert system. There are two interfaces to the outside world:

- a dialog controller with built-in help and explanation facilities, and
- interfaces to mass storage and gateways to other processors.

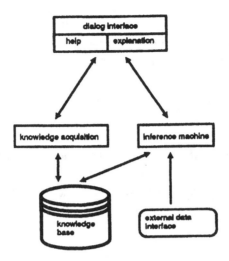

Figure 8: Simplified architecture of an expert system

The kernel consists of an inference machine and the knowledge base. It should be noted that the principal benefit of an expert system lies in its facilities to interact intelligently with its human dialog partner.

Evolution in expert system technology has widened the scope of available methods from mere rule-based representations to a broad range of integrated modelling concepts to cater to the needs of many fields of application. Nearly equally important are knowledge acquisition tools for design and maintenance of knowledge bases, to detect conflicts and contradictions, monitor performance, and to adjust existing rules. In some cases, expert systems are even capable of adjusting their knowledge base autonomously by induction, although this kind of "learning" has surely not found industrial acceptance yet.

Quite a few companies are now marketing expert system shells which are nothing less than comfortable programming languages for expert systems together with flashy user interface techniques. It may be doubted that any human expert has been really replaced by an expert system, since maintenance of a knowledge base might easily be more time consuming than doing the job right away [16]. However, systematic extraction of knowledge from the expert pays by granting more transparency in the expert's field of expertise and by much better availability, even at various locations.

AI TECHNIQUES AND REAL-TIME PROCESSING

Since intelligent behavior is perhaps mostly required in time-critical and dangerous environments, use of knowledge engineering techniques increased very swiftly in this field. As long as they are used only as surrogate for other methods of software engineering, there is hardly a difference to conventional systems. So this discussion should be confined to AI systems in a literal sense.

AI Techniques In Hard Real-Time Environments

A real-time system is defined to perform state transitions bounded in the temporal dimension. In a *hard* real-time system it can be assumed that violation of such bounds may invalidate operational consistency and even be a threat to the health and well-being of human operators.

Conventional approaches utilize hardware investment for speed-up and interrupt scheduling and priorization to meet the demands of hard real-time processes.

- For an AI system, speed-up can be achieved by parallelization; this holds for neural systems in particular, but to a certain degree also for rule-based and logic-based systems like PROLOG.
- Most of all available sub-symbolic models are predictable in their consumption of time. When using a symbolic system, behavior depends on the structure of the knowledge base: if processing is straight-forward and comparable to the evaluation of a decision table, it is certainly possible to determine an upper bound for execution time; if the knowledge base contains more sophisticated constructs involving recursion and exceptions, such bounds may be at least temporarily violated (see e.g., [17]). To keep behavior predictable, all modifications to the knowledge base must be monitored in this regard, which is rather complicated.
- Interrupt scheduling can be imposed on knowledge-based systems on the condition that all global variables (i.e.,all or part of the knowledge base) are consistent (see e.g.,[18]). However, timely task execution can still not be guaranteed.

So the final question is, 'Can an AI system be useful in a hard real-time environment'? Taking into account the fact that much of the "intelligence" expressed by an AI system is directed towards a human operator, and human interaction, in turn, cannot be part of a time-critical operation, the answer should be 'No', except for systems where no human interaction is intended, in particular sub-symbolic configurations. Many knowledge-based systems in the real-time world simply use some structural techniques, and, for the sake of proper operation, forsake their flexibility. The term "real-time expert system", which is pretty far-spread in literature, is a contradiction in terms and thus not worthy of further discussion.

Further Applications

If we revert to loose coupling of knowledge-based systems, even expert systems, to real-time processes, there is a broad range of possible and useful applications (cf. e.g., [19,20,21]). The next paragraphs discuss the most important functions to benefit from AI technology.

- *Process Parametrization.* the complicated and tedious taks of installing and maintaining a process configuration can well be supported by an expert system which incorporates factual knowledge and heuristics derived from previous installations and customizations.
- *Process visualization.* By maintaining a knowledge base on the processes and operator preferences, an AI system can provide a hand-tailored interfacing device between an otherwise conventionally programmed real-time system and a human operator. Thus, ease of handling, operator alertness, and system transparency can be improved.
- *Monitoring processes.* AI techniques are also helpful in condensing the flood of signals and identifying possible trouble spots for an early operator alert.
- *Process diagnostics.* In a crisis, an expert system in control of all data pertaining to processes, can prove a valuable assistant to operators by defining precisely the reason for a breakdown and suggesting how to proceed.

- *Process planning.* Expert systems are already widely in use for planning of processes, mainly in manufacturing systems. They allow interactive simulation of hypothetical schedules, identify bottlenecks, and optimize according to varying goals.

SUMMARY

AI as an independent field of knowledge is in deep crisis, at least in the U.S.A. where massive application of pure knowledge processing has been intended in the military sector and now been partially cancelled (cf. [22]). Less so in the European market, where many people still believe their own hype. The problem is that

... there's been great progress in other areas of computer science while AI is still saying, 'Give us more time'. How long should we wait? They aren't going to make any sudden breakthroughs. Let's face it, the stuff isn't working. People have invested lots of money, but Wall Street doesn't throw money away forever. [23]

One way to counter this problem is to push stakes even higher, to nearly metaphysical level, and to announce that AI is actually intended to beat death by creating the technological foundations for full downloads of human minds to computers. These might be the aberrations of elder scientists or desperate attempts to obtain fresh funding... Anyway, the more sensible solution is to abandon the claim that AI could substitute and paradigmatically improve all conventional approaches and to retreat into sectors where the merits of AI are not challenged. The immense recent progress in software engineering would have been unthinkable without the concepts provided by active research in the area of AI.

This holds as well for AI applications in real-time environments. There may be no truly intelligent real-time systems in the works, but conceptual influx from AI has brought and will bring less costly, easier maintainable, and more flexible systems, which offer comfortable handling, will run fast and reliable - in a nutshell: which are more beneficial to all of us.

REFERENCES

[1] P. H. Winston: "Artificial Intelligence", Reading MA.

[2] P. Kahn: *oral communication*, München, 1991.

[3] D. R. Hofstadter: "Gödel, Escher, Bach: An Eternal Golden Braid", New York NY, 1980.

[4] N. Graham: "Artificial Intelligence", *Blue Ridge Summit*, PA., TAB, 1979 (Eds.): The Foundation of Artificial Intelligence, Cambridge, 1990.

[5] N. J. Nilsson: "Principles of Artificial Intelligence", Berlin, 1982.

[6] H. Stoyan: "Programmiermethoden der Künstlichen Intelligenz", Vol. 1, Berlin, Springer-Verlag, 1988.

[7] W. F. Clocksin, C. S. Mellish: "Programming in Prolog", New York NY, 1981.

[8] J. McCarthy: "Recursive Functions of Symbolic Expression and Their Computation by Machine", *CACM 4*, 1960.

[9] A. Church: "The Calculi of Lambda-Conversion", *Ann. of Math. Studies*, Vol. 6, Princeton NJ, 1941.

[10] M. Minsky: "A Framework for Representing Knowledge", in: *Winston, P.H. "The Psychology of Computer Vision"*, New York NY, 1975.

[11] A. C. Kay: "The User Interface As A Theatrical Mirror", in: *Wedekind, H, Kratzer, KP (Eds.) "Büroautomation"*, Teubner, Stuttgart, 1985.

[12] A. Goldberg, D. Robson: "Smalltalk-80: The Language and its Implementation", *Reading MA*, 1983.

[13] R. P. Lippmann: "An Introduction to Computing with Neural Nets", *IEEE ASSP Magazine*, 1987.

[14] D. E.Rumelhart and J. L. McClelland: "Parallel Distributed Processing: Explorations in the Microstructure of Cognition", Cambridge MA, 1986.

[15] K. P. Kratzer: "Neuronale Netze: Grundlagen und Anwendung", München, 1990.

[16] K. Kurbel: "Entwicklung und Einsatz von Expertensystemen", Berlin, 1989.

[17] J. F. Muratore, T. A. Heindel, T. B. Murphy, A. N. Rasmussen, R. Z. McFarland: "Real-Time Data Acquisition at Mission Control", *CACM 12*, 1990.

[18] C. Lischka, H. Voss: "TEX-1: Echtzeitfähige Expertensysteme in der prototypischen Anwendung", in: *Brauer, W, Freska, C (Eds.) "Wissensbasierte Systeme"*, Berlin, 1989.

[19] P. F. Elzer: "GRADIENT - Ein Schritt in Richtung intelligenter Prozessüberwachung", in: *W. Brauer, C. Freska (Eds.) "Wissensbasierte Systeme"*, Berlin, 1989.

[20] P. Mertens: "Expertensysteme in der Produktion", München, 1990.

[21] R. Soltysiak: "Wissensbasierte Prozessregelung", München, 1989.

[22] S. E. Dreyfus: "Expert Human Beings, Competent Systems, and Competent Neural Networks", in: *W. Brauer, C. Freska (Eds.) "Wissensbasierte Systeme"*, Berlin, 1989.

[23] G. Martins: "Artificial Intelligence Smarting from Unfulfilled Promises", San Francisco Examiner 12/21/1986, San Francisco CA, 1986.

[24] P. Harmon: "Expertensysteme in der Praxis", München, 1987.

[25] F. Hayes-Roth, D. A. Waterman, D. B. Lenat: "Building Expert Systems", London, 1983.

[26] M. A. Ellis, B. Stroustrup, "The Annotated C++ Reference Manual", *Reading MA*, 1990.

16

RECOMMENDATIONS FOR A REAL-TIME SYSTEMS CURRICULUM

Wolfgang A. Halang
Rijksuniversiteit te Groningen
Vakgroep Informatica
Postbus 800
NL-9700 AV Groningen
The Netherlands

INTRODUCTION

The objective of courses for the education of real-time computer and control systems engineers is to discuss topics and problems encountered when developing embedded systems for hard real-time environments. With the latter term, industrial, scientific, and military areas of application are meant which are characterized by strict time conditions that must not be violated under any circumstances. In contrast to this, commercial systems e.g., for automatic banking and airline reservations, only need to fulfill soft real-time requirements, i.e., although they are designed with the objective of fast response time, the user may expect the completion of his transactions within somewhat variable time frames.

The significance of real-time systems is rapidly growing. The vast spectrum of these systems can be characterized by just a few examples of increasing complexity: controllers in washing machines, air-traffic-control and air-defence-systems, control and safety systems for nuclear power plants, and finally future space borne systems like the Space Station. The importance of such systems for the well-being of people requires considerable effort in research and development of highly reliable real-time systems. Furthermore, the competitiveness and prosperity of nations now depends on the early application and efficient utilization of computer integrated manufacturing systems (CIM), of which real-time systems are an essential and decisive part. The above arguments underscore why it is timely and important to take a closer look at the education of real- time systems engineers.

Observing the development of real-time systems for some 18 years, the author has no doubt that European researchers and application engineers are leading in this field. This is rather surprising, because much larger military and space flight projects have been carried out in other countries, which are very demanding with respect to real-time data processing. This leadership may be due to the traditionally strong position of the Europeans in *civil* industries. Indeed, electrical, chemical and control systems engineers were pioneering this field during the last 20 to 25 years. The first and the majority of high-level real-time languages were defined and developed in Europe. European research on real-time operating systems, on special purpose process control peripherals, and on safety engineering methods is also more advanced than comparable efforts in other parts of the world. This can be related to the greater attention the field gets in European institutions of higher education, i.e. universities and polytechnics. Comparing job offerings for computer science and engineering faculty, for example, it is evident that corresponding departments in North America are not too interested (yet) in finding people with a background in real-time and embedded systems. According to a recent study [19], there are only between 5 and 10 university research groups in the United States working on issues of real-time computing. Finally, in Europe, software tools for supporting the entire development process of real-time systems, from hardware configuration and software requirements specification to code generation and documentation, are already in widespread use. In contrast to this, there seem to prevail ad hoc methods [21] of real-time systems realization in the United States. Another measure for the state of the art is the fact that in Japan 85% of the real-time software is still written in assembly languages compared to about 60% in Germany. These estimations are based on statements made in several international conferences.

In order to catch up with the European leadership in research, development, and application of real-time systems, it is important to provide the best possible education to future systems engineers. The thus specialized graduates will be able to effectively strengthen the industries in their countries by innovations and so contribute to the competitiveness of their products and services on the world market. In the sequel, an outline of a syllabus recommended for the education of real-time computer and control systems engineers is given.

OUTLINE OF AN EDUCATIONAL PROGRAM

Prerequisites and Relation to Other Courses

The purpose of this course is to provide an additional qualification for senior undergraduate or master students of technical subjects such as chemical, electrical, and mechanical engineering, but also of computing science. In order to cope with the heterogeneous knowledge background of the participants, the course covers its area completely. Hence, there are some overlaps with other courses, which some of the students may already have attended. The overlapping material, however, will always be presented with a distinct real-time systems specific emphasis. Besides a good basic knowledge in science and engineering, the only prerequisite required is that the students are proficient in an Algoloide programming language such as Pascal or C.

Syllabus for Basic Courses

At the beginning of a course or a series of courses on real-time systems, there should stand a *historical survey* and an *overview about the importance of the subject for industrial applications*. Then, the *basic requirements with regard to the time behaviour,* viz. timeliness, simultaneousness and predictability, and fundamental concepts like real-time mode of computer operation, process types, process coupling, and interrupt systems are introduced, and the time conditions encountered in process control environments are classified. The discussion of the two methods at hand to achieve timely and (quasi-) simultaneous behaviour of real-time software, i.e. the *cyclic executive approach and asynchronous multitasking,* concludes the introduction to the material.

Initially, some notions are shortly summarized, which are already treated in courses on operating systems, but that may still be unfamiliar for students of other disciplines such as chemical engineering. This includes the *task* as the concept of a parallel program thread in a multiprogramming environment. Further topics in this context are *task scheduling operations* and *state models, synchronization primitives* for organizing the cooperation of tasks, and the *timing of task executions.* After having discussed the mentioned subjects, *structured real-time program-*

ming is treated. Initially, this can be carried out in a language-free graphical form, as introduced in [5], in order to concentrate on the clarification of the intrinsic concepts and on the representation of timing in concurrent systems.

The next major part of the course is dedicated to *programming in high-level real-time languages*. It starts with compiling the requirements of such languages, followed by the different lines of development these languages have taken, viz. extension of existing mathematically oriented languages by real-time language features, definition of new high-level languages, and specialized languages and program generators aimed towards meeting the programming needs in specific application areas. Subsequently, the major languages in this field are introduced, compared with each other, and evaluated [1,2,4,7,10,15,18]. Then, as a case study, one of these languages must be studied more thoroughly. By carrying out a number of programming exercises, the students are to gain proficiency in this language and especially in using its real-time and process control elements. The selection of a real-time programming language mainly depends on the availability of appropriate programming systems for educational use, but also on other, sometimes irrational factors. According to [7,18] and Chapter 4.1, LTR [15], PEARL [4], and Real-Time Euclid (cf. Chapter 2.2 and [22]) appear to be the best-suited real-time languages, both from the viewpoints of teaching clean, structured concepts and of expressive power for timing behaviour and for applications in industrial automation. Unfortunately, LTR has been phased out in France, its country of origin, and replaced by Ada, and Real-Time Euclid did not leave the research laboratory. Ada and PEARL are the only high-level real-time languages readily applicable in industrial control environments, since they have been implemented on a wide range of computers including the major 16 and 32 bits microprocessor series, on which most contemporary process control computers are based. Owing to its simple syntax, which resembles Pascal or other Algoloide languages, for students with some background in programming, the teaching of PEARL can be kept very short, since it is sufficient to concentrate on its real-time specific features. Thus, the language can be mainly covered within one class. Furthermore, PEARL has an additional advantage, because recently a third version was introduced [4], which allows the programming of networks and distributed systems fully within the framework of the language: the various program modules can be assigned to the nodes, logical connections can be established, and transmission protocols specified. Moreover, it provides language facilities to detail the dynamic reconfiguration of a system for the error case. For a more detailed justification of the selection of PEARL as the real-time language most appropriate for teaching purposes, the reader is referred to Chapter 3.1 of this book, where it is compared with its major competitors and especially with Ada. One main objective of this part of the course should be to stress that assembly language programming can be renounced, since high-level real-time programming systems have reached a stage which makes their usage feasible for even the most time-critical applications. This can be shown by considering experimental data [16], which reveal that assembly coding can improve the run-time and storage efficiencies of e.g. PEARL programs only by single digit percentages.

After having become familiar with real-time programming, *real-time operating systems* require a significant amount of attention. The coverage commences with the functional description of state of the art real-time operating systems along the lines of the recommendation [3]. Special emphasis must be placed on an in-depth comparative treatment of the critical topics of deadlock handling, particular prevention, synchronization concepts, memory management, and priority and advanced - i.e. time driven - task scheduling strategies for hard time conditions. In contrast to standard operating system courses, the consideration of each of the mentioned topics is to focus here on the predictability aspect and that execution time bounds can be guaranteed. The students can gain a very profound knowledge of this subject by carrying out a project of designing and, during laboratory sessions, implementing a small real-time operating system, which should contain the functions for interrupt, time, task, and processor administration, an input/output driver for a slow device, and should provide one or the other advanced feature.

Turning to *hardware*, the system structures of real-time computers are treated. This includes local and global bus systems, hardware architectures, especially with regard to achieving fault-tolerant systems, as well as centralized, decentralized, hierarchical, and distributed configurations and their suitability for certain applications. Special emphasis is to be put on the objectives for employing distributed automation structures which match the hierarchical organisation of entire industrial plants, and the criteria for suitable communication structures in such systems.

The main topic of *process interfacing* and *peripherals* needs to be introduced with the classification of processes, noise, and signals, and some material on sampling, signal quality, and signal conditioning. The influence of noise must be stressed and a variety of measures for the prevention and reduction of electric disturbances are discussed and studied in laboratory sessions. Generally, this material is very extensive and usually covered in detail by other electrical engineering courses. The purpose of its short discussion here is only to create awareness for this area of possible major problems in real-time systems design. The coverage is mainly directed to computing science and other engineering students who do not attend electrical engineering courses. Then, the structure and principle of operation of process sensors and actuators and the main components of process interfacing are discussed in class and applied in experimental setups, viz. analog, digital, and impulse interfaces, multiplexers, various types of analog-to-digital converters, sample/hold-amplifiers, timers, data acquisition subsystems, and interfaces to certain bus standards like IEEE 488 and CAMAC. The inherent noise suppression properties of integrating analog-to-digital converters and the flexibility of microcomputer based programmable process input/output units are to be pointed out. In this part of the course, the students are required to write some specific device driver programs, linking their knowledge of process peripherals with the one of real-time operating systems.

As the next main topic, *system reliability*, especially of the software, deserves broad attention. The discussion commences with clarifying the terms "reliability" and "safety", with a characterization of errors and failure situations, and with introducing the concepts of *redundance* in its various forms and of *diversity*. With regard to the hardware, measures to achieve fail-safe behaviour are presented. Then, the students are trained to develop reliable software by writing robust, fault-tolerant programs of low complexity observing rigorous development disciplines and by utilizing design tools and diverse approaches. To prove the reliability of programs, methods for software quality assurance and software verification methods suitable for real-time environments are described. In this context, procedures for real-time error diagnostics and recovery are also studied. Very important is the discussion of measures to improve the security of realized systems and of methods to carry out the safety licensing of real-time systems.

The final part of the course or course-series is dedicated to *integrated development environments*, i.e. software tools, for the realization of embedded and real-time automation projects, which became necessary by the need to jointly design integrated hardware and software systems. They allow the realization of independent structuring and description of a system and for the specification of the hardware and software structures as well as for the specification of timing constraints. By taking also hardware and timing aspects into consideration, these tools distinguish themselves from sole software specification and CASE tools, which are covered in computer science courses. Here, software tools are introduced which support all phases of engineering work, viz. requirements engineering, system design, implementation, verification, and validation, and provide automatically the accompanying documentation. In particular, on the one side software design objects such as modules, procedures, tasks, data, and interfaces and on the other side hardware design objects like components, signals, bus connections, and computers themselves are described, and most importantly, the relations between the software and hardware design objects, i.e. the relations between the logical and the physical structure of a system. Thus, the main difficulties of using separate design tools for the development of software and hardware components can be overcome, which result from the enforced early and final decomposition in software and hardware parts and from the lack of being able to analyze relations between the software and hardware structure. Such integrated project-support-systems are fully computer supported. Using a specification language and a model concept suitable for the considered project, which is selected out of a number of available concepts, the system engineer describes the problem, the objectives, and the methods for realizing the problem solution. The knowledge based environments allow for the stepwise refinement of a design and, thus, support the model of design levels.

It appears that EPOS [6,12,13] is the most advanced of the available integrated development support systems. Besides requirements engineering and system design, it also supports project management and product quality assurance functions. The corresponding specifications are collected in a common project database, from where they are provided as inputs for a number of tool systems. The latter analyze the problem descriptions and specifications in order to find consistency and design errors, provide a host of different documentation facilities, support the management while planning and carrying through a project as well as with version and quality control of software, and finally generate program code in high-level languages directly from the specifications. EPOS allows the user to employ different design methods and different views of a problem. There is also a tool to transform these views automatically into one another. The engineering and project management oriented support system EPOS is used by a large number of companies for several years now and is also being intensively utilized in educational institutions for semester and master thesis projects. Thus, it has proven its suitability for being applied in the laboratory part of a course on real-time systems as described in this Chapter. In order to experience the full power of an integrated development support system, the practical laboratory work accompanying this part of the course must be very extensive. It is desirable that the system is then later also applied in the course of further project and thesis work.

Semester Schedule

The above outlined syllabus is suitable for a one-semester senior or graduate course. It was followed by the author who taught such a course at three universities and, in block form, as a continuing education short course. If a second semester is available, then the mentioned topics can be presented in more depth. This holds especially with respect to integrated project development environments, whose application can be practiced in a number of student projects to be presented later in class.

Owing to their increasing importance, the following four software related subjects are studied in the second semester: *real-time communications in distributed systems, database organization* and *programming under real-time conditions with special emphasis on distributed databases,* and *building of expert systems,* all of which are to observe hard deadlines, as well as the utilization of *schedulability analysis* tools. Aiming at guaranteeing predictable system behaviour, with the latter, worst case task execution times are to be calculated and it is to be evaluated before actual implementation whether a system will meet its timing requirements. Whereas suitable computers and standard peripherals are usually available off the shelf, the real-time system engineer often has to *design* and *build special purpose interface boards* to interconnect the technical external process to the computer system. This constitutes another topic for detailed study in the advanced course which must, of course, be supplemented by intensive laboratory experiments.

Accompanying Laboratory Work

A main feature of the here introduced course on real-time systems is that it is accompanied by extensive laboratory sessions providing practical experience about all topics discussed in class. As previously mentioned, the laboratory work incorporates the following main components:

- Writing of a number of application programs, e.g., performing process measurements and control, in the real-time language selected for the course.
- Development and implementation of a small operating system or at least a real-time kernel.
- Connection of the computer available to various devices and writing of the corresponding driver routines.

Suitable devices for a laboratory environment are models of continuous, discrete, and sequential processes. The investment to set up a laboratory with the mentioned facilities does not need to be high. Process models are available at low prices; a variety of very useful devices can even be obtained on the market for toys and hobby equipment. Real-time programming systems with appropriate operating systems as well as the software tools for requirements engineering are now available on personal computers of the AT class. Their prices are oriented at the hardware prices, i.e. they are also low. Both, the language PEARL and the requirements engineering tool EPOS, which were recommended in the last section, are available for personal computers at a nominal fee for educational institutions. The PEARL programming system is based on the operating system PORTOS, which is the first, and still only, commercially developed and distributed real-time operating system supporting deadline driven scheduling of tasks. PORTOS fully supports PEARL and is controlled by this language. Hence, it is practically invisible to the user, who does not need any training in order to apply this operating system.

Course Materials

As far as a textbook for the above outlined course(s) is concerned, there is no better one than [11], which just recently appeared in a completely revised second edition. It has to be complemented by manuals for the language and the other software tools used as well as by a laboratory handbook describing the function of the process models to be interfaced and controlled. The book [11] is written in German and has already been translated into Italian. It would be of great value to have this text translated into English, since, unfortunately, there is no good English language textbook for a real-time course yet. As additional reading, especially for the second course semester and in order to catch up with current research work, the new collection of survey articles [9], the tutorial text [20] and, naturally, the present book represent very good sources of information.

Areas for Graduate Research

The scientific research into real-time systems is presently intensifying, which provides a good opportunity for graduate students to participate in challenging projects. A (nearly) exhaustive list of actual research areas reads as follows:

- Conceptual foundations of real-time computing;
- Predictability and techniques for schedulability analysis;
- Requirements engineering and design tools;
- Reliability and safety engineering with special emphasis on the quality assurance of real-time software;
- High-level languages and their concepts of parallelism, synchronization, communication, and time control;
- Program transformations to ensure better guarantees;
- Real-time operating systems;
- Scheduling algorithms;
- Systems integration;
- Incorporating non-real-time components;
- Distributed, fault-tolerant, language and/or operating system oriented innovative computer architectures;
- Hardware and software of process interfacing;

- Communication systems;
- Distributed databases with guaranteed access times;
- Artificial intelligence with special emphasis on real-time expert and planning systems;
- Practical utilization in process automation and real-time control;
- Standardizations.

A Suitable Real-Time Operating System

Owing to its wide range of real-time features, most available or older real-time operating systems like Ready Systems' VRTX or Intel's iRMX are unable to support all functions of PEARL, especially with regard to timing control. Therefore, new operating systems had to be developed. One of these PEARL-oriented operating systems is PORTOS. In this section, we shall take a closer look at it, because it runs on personal computers and provides deadline driven scheduling as a unique additional feature, which is indispensable for a course as described here. This is because deadline based scheduling methods play an eminent role in current research and the concept of deadlines is more problem-oriented than the one of priorities.

As suggested by the acronym, PORTOS (POrtable Real-Time Operating System) is a portable and configurable operating system for dedicated real-time applications. PORTOS is available on all conventional microprocessor systems, such as IBM PC-XT, IBM PC-AT, and Intel SBC 286/310, and is used in embedded systems, which are based on the Intel Microprocessor Series.

PORTOS controls the execution of application programs in a hard real-time environment, it does not, however, support the development of these programs. The latter are primarily written in PEARL, but Ada, C, C/ATLAS, PL/M, and SYSLAN can also be used. PORTOS is highly configurable, i.e. it can be automatically adapted to special hardware configurations and, in its function spectrum and queue layout, it can be adjusted to the operating system support requirements of the user programs.

The configuration of PORTOS is automatically generated by the PEARL compilation system; for all others of the above mentioned programming languages, however, the configuration procedure is carried

Curriculum

First semester:
- History, importance and basic concepts of computerized process automation
- Real-time software enginneering
- Programming in high-level real-time languages
- Real-time operating systems with special emphasis on task scheduling
- Hardware architectures and distributed automation structures
- Process interfacing and peripherals
- System reliability and fault-tolerance
- Integrated project development support systems

Second semester:
- Advanced topics from the first semester's subject areas, esp. on the integrated project development support systems
- Real-time communications in distributed systems
- Organisation of (distributed) real-time databases
- Real-time expert systems
- Schedulability analysis
- Design and construction of special purpose process peripherals

out in a dialogue with the user. Depending upon the requirements of the application programs, PORTOS needs a memory area of 4 - 100 kB. One of the most important measures for the performance of a real-time operating system is the time required to react to external events. On an IBM PC-AT, PORTOS ensures the processing of interrupts within less than 50 microseconds.

PORTOS is structured in accordance with the layer model for real-time operating systems, which was developed in [3]. Hence, the machine dependent operating system components, such as interrupt handling, time management, process management, communication, synchronization, input/output, as well as memory and queue management, are realized in the PORTOS nucleus, whereas the hardware and language independent operating system services, such as file handling, user interface, and scheduling, are implemented in the PORTOS shell. A facility is provided, allowing the user to functionally extend each layer of the operating system in order to cope with special application requirements.

Hitherto, processor allocation to runable tasks was exclusively priority controlled. Essentially distinguishing it from all other commercially available real-time operating systems, PORTOS features deadline driven task scheduling [8,14,17], which is feasible in contrast to priority based scheduling schemes. Thus, under PORTOS, it is assured that as many tasks as possible meet their deadlines as specified by the user. Although deadline driven task scheduling has a number of advantages, including to match human thinking, PORTOS supports both scheduling methods for compatibility purposes with other real-time operating systems and languages. Both methods can be used simultaneously, with priority scheduling usually applied for tasks with weak real-time demands. A response time or a priority is assigned to each task upon its installation, i.e. within the framework of a PEARL task declaration or, for all other programming languages, when PORTOS is configured.

In PORTOS, the user provided response times take effect in the course of task (re-)activations and terminations. Upon task activation and continuation, a task's due date is derived from the response time and stored into the variable T.DUE of the corresponding task control block (TCB). The processor is then assigned to the task with the earliest due date. Upon termination of a task, a check is made whether its due date has been met. If this was not the case, a counter in the TCB is incremented. This event is signaled to the PEARL programmer to be dealt with by an exception handler, but is also available for evaluation by the PORTOS debugger, which will be outlined below.

For tasks with multiple sections, each of which having its own particular response time, dependent on the external technical process, POR-TOS provides a facility to change response times dynamically.

Finally, a few features provided by the test system of PORTOS are to be mentioned here. Its main purpose is to make transparent to the user, the dynamics of task behaviour. This facility is necessary to verify the execution of real-time software, not only in the sense of correct data processing, but also with regard to the requirement imposed by the dimension time, viz. *timeliness*. In the test phase, the PORTOS debugger is used for monitoring PORTOS objects such as tasks, queues, times, etc. If this feature is, to a certain suitable extent, also utilized in the operational phase of a software package, it can facilitate the post mortem error analysis when a system crash occurs. In addition to this debugger, there is also a PEARL test system, which, at PEARL source code level, allows to perform various tests and to take time measurements without influencing the time behaviour of the user program.

CONCLUSION

A comprehensive syllabus for a one or two semester course on real-time computer systems engineering for senior undergraduate or master's students of engineering subjects is recommended. The course(s) may serve as preparation for commencing thesis projects in a field of rapidly intensifying research activity. Corresponding areas of research topics are identified. The main themes covered in the course(s) are basic concepts of process automation, real-time software engineering and programming in high-level real-time languages, real-time operating systems with special emphasis on task scheduling strategies, hardware architectures and particularly distributed automation structures, process interfacing, system reliability and fault-tolerance, and, finally, fully integrated project development support systems. Accompanying course texts are mentioned, and the subjects and goals of laboratory exercises are established. It is pointed out, how such a laboratory can be developed with advanced, but low-cost, equipment and software tools. Special consideration is given to the question of selecting a suitable high-level real-time language to be discussed in class and to be used for practical exercises due to its essential role in the framework of the course(s) and for the application of the material for process control and automation purposes. PEARL is identified as the most powerful language for the real-time systems and automation engineer.

REFERENCES

[1]   "The Programming Language Ada Reference Manual", American National Standards Institute, Inc., ANSI/MIL-STD-1815A-1983, Lecture Notes in Computer Science 155, Berlin-Heidelberg-New York-Tokyo: Springer-Verlag 1983.

[2]   R. Barnes, "A Working Definition Of The Proposed Extensions For PL/1 Real-Time Applications", *ACM SIGPLAN Notices*, Vol. 14, no. 10, pp. 77 - 99, October 1979.

[3]   R. Baumann *et al.*, "Functional Description of Operating Systems for Process Control Computers", Richtlinie VDI/VDE 3554, Berlin-Cologne: Beuth-Verlag 1982.

[4]   "DIN 66253: Programming Language PEARL", German National Norm, Part 1 Basic PEARL, 1981, Part 2 Full PEARL, 1982, Part 3 Multiprocessor-PEARL, 1989, Berlin-Cologne: Beuth-Verlag.

[5]   P. Elzer, "Ein Mechanismus zur Erstellung strukturierter Prozessautomatisierungsprogramme", *Proceedings of the "Fachtagung Prozessrechner 1977"*, Augsburg, March 1977, Informatik-Fachberichte 7, pp. 137 - 148, Berlin-Heidelberg-New York: Springer-Verlag 1977.

[6]   P. Göhner, "Integrated Computer Support for the Development and Project Management of Software/Hardware Systems",
Proceedings of the 3rd IFAC Symposium on Computer Aided Design in Control and Engineering Systems, July/August 1985.

[7]   W.A. Halang, "On Real-Time Features Available in High-Level Languages and Yet to be Implemented", *Microprocessing and Microprogramming*, Vol. 12, no. 2, pp. 79 - 87, 1983.

[8]   R. Henn, "Feasible Processor Allocation in a Hard-Real-Time Environment", *Real-Time Systems*, Vol. 1, no. 1, pp. 77 - 93, 1989.

[9]   K.H. Kim (Ed.), "Advances in Real-Time Computer Systems", Greenwich, CT: JAI-Press 1991.

[10] W. Kneis (Ed.), "Draft Standard on Industrial Real-Time Fortran", International Purdue Workshop on Industrial Computer Systems, *ACM SIGPLAN Notices*, Vol. 16, no. 7, pp. 45 - 60, 1981.

[11] R. Lauber, "Prozessautomatisierung", Vol. 1, 2nd edition, Berlin-Heidelberg-New York-London-Paris-Tokyo: Springer-Verlag 1989.

[12] R. Lauber, "Development Support Systems", *IEEE Computer*, Vol. 15, no. 5, pp. 36 - 46, 1982.

[13] R. Lauber and P. Lempp, "Integrated Development and Project Management Support System", *Proceedings of the 7th International Computer Software and Applications Conference COMPSAC '83*, Chicago, November 1983.

[14] C.L. Liu and J.W. Layland, "Scheduling Algorithms for Multiprogramming in a Hard-Real-Time Environment", *JACM* Vol. 20, pp. 46 - 61, 1973.

[15] "LTR Reference Manual", Compagnie d'informatique militaire, spatiale et aeronautique, Velizy, October 1979.

[16] PEARL-Association, Brochure on the UH-RTOS System.

[17] H. Piche, "PORTOS-Nutzerhandbuch der PORTOS-Algorithmen - Antwortzeitgesteuerte Prozessorzuteilung", Report GPP/442/88, GPP mbH, Oberhaching, 1988.

[18] K. Sandmayr, "A Comparison of Languages: CORAL, PASCAL, PEARL, Ada and ESL", *Computers in Industry*, Vol. 2, pp. 123 - 132, 1981.

[19] J.A. Stankovic (Ed.), "Real-Time Computing Systems: The Next Generation", COINS Technical Report 88-06, Department of Computer and Information Science, University of Massachusetts, Amherst, 1988.

[20] J.A. Stankovic and K. Ramamritham (Eds.), "Hard Real-Time Systems", Tutorial, Washington: IEEE Computer Society Press, 1988.

[21] J.A. Stankovic, "Misconceptions About Real-Time Computing", *IEEE Computer*, Vol. 21, no. 10, pp. 10 - 19, October 1988.

[22] A. Stoyenko and E. Kligerman: "Real-Time Euclid: A Language for Reliable Real-Time System", *IEEE Transactions on Software Engineering*, 12, 9, 941 - 949, September 1986.

Remark: The software mentioned in this paper, i.e. a PEARL compiler, the real-time operating system PORTOS, and the requirements engineering tool EPOS can be obtained from

GPP mbH, Kolpingring 18a, 8024 Oberhaching bei München, Germany

Tel.: + 49-89-613041/Fax: + 49-89-61304294/Telex: 5 216 612 gpp

GLOSSARY OF REAL-TIME TERMINOLOGY

Atomic Action Indivisible program part. In reality: smallest uninterruptable programming module used to implement ->semaphores or ->bolts. Unfortunately, uninterruptability cannot be guaranteed if it is not a feature of the ->Realtime Operating System.

Bolts A generalized semaphore used to synchronize the access to resources which allow (exclusive) writing access as well as (shared) reading access.

Clock Device that adds the dimension time to the otherwise only logically arranged computer programs. The clock is used to measure time intervals and/or to synchronize processes. In distributed systems, a global clock is used for these purposes. Unfortunately, global clocks can only be synchronized within certain limits, typically a few microseconds.

Context Switching The switching of the processor from one activity to another. Context switching occurs as different processes are scheduled for execution.

Correctness Correct program execution should be achieved to guarantee ->deterministic behaviour which is of primary importance for real-time programs. Unfortunately, it is impossible to prove the correctness of a program. Therefore, the goal of software design must be to lower the risk to a defined level, depending on the importance of the application.

Determinism The action of a computer system that is completely forseeable and reproducible, a task that can only be accomplished fully in theory.

Embedded Systems (ES) An ES is an electronic system embedded within a plant or an external process. The external process may comprise both a physical/chemical/technical system (usually consisting of subsystems) and also humans performing some supervising or parameter setting tasks. Essential to the external system is that it obeys its own dynamics and does not allow complete restoration of a previous state in case of failure. Coupled with this property is the requirement that the ES has to respond to the external process within prescribed time constraints (->real-time requirement).

Exception   An exception is an error situation which may arise during program execution (e.g., division by zero, memory protection violation, etc). An exception may be raised in several programming languages so as to signal that the error has taken place. An exception handler is a potion of program text specifying the response to the exeption.

Fault-Tolerance   A method to ensure continued system operation in the presence of faults. Since it it not possible to write non-trivial programs fault-free and, since hardware components will have faults, it is crucial for the reliability of a system to be fault tolerant. It should be remembered that with the advent of Very Large Scale Integrated circuits (VLSI) and by using microprogramming techniques, the faults may already have been incorporated into the microprocessor hardware proper. Countermeasures are: introduction of redundant components both in software and hardware; ensuring →graceful-degradation or →exception handling.

Graceful Degradation   To ensure →fault-tolerance, failing system components (both in software and hardware) must be handled in such a way that the effect on the remaining system is only a partial degradation of system performance, not a total system crash.

Hard Real-Time Systems   Systems designed to meet given deadlines under any circumstances. Late results are useless and tardiness may result in danger and high cost. Typical examples are fly-by-wire-systems used to control all important elements of an aircraft.

Interrupt   Devices may want to interrupt normal program flow intermittently, e.g., to send data. This is done by requesting an interrupt. When the interrupt is acknowledged, the processor suspends the process currently executed (→ context-switching) and calls an interrupt-handler that services the requesting device. Upon completion of the interrupt handler, normal program execution continues, i.e., the processor is switched back to the process that was interrupted.

Interrupt latency   The time between the occurrence of an external → interrupt and the beginning of the execution of the interrupt handler. The interrrupt latency is a measure for real-time performance of a system.

Polling   Cyclic checking of an external flag within a program. The program stalls until the flag is set.

Semaphore Originally a person using flags for signaling purposes. Here: a simple mechanism to ensure mutual exclusion or synchronization of separate processes. A semaphore is an integer variable on which two operations ("P" and "V") are defined. Prior to accessing a device, a process must issue a call to the procedure implementing the P operation.

If the value of the semaphore is 0, the process has to wait until the value is greater than 0. If it is greater than 0, "P" decrements the value of the semaphore and the process continues to execute. After releasing the device, the process must call the procedure "V" which increments the semaphore, thereby freeing the device for access by other processes.

INDEX

442